Forrest Mims'
Circuit Scrapbook II

Forrest Mims'
Circuit Scrapbook II

Forrest M. Mims, III

HOWARD W. SAMS & COMPANY

A Division of Macmillan, Inc.
4300 West 62nd Street
Indianapolis, Indiana 46268 USA

© 1987 by Forrest M. Mims, III

FIRST EDITION
SECOND PRINTING—1988

International Standard Book Number: 0-672-22552-2
Library of Congress Catalog Card Number: 86-63070

Acquisitions Editor: *Greg Michael*
Editor: *Don Herrington*
Interior Designer: *T. R. Emrick*
Cover Graphics: *Lawrence Simmons*
Composition: *Impressions, Inc., Madison, WI*

Printed in the United States of America

Trademark Acknowledgments

Contents

Preface

How would you like to assemble a miniature laser system that fits in a shirt pocket? Would you like to experiment with state-of-the-art optical fiber sensors? Or would you prefer to experiment with piezoelectronics, solid-state heat pumps, radio control, ultrasonic sound, lightwave communications, or active filters? If you would like to add adventure to your projects, you might try a do-it-yourself aerial photography system. That project has provided my son and me hundreds of dramatic aerial photos and some interesting firsthand experiences. Like retrieving a downed kite and camera from a cattle-filled pasture, maneuvering a camera suspended from a balloon through tree branches, and helplessly watching a kite and camera plunge hundreds of feet to graze the waves of the Gulf of Mexico before swooping skyward again as if nothing had ever happened.

These topics are a sample of what's included in the pages that follow. This book is the latest in a series of compilations of columns that were originally published in *Popular Electronics*, *Computers & Electronics*, and *Modern Electronics*. The previous books in the series are *103 Projects for Electronics Experimenters* (Tab Books, Inc. 1981), *The Forrest Mims Circuit Scrapbook* (McGraw-Hill, 1983), and *Forrest Mims's Computer Projects* (Osborne/McGraw Hill, 1985).

As with previous books in this series, all of the columns compiled in this volume have never before appeared in book form. I have reviewed, checked for accuracy, and updated each column. For example, several examples of interesting correspondence have been appended to some of the columns.

The circuits in this book are my own design or were adapted from manufacturer's application notes and other sources. In every case, I have personally built, assembled, and tested each circuit to verify its operation. If a particular circuit or project strikes you as interesting, I hope you'll try it. You can use the circuit in its present configuration, or you can modify it for other applications.

If you have little or no previous experience in electronics and would like to learn some basics before assembling circuits in this book, I invite you to read *Getting Started in Electronics* (Radio Shack 1983) and Radio Shack's series of *Engineer's Mini-Notebooks*. I wrote each of these books with the electronics novice in mind.

If you enjoy the topics in this book, I hope you will read "Electronics Notebook," my monthly column in *Modern Electronics*. Incidentally, *Modern Electronics* was conceived and is edited by Arthur P. Salsberg. It is very much in the tradition of the former *Popular Electronics*, which shouldn't be surprising since Art served as the editor of that great magazine and its sequel, *Computers & Electronics*.

Finally, I would like to acknowledge the assistance of my family in preparing this book for publication. My wife, Minnie, and daughter, Vicki, spent many hours photocopying the original manuscripts and art. Without their help the deadline for this book would not have been met. Eric, my son, assisted during the preparation of some of the columns. He was particularly helpful during the numerous flight tests of the kite- and balloon-borne radio-controlled camera system described in Section Six. I very much appreciate their help. And I hope you enjoy building and experimenting with the projects in this book as much as I have.

FORREST M. MIMS III

Disclaimer

Reasonable care has been exercised with regard to the accuracy of the circuits and procedures given in this book. Variations in components, tolerances, and construction methods may cause the performance of circuits built in accordance with the instructions herein to differ. Therefore, the author and publisher assume no responsibility for the suitability of this book's contents for any application. Since they have no control over the use to which the information in this book is put, neither do the author and publisher assume any liability for any losses, damages, or personal injury resulting from its use. It is the reader's responsibility to determine if commercial use, sale, or manufacture of any device that incorporates information in this book infringes any patents, copyrights, or other rights. Readers are urged to carefully heed the safety notices given in the text.

Transistor and MOSFET Circuits

ONE

Transistor and MOSFET Circuits

Rediscovering the Transistor

With the availability of literally hundreds of different kinds of integrated circuits, it's common to design even the simplest circuit around one or more ICs when a transistor or two would suffice. For example, the usual way to drive a meter or trigger a relay in response to a changing voltage or current is to use an operational amplifier or comparator. Often the same function can be performed with one or two transistors.

Another common example is the pulse generator. Most designers use a timer such as the popular 555 or a pair of cross-coupled gates when a simple pulse generator is required. Often, however, a simple two-transistor circuit will perform the same function with fewer components.

Let's look at several practical examples of how a simple transistor circuit can provide some or all the functions of an integrated circuit version. I think you'll agree with me that simple transistor circuits still have an important role to play in electronics today.

Moisture Detection Circuits

If asked to design a moisture detection circuit, the typical design engineer will use an op amp or comparator as the principal active circuit element. Figure 1-1 shows a much simpler approach.

The ultrasimple circuits in Fig. 1-1 each use a single bipolar transistor as the active element. Most common silicon npn small signal transistors (2N2222, 2N3904, etc.) can be used.

The moisture meter can be used to measure the level of moisture in a flower pot or in garden soil. Probes can be made from nails or, even better, stainless steel wire. The circuit is calibrated by adjusting R2 for a meter reading of 1 milliampere when the soil moisture is at the desired level.

The moisture activated relay in Fig. 1-1 is a modified version of the moisture meter. The relay is actuated when the moisture level exceeds the level determined by the setting of R2. If you replace the sensor probes with a circuit board upon which has been etched an interlacing, comb-like grid, the circuit will be actuated when a rain drop bridges the gap between the two foil patterns.

It's interesting to compare these circuits with integrated versions. The most obvious difference is cost. A suitable transistor can be purchased for under 15 cents if you're willing to pay a dollar or two for a bag of a dozen or more.

Another advantage of the transistor version is simplicity. A transistor has only three connection leads, whereas connections must be made to at least five of the eight pins of a typical op amp. Furthermore, since most transistors are equipped with leads rather than pins, the transistor moisture detection circuits can be easily assembled on a perforated board the size of a postage stamp.

Finally, at least one of the transistor circuits can be powered

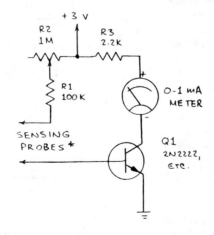

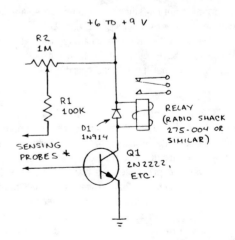

Fig. 1-1. Transistorized moisture sensing circuits.

(A) Moisture meter.

(B) Moisture activated relay.

3

by a pair of penlight cells. (The relay circuit requires more voltage to drive the relay.) Although CMOS op amps that can be powered by as little as a volt or two are available, they are far more costly than a single transistor.

A Two-Transistor Metronome/Tone Source

Figure 1-2 shows a simple metronome circuit that, together with its two transistors and a speaker, includes only six components. Ordinarily the ubiquitous 555 timer is used to make a simple circuit like this. However, the 555 version also requires six components (assuming it, too, employs a resistor in series with its frequency control potentiometer).

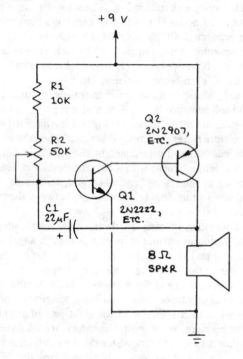

Fig. 1-2. Simple but effective two-transistor metronome.

Many common silicon switching and small signal transistors can be used for Q1 and Q2 in Fig. 1-2. The click rate can be adjusted by adjusting R1 or changing the value of C1. Since the transistors have leads rather than pins, it's easy to build this circuit on a small perforated board.

FET Electrometer

The electrometer in Fig. 1-3 is a good circuit to try on a dry winter day. Though it uses only four components, its meter will indicate the presence of static charges up to several feet away. The circuit I tried, for instance, can detect from several feet away the presence of a plastic comb which I charged by stroking through my hair. Even *very* small movements of the charged comb will cause the meter needle to respond in kind. Potentiometer R2 allows the circuit to be calibrated.

You can make a permanent version of the electrometer to check for the presence of static electricity near computers and

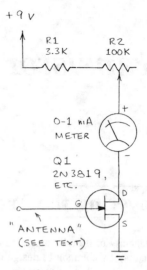

Fig. 1-3. Ultrasimple FET electrometer.

other devices that can be affected by electrostatic discharges. For best results, install the circuit in a plastic box. A circuit board will not be necessary, since the leads of the FET and R1 can be soldered directly to the meter and the potentiometer. Q1 should be placed near the top portion of the box. A short, stubby "antenna" wire should extend from the top of the box.

More sophisticated electrometers can be made with CMOS op amps. But they are more costly, trickier to design, and harder to construct. For a simple indication of the presence of static electricity, the ultrasimple single FET circuit in Fig. 1-3 may be a better choice.

An FET Timer

When faced with the need to design a timer, I inevitably use a 555 or its CMOS counterpart, the 7555. For brief timing durations of a few minutes, the simple FET circuit in Fig. 1-4

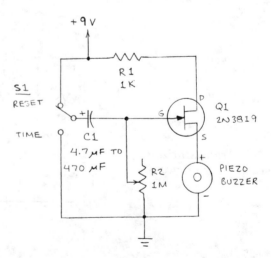

Fig. 1-4. A simple FET timer circuit.

may be a better choice. Certainly it's simpler, easier to build, and less expensive.

In operation, when S1 is switched to RESET, C1 is charged to near the power supply voltage through R2. When S1 is switched to TIME, C1 is slowly discharged through R2. Eventually, the voltage on C1 falls low enough to allow Q1 to switch on, thus allowing the piezoelectric buzzer to sound.

Power MOS Timers

VMOS, TMOS, DMOS, and other MOSFET power transistors can be used to make more effective, longer duration timing circuits than the simple FET version in Fig. 1-4. Thanks to the ultrahigh-impedance input of the MOSFET, time delays of up to half an hour or even longer are possible. And, again, the necessary circuitry is simpler than an IC version using the 555, 7555, or similar timer chips.

Figure 1-5 shows two simple timers in which a power MOSFET (Q1) plays the key role. Q1 can be a VN10, VN67, IRFD-1Z3, or other common power MOSFET.

The circuit in Fig. 1-5A is an "Off After Delay" timer. In operation, the piezoelectric buzzer is normally off. When S1 is momentarily closed, however, C1 is charged, thus switching on Q1 and the buzzer. When C1 eventually discharges through natural leakage paths, Q1 switches off and silences the buzzer.

Very long time delays are possible when C1 has a high capacity. For best results, use a capacitor having a very low loss dielectric. To control the timing cycle, you can add a resistor across C1 (R1). R1 will provide a discharge path for the charge on C1.

Figure 1-5B is a modified version of the circuit in Fig. 1-5A. Here a bipolar transistor (Q2) inverts the switching status of Q1, thus providing an "On After Delay" operating mode. In this case the piezoelectric buzzer sounds *after* the time delay is complete.

Though I used a piezoelectric buzzer in both circuits in Fig. 1-5, you can instead use a relay, small lamp, motor, portable radio, or other device. In any case, do *not* exceed the power rating of Q1 in Fig. 1-5A or Q2 in Fig. 1-5B. The latter circuit includes a series resistor (R3) to limit current through Q2 and the piezoelectric buzzer. If it's necessary to reduce the drive current in the circuit in Fig. 1-5A, insert an appropriate series resistor between the positive supply and the controlled device.

A Multifunction Two-Transistor Oscillator

Some integrated circuits, particularly the 741 operational amplifier and the 555 timer, have for many years been considered standard devices by engineers and experimenters alike. Because these and other ICs are so exceptionally versatile, it's easy to overlook simple circuits made from discrete components that are equally versatile and sometimes simpler.

One of the best examples of a highly versatile nonintegrated circuit is the two-transistor oscillator shown in Fig. 1-6. This circuit is designed to function as a code-practice oscillator when connected to a telegraph key, speaker, and battery. However, it has numerous other applications, some of which have greatly influenced my career as an electronics experimenter. Although there's little chance this circuit will affect anyone else's career, perhaps it will lead experimenters to reconsider the value of discrete transistors in their circuits.

Back to Basics

For some twenty years, Radio Shack stores sold an assembled version of the code-practice oscillator circuit in Fig. 1-6 for

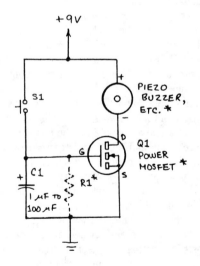

(A) Off after time delay.

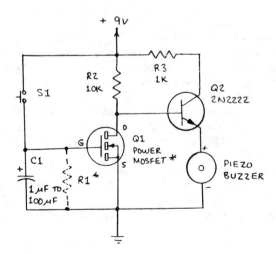

(B) On after time delay.

Fig. 1-5. Two power MOSFET timer circuits.

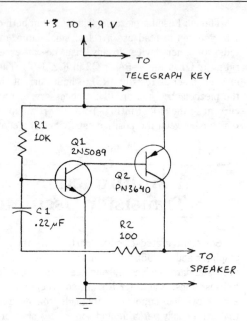

+3 TO +9 V

TO TELEGRAPH KEY

R1 10K

Q1 2N5089

Q2 PN3640

C1 .22 μF

R2 100

TO SPEAKER

Fig. 1-6. Radio Shack code practice oscillator circuit.

a few dollars (Cat. No. 20-1155). During my senior year at Texas A&M University in 1966, I used one of those circuits to supply pulses to some of the first commercially available infrared- emitting diodes. I used those pulsed infrared emitters as optical sources for miniature travel aids for the blind and in lightwave communications experiments.

Later in 1966 I used the basic circuit in Fig. 1-6 to drive a silicon solar cell. The cell emitted pulses of infrared which were detected by a second, identical solar cell nearby. I also used a miniaturized version of the circuit to drive an infrared- emitting telemetry transmitter which I launched in small rockets. A light-sensitive cadmium sulfide photocell varied the pulse rate of the infrared transmitter, thus providing an indication of the rocket's roll rate. In 1968, I modified the circuit slightly so that it would cause a small incandescent lamp to emit brilliant flashes of light. I built a series of miniature flashers to assist in the recovery of dozens of night- launched model rockets that carried an experimental guidance system.

These experiments with the basic circuit in Fig. 1-6 had consequences more far ranging than I could ever have imagined. The rocket light flasher became the subject of my first magazine article and led directly to my decision to become an electronics writer. Ed Roberts helped with some of those night launchings. In 1969, Ed and I joined with two friends to form MITS, Inc. Though our first product was a model rocket light flasher, MITS is best remembered for the Altair 8800, the first personal computer.

Designing travel aids for the blind, a project which still occupies a good deal of my time, led to a series of articles and books about light-emitting diodes and lasers. The experiments in transmitting infrared pulses between two identical solar cells led to similar work with LEDs. In 1973, after I used a pair of

identical LEDs to transmit audio in both directions through both the air and an optical fiber, I submitted an invention disclosure to Bell Labs. After agreeing to pay for the suggestion if they used it, Bell Labs rejected it as being impractical. Five years later, however, Bell Labs announced it had developed an optical telephone that incorporated the core of my suggestion. Since *Business Week* said the new phone would "dramatically alter the basic nature of the phone network" (December 4, 1978), I spent several months asking Bell Labs to pay for the use of the suggestion. They refused, I sued, and they eventually settled out of court.

After more than 20 years Radio Shack discontinued the code practice oscillator circuit in Fig. 1-6. For sentimental reasons, I was sorry to see the end of that product. On the other hand, the basic circuit can be assembled in a minute or so on a plastic breadboard. I've experimented with an assortment of applications for the circuit and many of them are covered in the following. First, let's review the circuit's operation. Referring to Fig. 1-6, when the key is closed C1 begins to charge through R1, R2, and the speaker. Both Q1 and Q2 are initially off. Eventually, the charge on C1 becomes high enough to switch Q1 on. Q1 then switches Q2 on.

When Q2 is on, the speaker is connected directly across the power supply through Q2's emitter-collector junction. Meanwhile, C1 discharges to ground through the base-emitter junction of Q1. When the charge on C1 falls below that necessary to keep Q1 switched on, Q1 switches off. Q2 then switches off, and current is no longer supplied to the speaker. The charge-discharge cycle repeats, and the result is a series of audible pulsations from the speaker. If the charging time is made brief by making C1 small, the speaker will emit a high-pitched tone.

Introducing the Circuits

Note that Q2 in Fig. 1-6 is a pnp transistor. For applications like light flashers where low-switching impedance is desired (for maximum current), it's best to use an npn transistor switch since its "on" resistance is less than that of a pnp transistor. The basic circuit can be modified for this purpose by rearranging it to be a mirror image of the circuit in Fig. 1-6. This version of the basic circuit is used in each of the circuits that follow. It can produce fast-rising and falling current pulses having a peak amplitude greater than 1 ampere. And it will oscillate when the power supply voltage is only about 0.7 volt.

Each of the following circuits specifies a 2N2907 for the pnp transistor and a 2N2222 for the npn transistor. However many different pnp and npn transistors can be substituted for these units. For best results, use silicon transistors designated for switching applications. You may wish to use a power transistor for the npn unit in applications that switch lots of current (e.g., driving incandescent lamps).

To monitor the operation of the circuits, break the connection between the emitter and ground of Q2 and insert a small resistor in the gap. Connect an oscilloscope across this resistor to monitor the pulsed output from the circuit. If this monitoring resistor has a value of 1 ohm, it will have very little influence on the circuit operation. If a circuit fails to operate, disconnect the power while looking for the problem. Otherwise, Q2 may stay turned on and both it and any device with which it is in series may be damaged or destroyed by the excess heat thereby generated.

Incidentally, the very fast rise and fall times of the basic circuit can produce interference on a nearby radio. Be sure to keep this in mind while testing the circuit.

Audio Oscillator Circuits

Figure 1-7 shows a mirror image of the circuit in Fig. 1-6 configured as an audio oscillator. R3 controls the circuit's oscillation frequency. When C1 is 0.01 microfarad, the circuit generates a tone. The tone slows to a series of distinct clicks or pocks when C1 is increased to 1 microfarad. R4 limits current through the speaker and thereby reduces the volume of the sound from the speaker. The volume can be increased by increasing the power supply voltage. R1 controls the time Q2 remains on during each cycle. When the pulse duration is brief (R1's value is reduced), then the speaker volume is reduced.

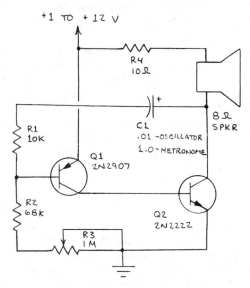

Fig. 1-7. Audio oscillator/metronome.

Figure 1-8 is an audio oscillator whose frequency is determined by the intensity of light that strikes a cadmium sulfide photocell (Radio Shack 276-116 or similar). When the light level is increased, the resistance of the photocell falls, thus increasing the frequency of oscillation.

A unique feature of the circuit in Fig. 1-8 is the substitution of a piezoelectric alerter (Radio Shack 273-064 or similar) for C1 in Fig. 1-7. The inherent capacitance of the alerter permits the circuit to oscillate. At the same time, the alerter emits an audio tone, thereby eliminating the need for a separate speaker as in Fig. 1-7.

Lamp and LED Flashers

Figure 1-9 is the circuit for an incandescent lamp flasher similar to one I have often used to track and recover night-launched model rockets. This circuit generates from one to two flashes each second. The rate can be altered by changing the

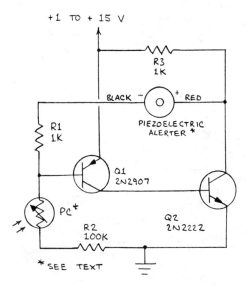

Fig. 1-8. Light-dependent oscillator.

value of C1 or R2. Use care when first operating the circuit or when altering its flash rate. If the lamp stays on continually without flashing, the heat generated by the heavy current flow will quickly overheat and possibly damage or destroy Q2. If Q2 becomes warm, substitute an npn power transistor equipped with a suitable heatsink.

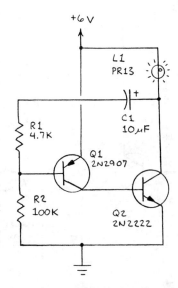

Fig. 1-9. Incandescent lamp flasher.

Figure 1-10 is an LED flasher circuit. With the values shown and when the power supply provides 9 volts, the circuit delivers a 1-millisecond long pulse twice each second. The pulse has an amplitude of 600 milliamperes. I have used this circuit to drive an ultrabright Stanley H2K red LED. The resulting flashes are too bright to observe at close range. (See "Super Bright LEDs" in Section Four for more information about the remarkable H2K.)

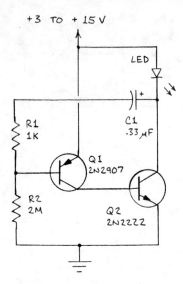

Fig. 1-10. LED flasher.

Figure 1-11 is a variation of the circuit in Fig. 1-10 in which the npn transistor used for Q2 has been replaced by a power MOSFET. Though I used a VN67, any n-channel MOSFET should work. With the values shown, the circuit delivers a 1 millisecond pulse to the LED every 80 milliseconds (a flash rate of about 12.5 Hz). The peak current through the LED is 500 milliamperes, and the pulses are very square.

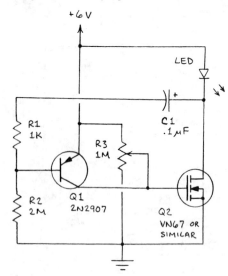

Fig. 1-11. Flasher circuit using power-MOSFET driver.

For the circuit in Fig. 1-11 to oscillate, potentiometer R3 must be properly adjusted. After the desired operating mode is achieved, the resistance between the rotor terminal and the two stationary terminals can be measured and a pair of fixed resistors substituted for the pot.

I have used the circuit in Fig. 1-11 to drive both near- infrared and red LEDs. For very high pulse current through the LED, use a low on-resistance MOSFET (under 1 ohm). If you can't find a low on-resistance MOSFET, two or more standard power MOSFETs can be connected in parallel to reduce the resistance of the current path through the LED. Be sure to avoid driving the LED above the current level for which it is rated. Otherwise, it will be degraded or even destroyed.

Figure 1-12 shows a pair of two-flash-per-second flasher circuits that are activated by the presence or absence of light at a phototransistor (Q3). In Fig. 1-12A, the flasher circuit is disabled when Q3 is dark. Light at the active surface of Q3 switches Q3 on and permits the flasher circuit to function.

The circuit in Fig. 1-12B can be used as a warning flasher that operates only at night. The circuit is disabled when light switches Q3 on, thereby clamping Q1's base to the positive supply. When Q1 is dark, the flasher operates normally. The circuit consumes only about 1 microampere when Q3 is illuminated and the supply provides 5 volts. When the supply provides 12 volts, the standby current drain increases to about 4 microamperes.

Incidentally, the dark-activated function can be accomplished by means of a cadmium sulfide photocell instead of a phototransistor. One way is to remove Q3 and connect a CdS photocell between the positive supply and Q1's base. Alternatively, connect the CdS photocell across C1. There is no reason why the light/dark activation methods described here cannot be used with any of the other circuits in this column. For instance, a phototransistor or photocell can be used to make a light- or dark-activated tone generator.

Figure 1-13 shows one way to control the basic flasher circuit by means of an external logic signal. The LED in an optocoupler is connected to the output of a TTL logic gate. When the gate output is low, the LED receives current and illuminates the phototransistor in the optoisolator. This permits the flasher circuit to function. When the output of the logic gate is high, the LED is extinguished, the phototransistor is dark, and the flasher circuit does not operate. The basic technique shown here can be applied to the other circuits in this column.

DC-DC Upconverter

A dc-dc upconverter can be made by replacing the LED of the previous circuits with the 8-ohm winding of an 8:1000-ohm miniature output transformer. For each current pulse through the winding, a high-voltage pulse is induced in the secondary winding. This voltage can be rectified and stored in a capacitor or used to drive low-impedance loads such as a piezoelectric bimorph tactile stimulator.

Figure 1-14 shows how this kind of circuit can flash a neon lamp. Though I used a Radio Shack 273-1380 output transformer, other transformers having a similar turns ratio should also work. D1 rectifies the high-voltage pulses from T1 and C2 accumulates the voltage. C2 discharges through the neon lamp when the charge stored in C2 reaches the lamp's firing voltage. The charge-discharge cycle then repeats.

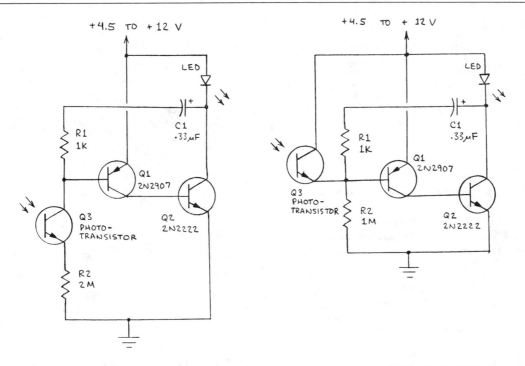

(A) Light-activated flasher.

(B) Dark-activated flasher.

Fig. 1-12. Light and dark activated flashers.

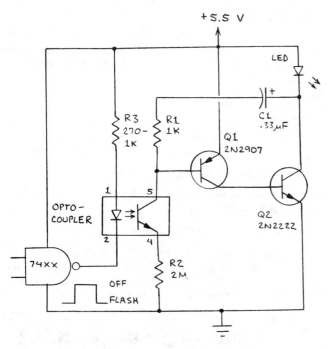

Fig. 1-13. TTL gated LED flasher.

The circuit in Fig. 1-14 can produce surprisingly high output voltages at relatively low current drains. The results I measured for the prototype circuit are given in Table 1-1.

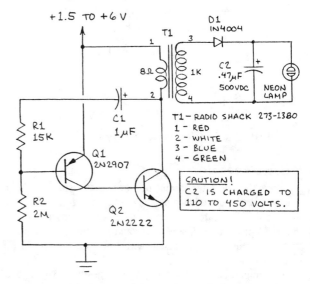

Fig. 1-14. A dc-dc upconverter.

Note in Table 1-1 how the output from the circuit saturates when the supply voltage exceeds 4.5 volts. Be sure C2 is rated for the expected voltage level.

Relay Controller

Adding a relay to the basic oscillator circuit permits it to supply a continuous stream of current pulses to a high power

Table 1-1. DC-DC Upconverter Readings

Supply Voltage	Current Drain	Output Voltage
1.5 V	0.6 mA	110 V
3.0 V	1.6 mA	320 V
4.5 V	2.2 mA	450 V
6.0 V	2.6 mA	450 V

device. Figure 1-15 shows an experimental circuit that does just that. Note that the supply should range from 5 to 6 volts for consistent results. Potentiometer R1 and C1 control both the time interval the relay is closed per pulse and the rate the pulses are applied. Potentiometer R2 controls only the pulse rate. Switching cycles of a few seconds can be achieved by careful adjustment of the various components.

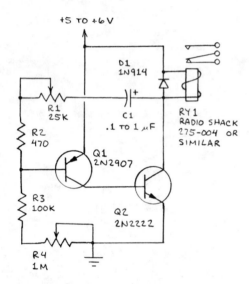

Fig. 1-15. Ultrasimple relay controller.

Pulse Generator

Since the oscillator circuit produces fast rising and falling pulses, it is well suited as a pulse generator. Figure 1-16 shows how a pulse generator can be implemented by inserting a 50-ohm resistor between Q2's emitter and ground. Table 1-2 summarizes operation of the circuit with the values shown in Fig. 1-16 when the supply was 12.5 volts.

The amplitude of the output pulse ranges from 10 volts (C1 = .001 μF) to 11 volts (C1 = 0.1 μF). The rise time for all

Table 1-2. Operation of Pulse Generator Circuit

C1	Pulse Duration	Maximum Pulse Rate
.001 μF	5 μsec	1500 Hz
.010 μF	22 μsec	225 Hz
.100 μF	200 μsec	23 Hz

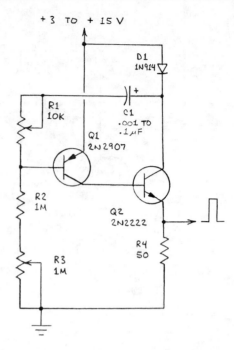

Fig. 1-16. Adjustable pulse generator.

values of C1 is a very fast 10 nanoseconds (measured at 10–90 % points).

Going Further

The circuits shown in Figs. 1-7 through 1-16 merely illustrate the wide range of applications for the basic two-transistor oscillator shown in Fig. 1-6. Here are some other applications you might want to explore:

Continuity Tester

Burglar Alarm

Low-Power Radio Frequency Transmitter

Pulse-Amplitude Modulator

Pulse-Duration Modulator

Monostable Multivibrator

Sound-Effects Generator

Infrared Tone Transmitter/Beacon

Finally, though to my knowledge the basic oscillator circuit isn't available in integrated form, you can easily assemble its components on a 14-pin dual in-line header. Several companies sell such headers along with plastic covers.

Experimenting with MOSFET Power Transistors

In this era of increasingly complex integrated circuits, one of the most important semiconductors is a family of comparatively simple power transistors. Collectively known as power

MOSFETs, these transistors are finding widespread application in power supplies, high current pulse generators, analog gates, and high performance audio and wideband amplifiers. A particularly important use is as high current drivers for microprocessors and other logic circuits designed to control heavy duty loads.

Some of the key advantages of power MOSFETs include ultralow on resistance, ultrahigh input impedance, nanosecond switching times, high power capability, and both linear and analog mode operation. I will also describe practical circuits for a linear MOSFET lamp dimmer, a pulse-modulated lamp dimmer, and a variable rate lamp flasher.

First, we'll examine in some detail the design and operation of a basic common-source MOSFET amplifier. If you prefer to experiment with real circuits rather than predict their performance on paper, be sure to read the entire discussion anyway. You may be pleasantly surprised to find that experimenting with an actual circuit gives better results than designing a circuit on paper.

We'll also experiment with both unidirectional and bidirectional MOSFET analog switches. And we'll conclude with a brief look at a MOSFET high power variable resistor.

Where To Find MOSFET Power Transistors

When I first wrote about MOSFET power transistors, Radio Shack was the only major hobby dealer that carried the new devices. In response to my question why other hobby electronics dealers don't carry power MOSFETs, a representative from a major mail order components supplier informed me there have been few customer requests for such devices. He also observed that prices are beginning to fall and his firm is taking a close look at several MOSFET devices for possible inclusion in an upcoming catalog.

In the meantime, Radio Shack remains one of the few major hobby sources of MOSFET transistors. You can also try industrial distributors who represent MOSFET manufacturers. In addition to Siliconix, major domestic manufacturers of MOSFET

transistors include International Rectifier, Intersil, and Motorola.

Operating Precautions

Figure 1-17 shows the pin outline for two typical MOSFET transistors and lists some of their key specifications. Additional information is provided in the data sheets for both devices.

Though the drain-source channel of a MOSFET transistor may safely handle very high currents and voltages, the gate connection retains the vulnerability of any MOSFET device to electrostatic discharge damage. Some power MOSFETs include a protective zener diode between the gate and the source. The diode protects the input from static electricity, but it can also impair the performance of the device. Therefore, some manufacturers no longer include the protective diode.

To avoid electrostatic discharge damage, you should handle MOSFETs like any other MOS transistor or IC. Be sure to store loose transistors in conductive foam.

If you use power MOSFETs in high power applications, be sure to observe all appropriate temperature and power ratings. In some cases a heatsink may be necessary. See the manufacturer's specifications for detailed information.

A Basic MOSFET Amplifier

Figure 1-18 shows a basic MOSFET common-source amplifier. The amplifier is so named because Q1's source is common to both the input and the output of the circuit. It is therefore the MOSFET counterpart to the bipolar transistor common-emitter amplifier.

In operation, R1 and R2 form a voltage divider that biases Q1's gate to a point where the drain-source voltage (V_{DS}) is half the supply voltage (V_{DD}). The required gate voltage (V_{GS}) can be measured with the help of a test circuit or it can be found by referring to the family of curves that shows the output characteristics for individual power MOSFETs as a function of drain current (I_D) and V_{DS}.

SPECIFICATION	VN10KM	VN67AF
MAXIMUM DRAIN-SOURCE VOLTAGE	60 V	60 V
MAXIMUM DRAIN CURRENT	0.5 A	2.0 A
DRAIN-SOURCE ON RESISTANCE	5 Ω	3.5 Ω
TURN-ON DELAY TIME	2 nsec	2 nsec
RISE TIME	5 nsec	2 nsec

Fig. 1-17. Pin outlines and key specifications for VK10KM and VN67AF.

VN10KM

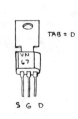

VN67AF

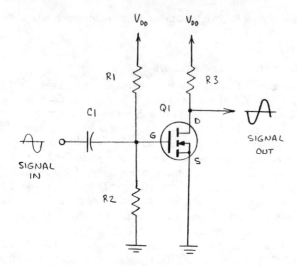

Fig. 1-18. MOSFET common-source amplifier.

Let's assume we wish to use the circuit in Fig. 1-18 as an audio tone amplifier that directly drives a small 8-ohm speaker. If V_{DD} is 9 volts and if the speaker is rated at 2 watts, then the maximum forward current through Q1 and the speaker (I_D) is, from Ohm's law, the power divided by the voltage (2/9) or 222 milliamperes.

Incidentally, knowing I_D and V_{DD}, we can now apply Ohm's law to find the necessary resistance for R3. Discounting the channel resistance of the MOSFET (typically 0.5 to 5 ohms when fully on) to provide a safety margin, it is V_{DD} divided by I_D (9/0.222) or 40.5 ohms.

Now that we know I_D and V_{DS}, we can refer to the manufacturer's output characteristics curves for Q1 to find the required V_{GS}. For Siliconix's VN10KM, V_{GS} is typically about 3.5 volts when I_D is 222 milliamperes and V_{DS} is 4.5 volts. The output characteristics curves are printed in Siliconix's *VMOS Power FETs Design Catalog*.

Knowing the V_{GS} required to bias Q1 so that V_{DS} is half V_{DD} means that values for R1 and R2 can now be selected. Since V_{GS} is 3.5 volts and V_{DD} is 9 volts, in this case V_{GS} is 0.39 V_{DD}. Therefore, the resistance of R2 should be 0.39 (R1 + R2). Assuming we wish to keep the circuit's input resistance high, a reasonable approximation using standard resistance values would be to use 750,000 ohms for R2 and 1,200,000 ohms for R1. This will provide a V_{GS} of 3.46 volts.

We can calculate the voltage gain (A_v of the basic amplifier by multiplying load resistor R3 times Q1's transconductance (g_{fs}). The typical g_{fs} of Siliconix's VN10KM is 200 milliohms. Therefore, A_v is 40.5 × 0.2, or 8.1.

Though a voltage gain of 8.1 may seem very small, the amplifier's power gain can be *considerably* higher. For example, assume the input signal is a 1-volt peak-to-peak sine wave originating from a source having an output impedance of 10,000 ohms. The equivalent input power (P_i) of this ac signal is found by dividing the square of the signal's rms voltage by the source's resistance. The rms value of the signal is 0.3535 times its peak-to-peak amplitude (or 0.707 times the peak amplitude). For the values given previously, P_i is $(1 \times 0.3535)^2/10,000$ or 12.5 microwatts.

The output power (P_o) is found by dividing the square of the rms output voltage by the load resistance (R3). Since the voltage gain (A_v) of the amplifier is 8.1, then the output voltage is 1×8.1-volts peak-to-peak. Therefore, P_o is $(8.1 \times 0.3535)^2/40.5$, or 0.202 watt. The power gain is P_o/P_i, or 16,195.

Incidentally, since the speaker in Fig. 1-19 is directly coupled to the MOSFET transistor, it receives a dc bias even with no input signal. The resultant displacement of the speaker's cone will cause distortion of high-level audio signals. This distortion can be eliminated by inserting a transformer between the circuit and the speaker at R3. It may then be necessary to recalculate the circuit parameters.

For more information about predicting the performance of a common-source power MOSFET amplifier, see *Design of VMOS Circuits* by Robert Stone and Howard Berlin. (Howard W. Sams & Co., 1980). This excellent book provides detailed step-by-step design procedures in Chapter 4. It also contains a wealth of information about various MOSFET circuits. Incidentally, the examples given on pp. 39–40 of this book multiply peak-to-peak signal values by 0.707 instead of 0.3535. Although this gives incorrect values for P_i and P_o, it does not affect the example calculation of power gain.

A Real MOSFET Amplifier

If you enjoy working with numbers, the preceding discussion probably makes the design of a common-source MOSFET amplifier seem relatively straightforward. Fortunately for those of us who also enjoy experimenting with real circuits, the mathematical approach has a serious drawback since the predictions are based upon "typical" values of transconductance (g_{fs}) and gate voltage (V_{GS}).

Since the voltage gain (A_v) of the amplifier is the product of the load resistance (R3) and g_{fs}, we can rearrange the formula to solve for g_{fs} ($g_{fs} = A_v/R3$). A_v can be found by measuring the voltage at the input and output and dividing the latter by the former. Now g_{fs} can be easily determined for individual MOSFETs under specific operating conditions. Incidentally, g_{fs} is sometimes designated g_m.

V_{GS} can be found by injecting a sine wave into the amplifier while watching the waveforms at the input and output of the amplifier on the screen of a dual trace oscilloscope. The voltage divider network (R1 and R2) should be trimmed until the output waveform is a maximum amplitude, undistorted version of the input waveform.

Figure 1-19 shows a practical version of the amplifiers that works quite well. Note the addition of R4 to permit quick adjustment of V_{GS}. The data sheet specifies for the VN10KM used in the circuit minimum and typical values of g_{fs} (or g_m) of, respectively, 100 and 200 milliohms. The "typical" value gives a predicted A_v of 4.4 ($A_v = R3g_{fs}$).

I measured an A_v of only 3.0 when the speaker was shorted to leave a load resistance of 22 ohms. This corresponds to a g_{fs} of 140 milliohms. As you can see, using the *typical* data sheet value can be misleading.

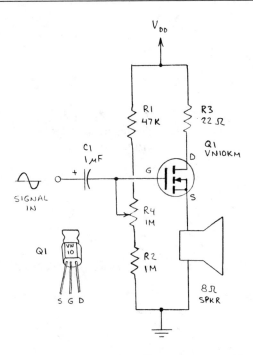

Fig. 1-19. Practical MOSFET amplifier.

Applications for the MOSFET Amplifier

The simple circuit in Fig 1-19 is well suited for use as a high input impedance small speaker driver. The input impedance can be increased by increasing R1 and R2 in the proper proportion to permit R4 to determine V_{GS}.

In audio applications a scope is not always necessary to adjust R4. Simply feed a tone or a voice signal into the input and listen to the speaker while adjusting R4 for maximum undistorted volume.

The circuit in Fig. 1-19 also makes an excellent LED driver for an amplitude modulated lightwave communications transmitter. Simply replace the speaker with an LED and increase R3's resistance to limit the current through the LED to a safe value. For maximum optical power output, select an AlGaAs of GaAs:Si near-infrared emitting diode.

To operate the circuit as an LED audio transmitter, connect a signal source or microphone preamplifier to the input. Then adjust R4 for best reception while monitoring the transmitted signal with a lightwave receiver. Use one of the circuits described in Section Five of this book or simply connect a photodiode or solar cell to the input of a transistor or IC amplifier.

The circuit in Fig. 1-19 works very well at higher frequencies. With the values shown and with the speaker removed, the frequency response is virtually flat to beyond a megahertz, the limit of my Heath function generator. When R4 is properly adjusted, the circuit faithfully reproduced 1-MHz sine and triangle waves. When a fast risetime (50 nanoseconds) 1-MHz square wave is fed into the amplifier, the output waveform experiences a delay of only 5 nanoseconds. The ringing that occurs at the leading and trailing edges of the output signal can be minimized by careful adjustment of R_{GS} and, at very high frequencies, careful, point-to-point wiring.

Finally, note that R4 will have to be readjusted if V_{DD} is changed. This can be troublesome if an unregulated battery power supply is used.

A MOSFET Unidirectional Gate

A MOSFET can easily be used as a one-way gate for a positive polarity, variable amplitude analog signal. Figure 1-20 shows how such a gate can be turned on or off by a CMOS gate. Any gate signal having sufficient amplitude to turn Q1 on can be used.

Since a power MOSFET can handle currents in excess of an ampere, the basic circuit in Fig. 1-20 is ideal for many different applications. It may not be well suited, however, for low distortion audio applications. Furthermore, if a waveform having both positive and negative components is applied to the gate in Fig.1-20, as much as half the signal will pass through the gate even when it is turned off.

A MOSFET Bidirectional Analog Gate

Siliconix's Application Note AN72-2 (Walt Heinzer, "VMOS— A Solution to High Speed, High Current, Low Resistance Analog Switches") describes a bidirectional MOSFET gate made from two VN88AF MOSFETs and a DG300 dual analog switch. Figure 1-21 shows a modified version of the Siliconix circuit that I assembled with two VN10KM's and a CMOS 4066 analog gate.

Fig. 1-20. A basic unidirectional MOSFET analog switch.

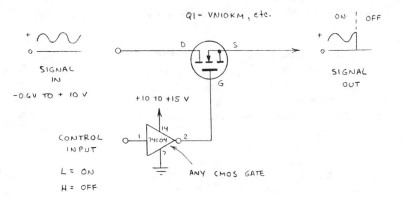

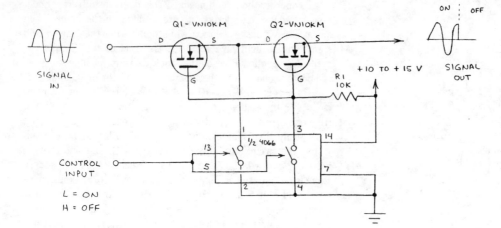

Fig. 1-21. A simple
bidirectional MOSFET
analog switch.

This circuit transmits ac analog signals at frequencies up to and exceeding a megahertz.

The circuit in Fig. 1-21 provides excellent input-output isolation in the off state when the output load is a low resistance (a few hundred ohms). When the output load is 10,000 ohms, about 4 percent (−30 dB) of the input signal appears at the output when the gate is off.

A MOSFET Variable Resistor

The drain-source channel of a MOSFET transistor can be considered a variable resistor when the drain-source voltage is about 3 volts. According to *Design of VMOS Circuits*, the book cited earlier, in this mode a power MOSFET "... exhibits a fairly linear inverse relationship between drain-source resistance and gate-source voltage. For the 2N6656, for example, its gate-source resistance can vary from about 2 ohms ($V_{GS} = 10$ volts) to essentially infinity ($V_{GS} = 0$)." (p. 90).

Figure 1-22 shows how a MOSFET power transistor can be used as a variable resistor having a much higher power rating than some miniature trimmer resistors. This circuit suggests many interesting applications, particularly since R1 can be replaced by temperature or light sensitive resistors.

A MOSFET Linear Lamp Dimmer

Figure 1-23 shows an ultrasimple lamp dimmer designed around a Siliconix VN67 power MOSFET. R2 controls the voltage on the gate of the VN67. This permits the VMOS FET to be operated from full off to full on, thus providing a linear light dimmer.

A Power MOSFET-Modulated Light Dimmer

The circuit in Fig. 1-23 works well, but it is inefficient since the MOSFET transistor dissipates considerable power even when the lamp is dimmed. An alternative approach is to drive the lamp with pulses from a simple CMOS oscillator using an arrangement such as the one in Fig. 1-24.

In operation, the oscillator switches the VN67 full on and full off. When the switching rate exceeds several tens of hertz, the lamp appears continuously on to the human eye. By altering the pulse rate, the lamp may be dimmed or brightened.

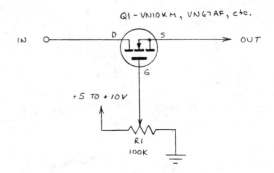

Fig. 1-22. A simple MOSFET high current
potentiometer.

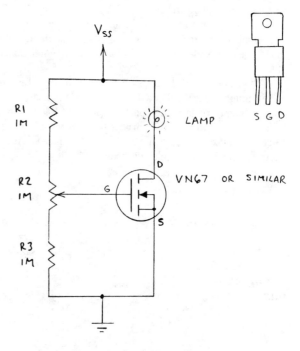

Fig. 1-23. Straightforward linear
dimmer circuit.

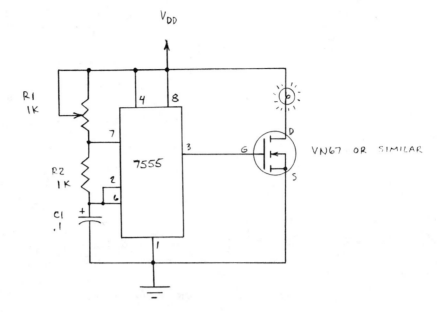

Fig. 1-24. MOSFET light dimmer circuit.

Other Applications for Power MOSFETs

The two circuits briefly presented here are merely representative of what can be done with power MOSFET transistors. Many other applications are possible. If you wish to experiment with the circuit shown in Fig. 1-24, try increasing C1 to a few tens of microfarads. This will convert the circuit to a very efficient lamp flasher. In any case, V_{DD} should not exceed the lamp's rated voltage unless the current pulses applied to the lamp are very brief. Figure 1-25 shows another way to make a MOSFET flasher.

A Multifunction MOSFET Oscillator

In this era of increasingly complex integrated circuits, it's easy to overlook the versatility offered by some very simple transistor circuits. Figure 1-26, for example, shows a multivibrator made with a pair of MOSFETs which has a wide range of useful applications.

To understand how this circuit operates, assume that initially C2 is discharged and C1 is charged to V_{DD}. Therefore, Q2 is off and Q1 is on. As C1 charges, the voltage at Q2's gate eventually rises to a point where Q2 is turned on. This causes C2 to be discharged through the path formed by Q2 and R3 until Q1 is turned off. The charge-discharge cycle then repeats, and the two transistors are alternately switched on and off.

The basic circuit has an oscillation frequency of approximately the reciprocal of 3.6 RC where R = R3 = R4 and C = C1 = C2. Under these conditions, the on and off times for Q1 and Q2 are equal. It's easy to produce nonsymmetrical operation where one transistor is on or off longer than the other simply by altering the RC time constant of one or both halves of the circuit.

Practical Applications

The circuit in Fig. 1-26 has two important advantages. First, the power MOSFET transistors are capable of driving directly such current demanding loads as incandescent lamps. Second, the almost infinite gate resistance of MOSFETs makes possible cycle times much longer than those obtained when bipolar transistors are used.

Figure 1-27 shows the most obvious application for a MOSFET multivibrator, a dual LED flasher. Here R1 and R2 limit current through the LEDs to a safe level. Note the inclusion of Q3 to provide an enable input that can be controlled by a TTL or CMOS signal. When Q3's gate is high, the oscillator operates. Otherwise the oscillator is disabled. Q3 can be omitted or replaced by an spst on-off switch if an enable input is not needed.

Another interesting addition to the circuit is potentiometer R5. Reducing its resistance *increases* the circuit's flash rate. If the resistance of R5 is reduced below about 1 kΩ, the circuit will cease oscillation. Therefore you may wish to insert a 1.5-kΩ fixed resistor in series with R5. Should the circuit cease to oscillate and then fail to restart when power is interrupted, it can be restarted by momentarily shorting one or both timing capacitors.

The circuit in Fig. 1-27 can be easily modified for different flash rates and nonsymmetrical operation by changing the values of the RC components. A particularly interesting application is to replace R3 and R4 with thermistors or cadmium sulfide photocells. The circuit can then be used to visually monitor temperature or light level differences in two locations.

For example, say you wish to match the temperature of two solutions of darkroom chemicals but you have misplaced your thermometer. First immerse the thermistors in each of the two solutions. If the temperatures are identical, the LEDs will flash on and off at equal time intervals. If, however, the temperatures are different, the flash rate will be uneven. Simply add ice to the warmer solution until the flash rate is even.

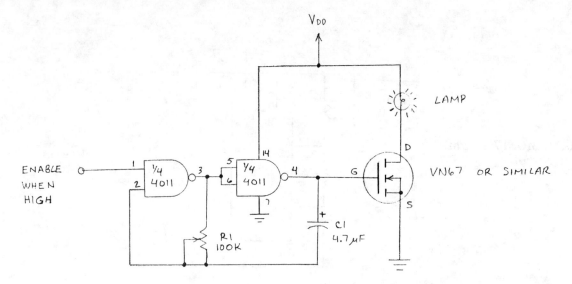

Fig. 1-25. Pulse-modulated light flasher circuit.

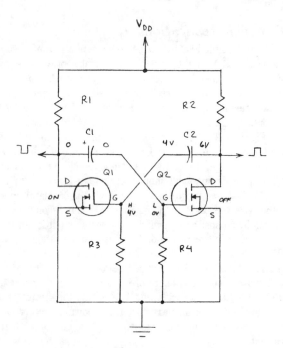

Fig. 1-26. Basic MOSFET multivibrator.

For this application to be successful, you should use glass bead thermistors. They are fragile, but they can be immersed. You will also have to devise some flexible leads to connect the thermistors to the circuit. Be sure to completely insulate the connection between the thermistors and the leads as moisture may cause erroneous results.

The circuit in Fig. 1-27 can also be used as a tone generator. Piezoelectric speakers can be connected directly across the LEDs. Or standard 8-ohm miniature speakers can be substituted for

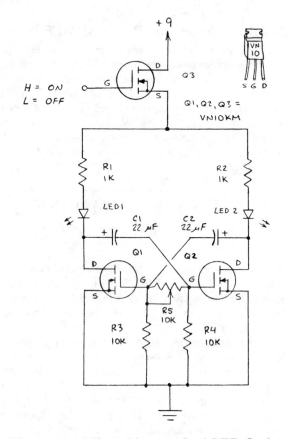

Fig. 1-27. Adjustable rate dual LED flasher with enable input.

the LEDs. You will need to reduce the resistances of R1 and R2 to about 100 ohms. And you can use one or two speakers depending upon your application.

If the device(s) you wish to drive affect adversely the operation of the circuit, you can always use additional MOSFET transistors as buffers. Simply connect their gate leads to the drain connection of Q1 or both Q1 and Q2.

Finally, to provide one cycle of operation, insert a capacitor between R4 and ground. Add a series connected normally open push button and 1.5-kΩ resistor across the capacitor. When the switch is closed, the circuit will operate. Release the switch, and it will cease operation after one cycle. Use any capacitance from 0.01 to 0.1 μF for the capacitor.

An Ultrasimple Power MOSFET Timer

One of the challenges of good circuit design is to accomplish the task at hand with as few components as possible. Power MOSFETs are often ideally suited for simplifying circuits since they are much easier to use than conventional bipolar transistors.

Istvan Mohos of Bergenfield, NJ has designed an exceptionally simple MOSFET circuit which has several practical applications. Istvan writes "A friend asked me to build a timer into his transistor radio. He kept falling asleep with the radio on and draining the 9-volt battery overnight."

Istvan solved his friend's problem with the simple circuit shown in Fig. 1-28. He explains "Pressing the miniature pushbutton switch charges the 1.5-μF tantalum capacitor to the supply voltage. The capacitor supplies hole charges to the gate of Q1, turning it fully on."

When Q1 is on, the 9-volt battery is connected to the radio through Q1. The timing cycle begins as C1 is slowly discharged through reverse-biased diode D1. Why a diode when a very high resistance resistor would work as well? Istvan explains he used the diode ". . . in place of a 200- to 300-megohm resistor I wasn't going to find anyway."

Eventually, the charge on C1 is too small to keep Q1 fully on. As the charge continues to leak through D1, Q1 is gradually turned fully off. The timing cycle can be reinitiated by again pressing S1.

Istvan tried various values for C1. "1.5 μF gave a timing period of 70 minutes," he writes. "During the last few minutes the FET operates in its linear region, slowly pinches off the current to the radio, and provides a built-in fade. Both the VN10KM and VN67AF MOSFETs worked well, but if the radio has a large supply bypass capacitor, the higher power VN67AF is the one to use to avoid troubles with excessive capacitive loading. My friend says he has never slept better."

I've breadboarded Istvan's circuit using various power MOSFETs and can report it works just as he describes it. The circuit is so small it's easy to see how Istvan was able to install it inside his friend's radio.

While experimenting with the circuit, several modifications came to mind. The most obvious is to increase the drive ca-

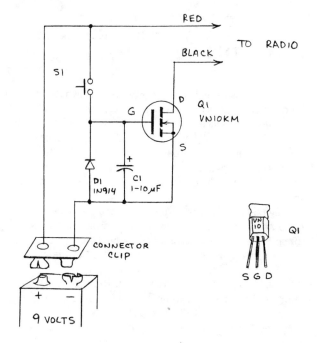

Fig. 1-28. Ultrasimple automatic radio turn off timer.

pability of the MOSFET by inserting a small relay in place of the transistor radio. This would permit the construction of an ultrasimple automatic shut-off switch for lamps and appliances. Connect the circuit to your car's headlights, for example, and they will provide light during the several minutes or so it takes you to get to your front door. They will then turn off automatically.

The circuit can control low-wattage, low-voltage lamps directly. An interesting aspect of direct control is the gradual dimming of the lamp shortly before it is extinguished. In this role it is important to observe the maximum power ratings for Q1.

If you experiment with Istvan's circuit, you may wish to try leaving out D1 altogether. The natural leakage of the capacitor will still provide timing operation. Depending upon such factors as humidity and the type of board the components are mounted on, the use of the diode may provide a more consistent timing cycle duration.

Be sure to experiment with various kinds of capacitors, also. Though tantalums are well suited for this application, even ceramic and other higher leakage capacitors will provide surprisingly long timing periods.

Finally, for precision timing purposes, don't overlook timing chips and binary divider chains designed specifically for that purpose. Coupled with crystal-controlled oscillators, these circuits can provide exceptionally precise timing cycles of up to weeks, months and even years. Of course, they cost more money, use lots of parts, and won't fit inside a transistor radio like Istvan's simple circuit.

A Digitally Programmable MOSFET Variable Resistor

Microprocessors and computers can be readily interfaced to power switching devices like SCRs, triacs, and MOSFETs. Although all these devices can be used for on-off switching, only the MOSFET can also be used in a variable resistor mode. This can be accomplished by setting the voltage V_{GS} at the gate of the transistor to points within the region which controls in a nearly linear fashion the resistance of the transistor's drain source channel.

A digital-to-analog converter provides a convenient means for allowing a computer to generate a variable voltage. Figure 1-29 shows a very simple, low-cost, 4-bit D/A converter that applies under digital control a variable voltage to the gate of a MOSFET.

The D/A converter is made from an R–2R resistor ladder network and a series of two op amps. Of course a single chip D/A converter such as the DAC801 can be used to provide higher resolution (8 bits or 256 voltage levels) and better accuracy.

How It Works

In operation, the D/A conversion is accomplished by the resistor ladder network. When all inputs are low, the network outputs 0 volts. When all inputs are high, it outputs nearly $+V$.

Intermediate binary inputs provide directly proportional output voltages.

The first 741 buffers, inverts, and gives dual polarity to the output from the ladder network. The second 741 provides a means for adjusting the baseline of the output voltage to ground, or above or below ground. This is achieved by adjusting R11. R11 therefore permits the output voltage applied to the gate of Q1 to be set to any desired point.

Testing the Circuit

Interfacing the circuit in Fig. 1-29 to a computer's data bus is best accomplished by interposing buffers between the data bus and the D/A converter. If you wish to test the circuit *without* a computer, the nibble generator circuit in Fig. 1-30 provides a convenient source of a stepped, automatically recycled binary count (0000 to 1111 and repeat). The count rate is controlled by R1. C1 can be increased to a few tens of microfarads for much slower count rates.

When the circuit is connected to a data bus or nibble generator that provides a repetitive series of ascending binary counts, you can test Q1's operation by connecting a small lamp (rated at V) at R_L. When R11 is properly adjusted, the lamp will respond to an ascending count by gradually brightening. It will then suddenly turn off and again begin to brighten as the cycle repeats.

Be sure to connect a voltmeter to pin 6 of the second 741 while performing the lamp test. You can then monitor V_{GS} while adjusting R11. For best results slow the count rate when you are making voltage readings.

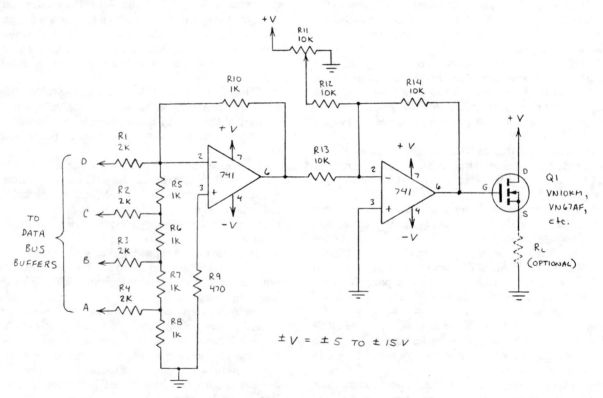

Fig. 1-29. Digitally programmable MOSFET variable resistor.

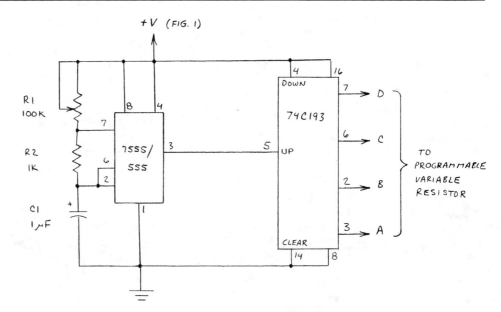

Fig. 1-30. Nibble generator for testing programmable variable resistor.

Going Further

This circuit has many interesting applications. For example, Q1 can be used as the frequency control resistor for a 555 tone generator circuit. Use the basic 555 clock circuit in Fig. 1-30 but omit R1 and connect Q1's drain to +V. Connect its source through a 50-kΩ trimmer resistor to pin 7 of the 555. Reduce C1's value to 0.01 μF. Connect a small speaker through a 100-ohm resistor from pin 3 of the 555 to +V. Adjusting R11, the tone generator trimmer and R1 of the clock will provide a repetitive series of stepped frequency tones.

TWO

Analog Circuits

- Jellybean Op Amps
- Experimenting with the Analog Comparator
- A Single-Chip Analog Building Block
- A Dual-Polarity Five-Volt Power Supply
- Experimenting with Low Power Integrated Circuits
- Event Failure Alarm
- An LM3905 Ap Note
- A Tunable Notch Filter
- Two 60-Hz Hum Filters
- An Easily Adjusted 60-Hz Hum Filter
- An Easy-to-Use Universal Active Filter
- A Fully Adjustable Pulse Generator
- Power Pulse Generator
- A Programmable Function Generator
- A Sound Effects Generator
- Bomb Burst Synthesizer
- An ICM7209 Ap Note

Analog Circuits

Jellybean Op Amps

The semiconductor industry makes many different kinds of operational amplifiers. To companies which specialize in high performance or precision op amps, however, mass-produced, inexpensive op amps are lumped into a catchall category and called *jellybeans*.

One of the most popular jellybean op amps is the 741/741C. This chip was introduced by Fairchild in 1968 as a successor to Fairchild's μA709, the industry's first widely accepted op amp on-a-chip. The μA741 was also intended to compete with National's new entry in the op-amp market, the LM301. Unlike either of these predecessors, the μA741 incorporated an on-chip compensation capacitor.

When the μA709, LM301 and μA741 were first introduced, they were considered major breakthroughs in semiconductor technology. They quickly became very popular with design engineers who found they could replace a handful of discrete transistors with a single chip.

Fairchild's 1965 data sheet for its μA709C, the commercial version of the 709, was entitled "High Performance Operational Amplifier." In those days, "high performance" meant a minimum input resistance of 50,000 ohms, a minimum voltage gain of 15,000, and a typical input offset voltage (V_{os}) of 2 millivolts.

The μA741C offered even better performance. The minimum input resistance rose to 300,000 ohms (2,000,000 ohms typical), the minimum voltage gain rose to 50,000 (200,000 typical), and the typical V_{os} fell to 1 millivolt. Like the A709C, the μA741C was designated a "high performance" operational amplifier.

Today both the 709 and 741 as well as other early op amps are still in widespread use. They're cheap, readily available, and literally hundreds of proved circuits are available.

Operational amplifier technology, however, has not stood still since the 741 was introduced thirteen years ago. There now exist a wide range of op amps with specifications far superior to those of the old standbys.

These new devices are called precision or high performance op amps to distinguish them from the 709, 741, 301 and other general purpose op amps, all of which are now considered jellybeans.

Precision op amps have much faster frequency responses than earlier chips. FET inputs can provide input resistance as high as 10^{12} ohms. And V_{os} can be as low as tens of microvolts.

Not every op amp offers a complete range of such exceptional performance ratings, but today's op amps are far more worthy of being termed high performance devices than were their predecessors.

Experimenters have avoided high performance, precision op amps because of their higher cost. Although jellybean chips like the 709 and 741 can be purchased for as little as 35 cents each, precision op amps have sold for as much as $10 or more.

In recent years, the superior characteristics of precision and high performance op amps have stimulated increased demand. Consequently, more manufacturers now make such chips and prices for better grade units have fallen.

When is it worthwhile to select a more costly precision op amp instead of one of the jellybean varieties? In the audio field alone, precision op amps can provide higher bandwidth, lower noise, more noise rejection, and better sensitivity than jellybean chips. All these advantages also apply to instrumentation amplifiers.

Probably the best way to appreciate the advantages of precision op amps over the jellybean variety is to see how they are used in actual circuits. Precision Monolithic Industries (PMI), a leading manufacturer of precision analog integrated circuits, has published a number of application notes for its line of precision op amps. AN-13, by Donn Soderquist and George Erdi, provides a detailed treatment of PMI's OP-07, a bipolar op amp which has an input offset voltage of only 25 microvolts. This ultralow V_{os}, which eliminates the nulling potentiometer required by economy grade jellybean chips, is achieved during manufacture by a one-time computer controlled adjustment of an on-chip trimming network.

Figure 2-1 is a high stability 10-volt reference which is described in AN-13. Though a 741 could be used in this circuit, relatively frequent recalibration of an offset trimmer potentiometer would be required because of the long term drift of the V_{os}. Long term drift of the OP-07 is only about 1 microvolt per month, about 1/100th the value of the 741.

Figure 2-2, which is also from AN-13, is a precision large signal voltage buffer with a worst-case accuracy of 0.005 percent. This high degree of accuracy is due to the ultralow V_{os} of the OP-07 and the total absence of external components. See AN-13 for more details.

You can find out more about precision and high performance op amps by contacting their manufacturers. Some of the leading manufacturers of precision op amps are Advanced Micro Devices, Fairchild, Harris Corporation, Intersil, Motorola, National

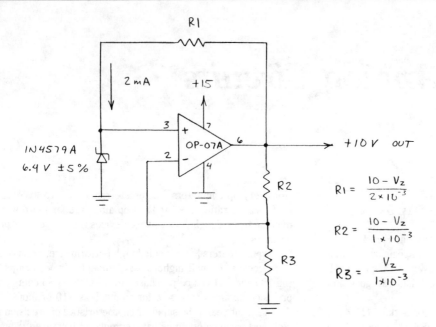

$$R_1 = \frac{10 - V_z}{2 \times 10^{-3}}$$

$$R_2 = \frac{10 - V_z}{1 \times 10^{-3}}$$

$$R_3 = \frac{V_z}{1 \times 10^{-3}}$$

Fig. 2-1. Precision 10-volt reference suggested by PMI.

Semiconductor, Precision Monolithic Industries, RCA, Raytheon, Signetics, and Texas Instruments. You will find the addresses of these and other companies mentioned in this book in the Appendix.

Several excellent books on operational amplifiers are also available, and Walter Jung has written two of them. *IC Op-Amp Cookbook* (Howard W. Sams & Co., 1975) covers everything you need to know about op amps in 591 pages. *Audio IC Op-Amp Applications* (Howard W. Sams & Co., 1987) is limited to audio uses for op amps.

The best organized book on all aspects of op amps is David Stout's *Handbook of Operational Amplifier Circuit Design* (McGraw-Hill, 1976). This book is more expensive than the Sams volumes. If it's beyond your budget, you can probably find it at a good technical library.

Experimenting with the Analog Comparator

In this digital age, analog (or linear) electronic circuits are sometimes considered obsolete. Of course nothing could be far-

ther from the truth. Indeed, analog circuits can perform many tasks for which digital circuits are totally unsuited. And, using just a few components, they can perform some tasks that would require highly complex digital circuits like programmable microprocessors.

One of the key analog circuits is the *operational amplifier.* This circuit is a two-input, differential amplifier that uses a feedback resistor from its output to one of its two inputs to control the voltage gain of the circuit. When the feedback resistor is omitted, even a very small input signal will cause the output of the amplifier to swing wildly from ground to the maximum possible positive or negative voltage extreme. When used in this fashion, the operational amplifier is considered an analog *comparator.*

The comparator has an amazing number of applications. Because of its two-state (on-off) mode of operation, many of its applications are digital in nature. In this discussion I'll explain how the comparator works and provide some sample application circuits with which you can experiment.

The Basic Comparator

Many different analog comparator integrated circuits are available commercially. Often, however, you can use a commonly

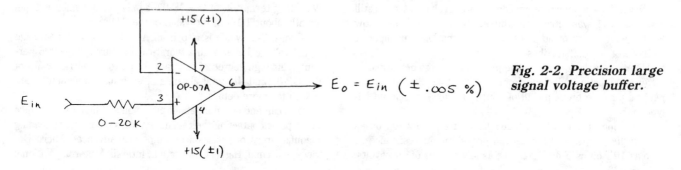

Fig. 2-2. Precision large signal voltage buffer.

available op amp such as the 741 in a comparator mode simply by leaving out the usual feedback resistor. Figure 2-3, for example, shows a basic comparator demonstration circuit made from a 741 and several resistors.

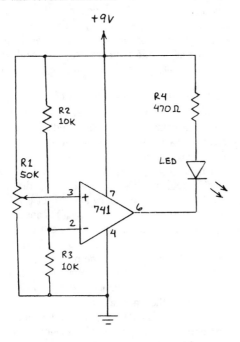

Fig. 2-3. Basic comparator demonstration circuit.

In operation, resistors R2 and R3 form a voltage divider that places half the supply voltage (4.5 volts) at the 741's inverting input. This voltage is called the *reference voltage*. R1 functions as an adjustable voltage divider that delivers a variable voltage to the noninverting input of the 741. This voltage is called the *input*.

When the amplitude of the input voltage is below that of the reference, the output of the 741 comparator is low (near ground). Therefore, the LED is switched on. When the input voltage rises above the reference, the output of the 741 suddenly switches on, rising to near the positive supply voltage. The LED is then extinguished.

If the input voltage is made very close to the switching threshold, the 741 may oscillate in an unstable fashion by rapidly and unpredictably switching on and off. But practically speaking, the comparator output is either full off (ground) or full on (near the positive supply voltage).

Note that the inputs of the comparator are designated inverting (pin 2) and noninverting (pin 3). You can reverse the operation of the circuit in Figure 2-3 simply by reversing the two inputs. Be sure to keep this in mind when you experiment with the following circuits.

Adjustable Light-Dark Detector

The basic circuit in Fig. 2-3 may seem simple, but it can be readily adapted for many applications. Fig. 2-4, for example, shows how to use the basic circuit as an adjustable light-dark detector. This circuit can be used to signal the arrival of dawn (or dusk) and to provide a warning when a refrigerator door has been left open. It can also be used as a simple break-beam object detector. Though the circuit uses a piezoelectric buzzer or alerter, an output relay can be included to control an external motor, lamp or other device.

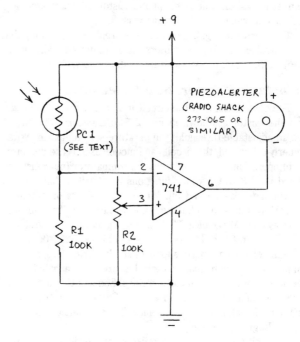

Fig. 2-4. Adjustable threshold light-dark detection circuit.

The circuit's light detector (PC1) is a low cost, but highly sensitive, cadmium sulfide photoresistor. The circuit activates the alerter when the photoresistor is illuminated by even a very low light level. After a simple modification is made, the circuit will trigger the alerter when the photodetector is dark. In either case, the circuit consumes only about 0.5 mA in its standby mode and about 4.5 mA when the alerter is sounding.

Comparing the two circuits, note that the photoresistor in Fig. 2-4 has replaced R2 in Fig. 2-3. Therefore, the photoresistor and R1 in Fig. 2-4 form a light-dependent voltage divider. Potentiometer R2, which forms a second adjustable voltage divider, permits the reference voltage at the noninverting input of the 741 to be alerted.

When the sensitive surface of the photoresistor is illuminated, its resistance is very low, typically a few hundred ohms. Therefore, the voltage appearing at pin 2 of the 741 can approach the supply voltage when the photoresistor is brightly illuminated. The 741 will switch on as soon as the voltage at pin 2 exceeds the reference voltage from R1 which is applied to pin 3. The alerter will then be actuated.

When the light level at the sensitive surface of the photoresistor is decreased, its resistance is increased. Indeed, the resistance may reach a million ohms or more when the light level is very low. When this occurs, the voltage at pin 2 approaches ground. In any case, when the light level falls to a point where the voltage at pin 2 falls below the reference voltage, the comparator will switch off. The trigger point, of course, can be conveniently altered simply by changing the setting of R2.

Incidentally, this operating mode can be reversed simply by exchanging the photoresistor and R1 in Fig. 2-4. The circuit then switches off when the photoresistor is illuminated and switches on when the photoresistor is dark.

The alerter in Fig. 2-4 can be easily replaced by a relay that can control external lamps, motors and other devices. The circuit in Fig. 2-5, which is described next, shows how.

Adjustable Temperature Detector

The photoresistor in the circuit in Fig. 2-4 can be replaced by a thermistor as shown in Fig. 2-5 to transform the circuit into an adjustable-threshold, temperature-sensing alarm. When properly calibrated, the circuit can function as a freeze detector.

In operation, the output from the comparator (pin 6) is connected via R3 to Q1 which functions as a switch that turns a low voltage relay on and off. When the comparator output is high, Q1 switches on and, in turn, allows current to flow through the relay coil. Q1 can be a 2N2222 or any general purpose silicon switching transistor. The relay is Radio Shack's 275-004.

Some electronics parts suppliers stock thermistors, and you can purchase them by mail order if they are not available locally. Check the ads in electronics magazines. Some of the many thermistor manufacturers include Keystone Carbon Company (Thermistor Division), Fenwal Electronics, Thermometrics, Inc., and Omega Engineering, Inc.

Many different kinds of thermistors are available. For best results, select a thermistor having a room temperature resistance of from 25 to 50 kilohms or so. I prefer to use glass bead

thermistors since they are very small and can be safely calibrated in water. But they are more expensive than other types of thermistors.

If the thermistor you select can be calibrated in water, you can easily adjust the circuit to trigger at the freezing point of water simply by inserting the thermistor in crushed ice or snow. You can set other calibration points with the help of a thermometer. Just adjust the temperature of a small cup of water to the desired point, insert the thermistor and calibrate R2.

Sine- to Square-Wave Converter

The sine wave is among the most important waveforms in electronics. The comparator is well-suited for transforming the ubiquitous sine wave into square- and other kinds of waves. As you can see by referring to Fig. 2-6, this manipulation of waveforms can be achieved with the simplest possible comparator circuit. This circuit can also be used to clip that portion of a signal which rises above or below any preset level.

In operation, the sine wave (or signal) is applied to the non-inverting input of the comparator. When the reference voltage applied to the inverting input is ground, the output of the comparator remains at ground until the positive (rising) voltage of the sine wave *exceeds* ground potential. The output then suddenly switches to its maximum positive value and remains there until the voltage of the wave falls to ground potential. The comparator then suddenly switches off. When the voltage falls below ground potential, the output voltage suddenly switches to its maximum negative value where it remains until the waveform voltage again reaches ground potential.

It should be obvious that this operating mode transforms a sine wave into a square wave. What is not obvious, however, is that the amplitude of the square wave at the output can be much higher than that of the sine wave at the input. This occurs when

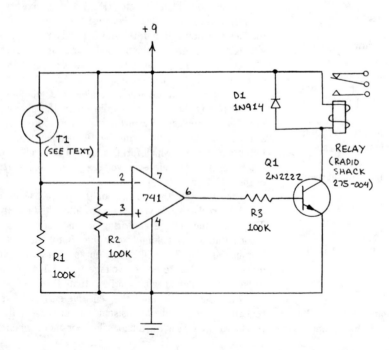

Fig. 2-5. Adjustable threshold temperature controlled relay.

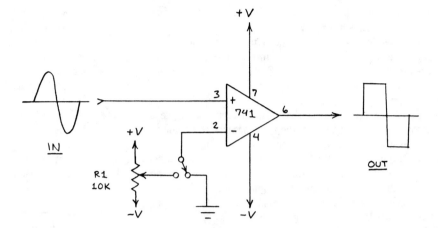

Fig. 2-6. Converting a sine wave into a square wave with a comparator.

the supply voltage exceeds the input voltage by about one volt or more. Therefore, it's important to adjust the supply voltage and possibly the amplitude of the input signal if true clipping of the sine wave is required.

The frequency response of the circuit in Fig. 2-6 depends largely upon the quality of the 741. The 741 I used in a breadboard version of the circuit had a peak response of 42.5 kHz at the −3-dB (half amplitude) points. Other operational amplifiers or comparators can provide a much higher frequency response.

Interesting effects can be had by connecting the noninverting input of the 741 to potentiometer R1 instead of ground. This permits the reference voltage and, consequently, the circuit operation to be altered. When, for instance, the reference voltage is *increased* above ground, the positive half of the output wave narrows and increases in amplitude while the negative half becomes broader and decreases in amplitude. The reverse occurs when the reference voltage is *reduced* below ground.

Potentiometer R1 also permits the shape of the output square wave to be transformed into either a positive or negative triangle wave having a clipped peak. If R1 is adjusted to provide a sharp peak, the comparator becomes unstable and oscillates.

The circuit in Fig. 2-6 will work when powered by a single polarity supply (pin 4 connected to ground instead of −V). However, the comparator will then respond only to the positive side of the incoming signal.

Incidentally, while experimenting with the circuit in Fig. 2-6, I applied a square wave to the input. The output was a trapezoid wave having sloping sides. When the amplitude of the input signal was adjusted to match that of the output, a substantial delay could be observed in the arrival of the maximum positive and negative excursions of the trapezoid.

For instance, when the frequency of the incoming wave was 10 kHz, the duration of both the positive and negative peaks of the incoming square waves was 50 microseconds. The positive peak of the trapezoid trailed the positive leading edge of the square wave by 20 microseconds. The negative peak of the trapezoid trailed the negative leading edge of the square wave by 32 microseconds.

This delay, which also occurs when other waveforms are processed by the circuit, has several possible applications. One is the conversion of a single-phase digital clock for a logic circuit into a two-phase clock.

Peak Detector

Often it's important to measure the maximum amplitude of an event such as rainfall, wind velocity, light level, temperature, revolution rate, and many others. If a transducer is available that converts an event to be measured into a proportional voltage, then the simple comparator circuit in Fig. 2-7 will detect

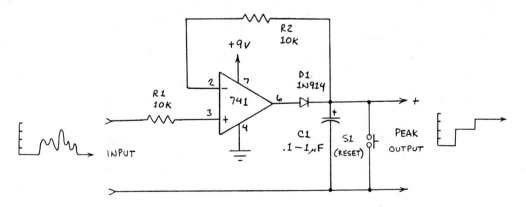

Fig. 2-7. Comparator peak voltage detector.

and store for several minutes the maximum amplitude of the signal from the transducer.

The circuit in Fig. 2-7 is called a peak detector. Digital circuits are available which can perform the same function, but they are far more complex and costly. Furthermore, they require an analog-to-digital conversion stage in order to measure the input voltage.

The peak detector in Fig. 2-7 is placed in operation by pressing S1 to discharge C1 and reset the system. Since this removes any charge stored in C1, the reference voltage coupled back to the inverting input (pin 2) of the 741 via R2 is 0. Any signal voltage applied to the input terminals of the circuit will immediately switch the 741 on since the signal voltage will exceed the reference voltage. C1 will then begin charging to the supply voltage through D1.

When the charge on C1 exceeds the input voltage at pin 3, the 741 immediately switches off and C1 stops charging. At this point, the amplitude of the charge stored in C1 equals the input voltage. If the input voltage rises above the voltage stored in C1, then the comparator will again switch on and C1 will again begin charging until the voltage level exceeds the reference. The comparator will then switch off.

As you can see, the peak detector automatically tracks the input voltage and stores its peak value. At any time a new cycle can be initiated simply by pressing S1 to discharge C1 and reset the system.

Reverse-biased diode D1 prevents C1 from discharging through the comparator. The circuit, however, is not perfect since C1 will not long retain its charge if it is not of high quality

or if a low-impedance voltmeter is used to monitor the level of its charge. The circuit will, however, hold a charge for several minutes or even much longer if a good quality, low-loss Mylar or polystyrene capacitor is used for C1. It's also important to monitor the output voltage with a high-impedance voltmeter to prevent C1 from being inadvertently discharged.

Going Further

The comparator applications discussed here are among the simplest. Many other applications are available, and you can find representative circuits in semiconductor application manuals and books about linear integrated circuits.

A Single-Chip Analog Building Block

An important multifunction analog integrated circuit is manufactured by Precision Monolithics Incorporated. The chip is designated the GAP-01, after *G*eneral-purpose *A*nalog *P*rocessor. It combines on a single chip two differential transconductance amplifiers, two low-glitch current mode analog switches, an output voltage buffer amplifier, and a precision comparator.

What distinguishes this unique chip from other multicircuit analog ICs, such as quad op amps, is that the outputs form the two differential amplifiers can be internally switched to the noninverting input of the output buffer. This is shown clearly in Fig.

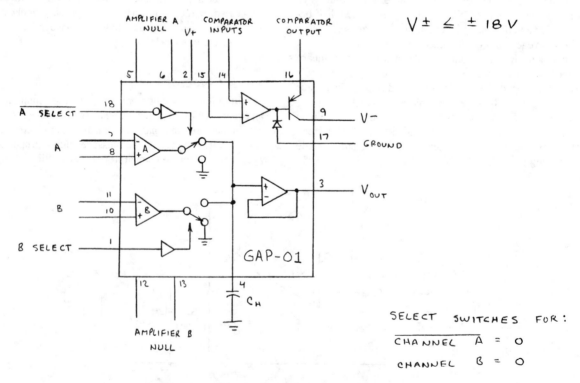

Fig. 2-8. Block diagram of the GAP-01.

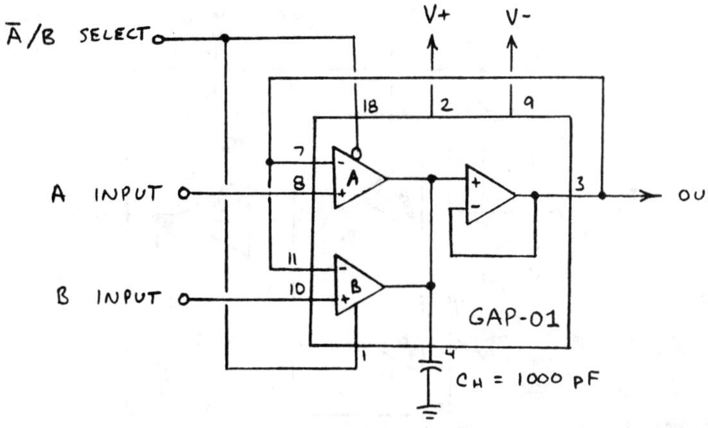

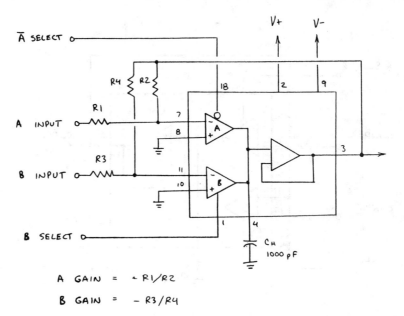

Fig. 2-9. Unity gain two-channel analog multiplexer.

2-8, a functional block diagram of the GAP-01. The external capacitor required for loop compensation (C_H) serves also as a holding capacitor when the GAP-01 is used in sample and hold applications.

The status of the two analog switches between the differential amplifiers and the output buffer is controlled digitally by means of logic signals applied to their respective control inputs (pins 18 and 1). This provides two programmable signal paths through the integrated circuit. The gain of each channel can be controlled by connecting a feedback resistor between the buffer's output (pin 3) and the appropriate amplifier input.

Unity Gain Analog Multiplexer

Figure 2-9 shows how the GAP-01 is connected as a two-channel analog multiplexer having unity gain. A low at the Channel Selector Input selects Channel A whereas a high selects Channel B. Applications for this circuit include such roles as selecting between two analog transducers (e.g., temperature and airspeed) in a remote telemetry system and selecting between two microphones in an audio system.

Incidentally, note that Fig. 2-9 employs a simplified diagram of the GAP-01 wherein each analog switch is pictorially merged with its respective amplifier. This diagram, which is used by PMI in a 12-page GAP-01 specification and application brochure, will be used in the circuits that follow.

Analog Multiplexers with Gain

Adding gain to the circuit in Fig. 2-9 is easily accomplished by inserting input and feedback resistors in one or both channels. For example, Fig. 2-10 shows how both channels are given feedback so they might function as adjustable-gain, inverting amplifiers. The gain of each channel equals the feedback resistance divided by the input resistance.

PMI's GAP-01 specification and application brochure shows two other ways to design an analog multiplexer with gain. The circuit in Fig. 2-11, for example, shows how to achieve both positive and negative gains.

If it is desired that both channels have the same positive voltage gain, the circuit in Fig. 2-12 can be used. Here a single voltage divider simultaneously sets the gain for both channels.

Fig. 2-10. Two channel inverting analog multiplexer with gain.

$$A \text{ GAIN} = -R1/R2$$

$$B \text{ GAIN} = -R3/R4$$

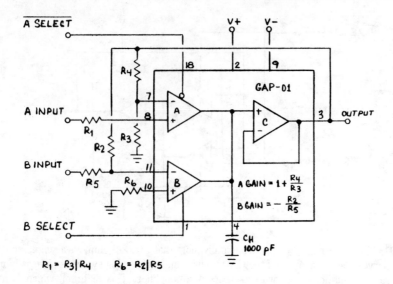

Fig. 2-11. Two-channel multiplexer with positive and negative gain.

A Continuously Switched Analog Multiplexer

In the previous three circuits, the GAP-01 can be considered a 2-line to 1-line analog encoder or multiplexer with individually controlled gain stages for each channel. In each case, the control signals that determine which channel is selected are CMOS and TTL compatible logic levels derived from external logic.

To more fully exploit the many possible applications for the GAP-01 operated as a two-channel multiplexer (or encoder), a fully adjustable switcher is required. The role is admirably filled by half of a 558 quad timer connected as a pulse generator having independently controlled pulse rate and pulse duration.

Figure 2-13 shows a circuit I've tried using a 558 in this free-running switcher mode. In operation, R5 controls the duration of pulses applied simultaneously to the two GAP-01 channel control inputs. R6 controls the pulse repetition rate. The

addition of a 10 kΩ voltage divider as shown in Fig. 2-13 permits the achievement of a constant duty cycle irrespective of the pulse rate.

For preliminary tests, set the gain of both amplifiers at 10 (e.g., R1 and R3 = 1 kΩ; R2 and R4 = 10 kΩ). Connect any desired signal to the inputs and monitor the output with an oscilloscope while adjusting R5 and R6 to vary the switching conditions.

A Dual Output Pulse Generator

The unused half of the 558 in Fig. 2-13 can be used to provide a dual output pulse generator having independently controlled pulse rates. This permits the construction of FSK (frequency-shift keyed) generators, twee-dell sirens, function generators, and tone burst sources using only two chips.

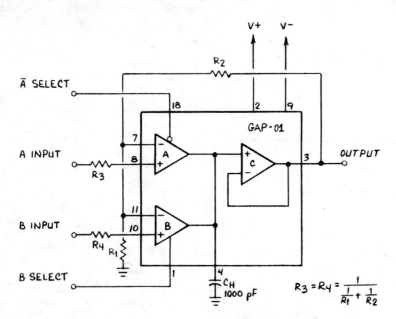

Fig. 2-12. Identical gain per channel two channel analog multiplexer.

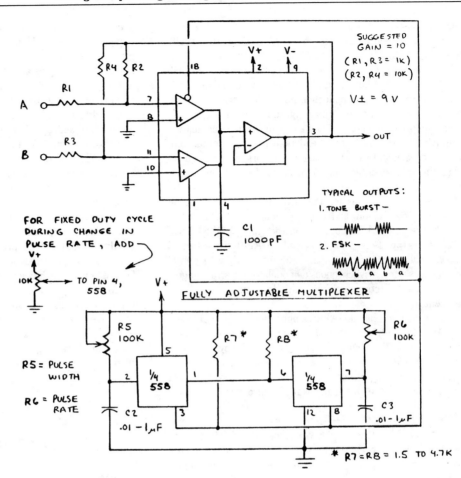

Fig. 2-13. Continuously switched two-channel analog multiplexer.

Figure 2-14 shows the circuit I used to implement these ideas. R9 and R12 control the pulse rates of the two pulse generators.

Incidentally, the 558 circuits in Figs. 2-13 and 2-14 may occasionally cease operating if an attempt is made to operate them beyond their maximum frequency limits. Should this occur, disconnect the power momentarily or ground pin 13 momentarily.

Learning More About the GAP-01

The most unique feature of the various analog multiplexer circuits with which we've experimented is their generic rela-

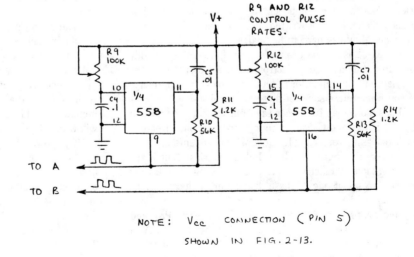

Fig. 2-14. Dual output pulse generator for circuit in Fig. 2-13.

tionship to so-called "active" devices. For example, if a filter with an op-amp gain block is an *active filter*, then the circuits in this column can be considered as *active multiplexers*.

Of course, there are many other applications for the GAP-01, and you can find out more about them by referring to the chip's specifications and application notes. It describes an absolute value circuit with polarity programmable output, a two-channel sample/hold amplifier, and a synchronous demodulation circuit. Also covered are recommendations for the external compensation capacitor, tips minimizing gain errors, information regarding the logic levels applied to the channel select inputs, and other useful topics.

Additional applications for the GAP-01 are covered in an article by David Gillooly and Paul Henneuse, both PMI engineers. The article, which appeared in the April 7, 1981 issue of *Electronics* (pp. 121–129), described a GAP-01 two-channel digital-to-analog converter, a four-quadrant multiplying digital to-analog converter, a synchronous demodulator for strain gages, and a photoelectric control circuit that's immune to the effects of ambient light.

Finally, five additional GAP-01 applications were published in the April 20, 1982 issue of *Electronic Products* (pp. 61–66). Finalists in a GAP-01 contest sponsored jointly by PMI and *Electronic Products* magazine, the circuits include a digital capacitance meter, a high performance digital audio system, and a signal conditioner for a linear position transducer.

Future Analog Building Block

Clearly the GAP-01 is a highly versatile analog building block. Yet the GAP-01 is exceedingly primitive when one contemplates what might be achieved by adding more amplifiers and digital switches. Many kinds of active filters, multiplexers, demodulators, sample/hold, and computational circuits could be implemented, all under digital control. I've begun experimenting with a variety of such circuits using discrete amplifier chips.

Where To Find the GAP-01 and Other New Components

Much of the mail I receive concerns problems readers have experienced in locating components for various projects. For that reason alone, I've not written about several of my favorite circuits since they use hard to find parts.

The GAG-01 is a good example of a chip that's not readily available from the mail order electronics parts dealers that cater to hobbyists and experimenters. Until it is, how do you purchase one if you want to try some of these experiments?

PMI has more than two dozen sales offices scattered around the United States and Canada. Check your telephone directory to see if one is in your city. If not, PMI components are sold by authorized distributors such as Bell Industries, Hall Mark Electronics, Pioneer Electronics, and others. Again, check your phone book to see if one of these distributors is located near you.

If you cannot find a sales office or distributor, write PMI and request a list of sales offices and authorized distributors. You might also want to request a copy of the GAP-01 specification and application brochure. Hint: A neatly typed letter written in a courteous, professional manner will go a long way toward expediting a reply. Unfortunately, some hobbyists and experimenters request far too much information, personal consulting advice, free samples and "anything else you can provide." This is the primary reason some electronics manufacturers do not even answer mail from individuals.

Incidentally, it is essential that your letter include your complete home address. If you don't have access to a typewriter, you should *print* your address as neatly as possible.

You can use this same procedure to find components made by virtually any company. The addresses of the manufacturers of any unusual components are given in the Appendix of this book. This listing will provide a starting point. Sometimes a price is given in this book. I did not give a price for the GAP-01 since by the time you read this, the price may have dropped somewhat. Furthermore, the price provided by PMI was for the chip when purchased in quantities of 100. This means the single chip price will be somewhat higher than the $7.50 posted by PMI.

Most components used in this volume are available from retail electronics parts stores and mail order suppliers like those that advertise in the various electronics magazines. It's wise to get on the mailing lists of as many companies as you can. Often they publish catalogs and flyers that list far more components than their magazine ads.

A Dual-Polarity Five-Volt Power Supply

TTL and low power Schottky TTL integrated circuits require a positive 5-volt power supply. Many op-amp and MOS chips can be powered by a dual-polarity 5-volt supply. And most CMOS circuits can be powered by 5 volts.

All these power supply needs can be met by a very simple supply assembled from a handful of components. Figure 2-15, for example, is a straightforward dual-polarity, 5-volt supply designed around a pair of fixed output, integrated voltage regulators.

The supply's positive output is provided by a 7805 5-volt regulator. This chip outputs from 4.8 to 5.2 volts at a current of up to one ampere. It requires a minimum input voltage of 7.3 volts, and it can withstand a maximum input of 35 volts.

An external heatsink is required for maximum power output. If the area of the heatsink is insufficient to dissipate the heat generated within the IC, an automatic thermal shutdown circuit will turn off the 7805 and prevent it from being damaged.

The power supply's negative output is provided by a 7905 −5-volt regulator. This chip outputs from −4.8 to −5.2 volts at a current of up to 1.5 amperes. It requires a minimum input of about −7 volts, and it can tolerate a maximum input of −35 volts. Like the 7805, the 7905 includes on-chip thermal shutdown circuitry.

Operation of the circuit in Fig. 2-15 is straightforward. Transformer T1 drops the line voltage to 12.6 V ac. It also provides isolation from the ac line. B1, a full-wave rectifier bridge,

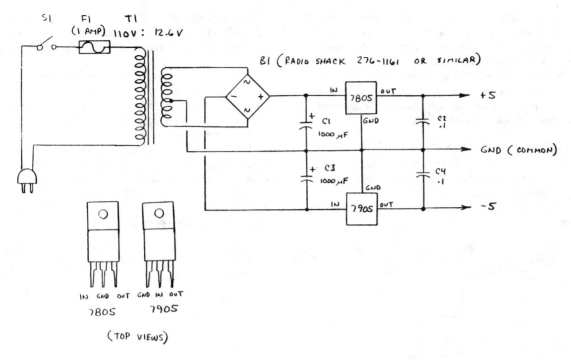

Fig. 2-15. Dual-polarity 5-volt power supply.

converts the low voltage ac from T1 into a series of dc pulsations.

Since the secondary of T1 is center tapped, it can be used to form a ground (0-volt). Both positive and negative dc pulsations can then be obtained by referencing the respective positive and negative outputs of B1 to ground.

The positive and negative excursions from B1 are applied to the inputs of, respectively, the 7805 and 7905 regulator chips. C1 and C3 serve as filter capacitors which smooth out the voltage pulsations from B1 into a reasonably steady dc.

Incidentally, if the connections between C3 and the 7905 regulator chip exceeds several inches in length, a capacitor having a value of at least 25 μF (aluminum electrolytic) or 2.2 μF (solid tantalum) must be connected from close to the chip's input to ground.

The 7805 can tolerate more distance between its input and the positive supply. If considerable separation is involved, connect a 0.22-μF capacitor from close to the chip's input to ground.

Though output capacitor C2 improves the transient response of the 7805, this chip will operate *without* it. The 7905, however, requires output capacitor C4 for operating stability.

Always exercise caution when building line operated power supplies! In particular, the connections between the power cord and the circuit should be secure and well insulated to provide short circuit protection. Likewise, the connections between the power switch (S1), the fuse (F1), and T1 should be well insulated.

For best results, assemble the supply on a perforated or etched circuit board and install it in an enclosure. It is essential to use an insulated strain relief when connecting a power cord to a circuit housed in a metal enclosure. Failure to follow these safety precautions can pose a dangerous or even fatal shock hazard.

Both the 7805 and 7905 are available in 10- and 15-volt versions (7812, 7815, 7912 and 7915). You can use these chips in the circuit in Fig. 2-15 to obtain higher output voltages so long as a transformer capable of supplying the required input voltage is used. Of course, the voltage ratings for B1 and all the capacitors will have to be adjusted upward. The 78XX and 79XX regulators can even be used to make *adjustable* output power supplies.

Experimenting with Low Power Integrated Circuits

Integrated circuits did not replace overnight those made from discrete components. Although it is true some of the delay in turning to ICs was resistance to new technology, early integrated circuits were often slower, more expensive, and required more power than their discrete transistor counterparts.

Today's ICs are far less costly than those sold a decade ago. They also operate at much faster speeds. Recently, considerable attention has been given the development of new ICs that consume much less power than their predecessors.

Though CMOS technology is the best known low power IC technology, other kinds of micropower integrated circuits have also been developed. Some combine conventional bipolar and MOS transistors. Others employ conventional bipolar fabrication in circuits carefully optimized for low power operation.

Though many one-of-a-kind low power linear (analog) and digital chips have been available for some time, several semiconductor companies have concentrated on the development of low power versions of existing chips. The new chips can usually be substituted directly for their power-hungry predecessors.

Low power versions of existing chips often cost more. But they pay their own way with the extension of battery life they provide. Keep this in mind as we look at several circuits before and after a low power IC has been substituted for a functionally similar or even identical conventional IC.

555/7555 Bouncefree Switch

Figure 2-16 shows a straightforward bounceless push button made by connecting a 555 timer or low power equivalent, the 7555, as a monostable multivibrator. Initially, C1 is short-circuited by a transistor in the timer. This forces the output (pin 3) to go low.

I have tried both 555 and 7555 versions of the circuit in Fig. 2-16. The 555 version worked perfectly the first time, but a problem developed when I exchanged the 555 for a 7555. The output remained high indefinitely. Or sometimes it seemed to switch from high to low for no explainable reason.

Since the low power 7555 is supposed to be a direct replacement for the standard 555, I at first could not understand the origin of this problem. Besides, I have used the 7555 in various other circuits with no problem whatever.

Then I recalled the trigger input of the 7555 is a MOS transistor having a very high input (gate) impedance. When such an input is allowed to float, as when the input switch is open, the lead connecting pin 2 to the switch acts as an antenna which couples stray signals directly to the trigger input transistor. This explains the erratic operation of the circuit.

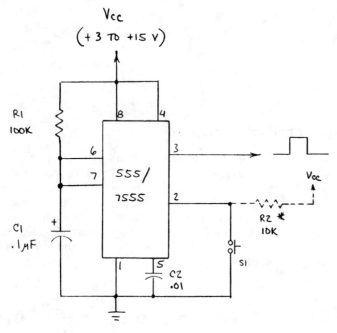

Fig. 2-16. Adjustable output bouncefree switch.

A negative pulse at the trigger input (pin 2) sets an internal flip-flop which turns off the transistor across C1. This causes the output to go high and allows C1 to charge through R1.

When the voltage on C1 reaches $\frac{2}{3}$ V_{cc}, an internal comparator resets the flip-flop, thereby turning on the transistor across C1. This discharges C1 and forces the output at pin 3 low.

The time constant of the circuit is $0.69 \times R1C1$. Therefore, when R1 is 100,000 ohms and C1 is 0.1 microfarad, the positive output pulse at pin 3 will have a duration of about 7 milliseconds.

Since the timing cycle does not begin until the trigger pulse is complete, momentary spikes such as those which accompany the closure of most mechanical switches will have no effect on the circuit's operation. They will merely reset the one-shot.

I solved the problem by inserting a 10-kΩ resistor between the trigger input and V_{cc}. This resistor forces the trigger input high when the switch is not depressed. It can be used with the 555 version, also. But my 555 version did not trigger on stray signals.

As you can see, the 7555 is not necessarily a direct substitute for the 555 in all applications. Its use, however, does not always mean additional components are required.

For example, in the free running astable mode, the 7555 does not require a decoupling capacitor between the control voltage input (pin 5) and ground. Furthermore, the 7555 has a wider operating voltage range (2 to 18 volts) and can operate at a speed of up to 500 kHz. It also requires much smaller trigger, threshold, and reset currents, typically 20 picoamperes.

How do the supply currents differ? The standby current drain of the circuit in Fig. 2-16 is 6.7 milliamperes when a 555 is installed and only 0.16 milliampere when a 7555 is used. When the output is triggered high, the drain of the 555 version falls to 5.5 milliamperes and that of the 7555 version rises to 1.1 milliamperes.

. Obviously, the very low current consumption of the 7555 version makes the circuit in Fig. 2-16 well suited for battery powered operation. The 555 version consumes 42 times more current in the standby mode!

Though the 7555 typically costs twice the price of the 555, I've had very good success using it in a wide range of circuits. I plan to use it in all battery powered circuits which require 555 timer chips.

The 7555 is made by Intersil and Exar. It's available from Jameco, Radio Shack, and other companies that advertise in the various magazines.

567/XR-L567 Tone Decoder

Figure 2-17 shows a basic tone decoder circuit which uses either a 567 or XR-L567, its low power counterpart, to detect a signal having a frequency of about 10 kHz. The L567 achieves low power operation by means of high value resistors. Conventional on-chip resistors cannot have high values without taking up considerable space. The L567 employs ion implantation, a method of selectively disrupting the silicon surface with an ion beam, to produce very small resistors having high resistance.

Current consumption of the L567 circuit in the standby mode is a very low 0.97 milliampere when V_{cc} is 7 volts. When a signal is detected and the LED is glowing, the current drain rises to 6.8 milliamperes.

The standby 567 consumes much more current. During standby the drain is 10.4 milliamperes. This rises to 22.5 milliamperes when the LED is on.

Though the circuit in Fig. 2-17 works well, I've not been able to get the L567 version to respond to a 1 kHz signal. For this experiment, R1 was changed to 10 kΩ, C1 to 0.1 µF, C2 to 2.2 µF, and C3 to 1 µF. Though the 567 version of the circuit responded well (at about 1.1 kHz), the L567 version failed to respond. Many applications for this chip involve very low signal frequencies, so it is therefore imperative that it work as well as its high power counterpart.

The L567 is made by Exar. It costs about 50% more than the 567. Contact an Exar distributor for current pricing and availability.

Low Power Op Amps

Many different low power operational amplifiers are available. RCA's CA3440 BiMOS (bipolar/CMOS), for example, dissipates only 500 nanowatts. RCA also makes a low power version of the 741. Designated the CA3420, the chip operates from as little as ± 1 volt and draws only 350 microamperes. National's LF441 is a low power version of the LM148 which draws only 150 microamperes.

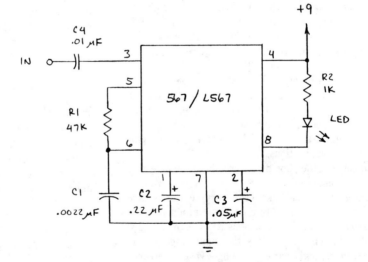

Fig. 2-17. A 10-kHz tone decoder.

I've had mixed results using the L567. The circuit in Fig. 2-17 works fine with either the 567 or L567. No component changes are necessary.

The frequency response of the two versions, however, is slightly different. The 567 version turns on the output LED when the input signal is about 9.13 kHz. The L567 actuates the LED when the input signal is about 9.1 kHz. This means *exact* substitutions may not be possible since trimming of some external components may be required before the L567 version will respond to the same frequency as the 567 version.

One of my favorite low power op amps is the LM108/208/308. This family of op amps will operate from a supply voltage range of from ± 2 to ± 20 volts. The maximum supply current is 300 microamperes.

The high input resistance of this op-amp family (70 megohms for the LM108/208 and 40 megohms for the LM308) allows its use in applications normally filled by FET amplifiers.

Figure 2-18 shows a high performance photodiode amplifier designed around an LM308. Though many different op amps can be used in this circuit, including the popular 741C, the LM308 provides superior gain (up to 300 volts per millivolt), low noise operation (about 300 microvolts at 10 kHz), and reasonably good frequency response (a unity gain frequency of nearly 1 MHz).

The circuit in Fig. 2-18 consumes 230 microamperes when the supply provides 6 volts. When a 741C is substituted for the

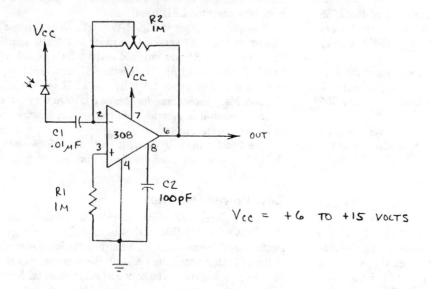

$$V_{cc} = +6 \text{ TO } +15 \text{ VOLTS}$$

all outputs are low. This is only a fifth the current required by the TTL 7400.

TTL LS costs more than standard TTL. But the low power consumption is worth the added price. Battery powered operation becomes a real possibility for many TTL LS circuits.

Fig. 2-18. Op-amp photodiode amplifier.

LM308, the current consumption rises to about 600 microamperes.

The noise level also rises when a 741C is used. Both the frequency response and the gain fall. Obviously the LM308 is superior to the more economical 741C in this and many other applications. The 741C, however, does not require an external frequency compensation capacitor (C2).

Incidentally, C2 can be connected to the LM308 in one of two ways. One enhances the chip's frequency response and the other reduces noise coupled from the power supply. See the LM108/208/308 for specific information.

The circuit in Fig. 2-18 is designed to receive pulsed light-wave signals. For dc applications, such as detecting starlight, remove C1 and connect the photodiode directly to the inverting input (pin 2) of the op amp.

In either mode, the circuit functions as a current-to-voltage converter with gain. The output voltage is the product of the photodiode current and the feedback resistor (R1). When used with an LM308, this circuit provides the input stage for an exceptionally sensitive, high performance lightwave receiver.

TTL vs TTL LS

TTL is well liked for its fast switching speed (10 nanoseconds propagation delay per gate). But TTL is power hungry. The four gates in a 7400 Quad NAND gate package, for example, consume a total of 4 milliamperes when all four outputs are high. The drain rises to 12 milliamperes when all four outputs are low.

Low power Schottky TTL preserves the very fast switching speed of conventional TTL but provides much lower current consumption. The 74LS00, for example, consumes a total of 0.8 milliampere when all outputs are high and 2.4 milliamperes when

CMOS

The advantages of CMOS have been proclaimed many times by this author. Consider, for example, the 74C00, the CMOS equivalent of the 7400/74LS00 described previously. This gate package typically consumes a miniscule 0.01 microampere and a maximum of 15 microamperes. What's more, CMOS ICs can operate over a supply voltage range of from about 3 to 15 volts. CMOS is also characterized by its high noise immunity.

Summing Up

Lower power chips represent one of the most important trends in today's integrated circuit technology. In the coming years, you can expect to see much more emphasis on CMOS and less on TTL and even TTL LS. You can also expect to see more low and micropower versions of many kinds of linear ICs.

Event Failure Alarm

There are many applications for an alarm which, absent some corrective action, sounds a warning a predetermined time after some event has taken place. Automobile seat belt alarms are a common example, but there are many others. Here are several:

1. An open door alarm for refrigerators and freezers which sounds, say, thirty seconds after the door is opened if the door has not first been closed.

2. A check list reminder which sounds an alarm when one or more actions are not taken within a predetermined time frame.

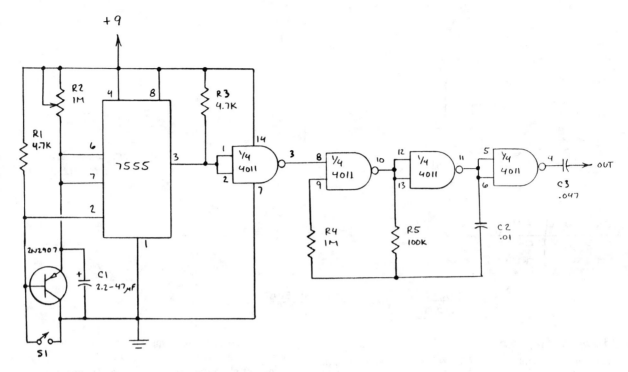

Fig. 2-19. Event failure alarm.

3. A delayed action alarm which ignores momentary faults, or even those lasting up to a minute or so, but which otherwise function normally.

4. A timer or quick-reaction tester for games and toys.

A Practical Event Failure Alarm

Figure 2-19 is the circuit for a straightforward two-chip event failure alarm. The 7555 timer is connected as a missing pulse detector, and the 4011 quad NAND gate serves as a tone generator.

In operation, the 7555 enters a timing cycle when power is applied to the circuit. The duration of the cycle is determined by R2 and C1. The circuit may be reset at any time by closing S1. This turns on Q1 which, in turn, discharges C1. If S1 is *not*

closed prior to the completion of the timing cycle, the 7555 output goes low, thus enabling the tone generator.

Only two of the gates in the 4011 are required for the tone generator. One of the spare gates is used to invert the enable signal from the 7555 output (pin 3). Pullup resistor R3 allows this gate to be interfaced directly with the 7555. The final spare gate in the 4011 provides a buffer between the tone generator oscillator and an external transducer or amplifier.

Though the circuit I prototyped incorporates a 7555, you can use a standard 555 timer if you prefer. The chief advantage of the 7555, which is a low power version of the 555, is its very low power consumption. You may also substitute a fixed resistor for R2 when you arrive at a suitable delay time. Remember that C2 also influences the delay time. Increasing the capacity of C2 increases the delay time.

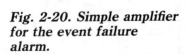

Fig. 2-20. Simple amplifier for the event failure alarm.

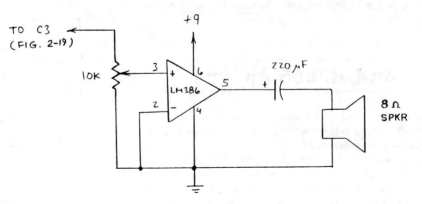

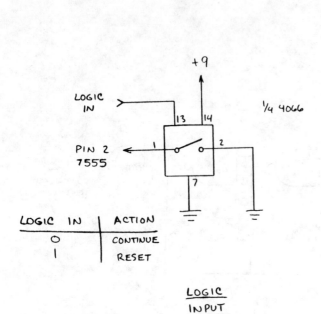

LOGIC IN

¼ 4066

PIN 2
7555

LOGIC IN	ACTION
0	CONTINUE
1	RESET

LOGIC
INPUT

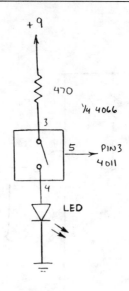

¼ 4066

PIN3
4011

LED

VISUAL
INDICATOR

Fig. 2-21. Adding a logic input to the event failure alarm.

Adding an Amplifier

Though the circuit in Fig. 2-19 will drive a small 8-ohm speaker at low volume, much better results are obtained by first amplifying the tone signal. Figure 2-20 shows a very simple power amplifier designed around a low cost LM386 and little else. Potentiometer R1 controls the input signal level and therefore functions as a volume control.

Adding a Logic Input

Many new applications for the basic circuit in Fig. 2-19 would become available if the alarm could be reset under digital control. A simple way to accomplish this is to replace S1 with one of the analog switches in a 4066 as shown in Fig. 2-21. When the input is low, the analog switch is *open*. When the input is high, the resistance of the switch falls from about 10^9 ohms to a few hundred ohms or less. In this circuit, this resistance is low enough to simulate a mechanical switch.

Adding a Visual Indicator

Figure 2-21 also shows how to add an LED to the circuit. In operation, the LED is normally off. When the delay time is up, the LED glows at the same time the alarm sounds.

An LM3905 Ap Note

Though less well known than its 555 predecessor, the LM3905 is a precision timer with a host of applications. Capable of operating from unregulated supplies of from 4.5 to 40 volts, the LM3905 and its LM122/LM222/LM322/LOM2905 counterparts can provide constant timing periods ranging from microseconds to hours. Variations in the supply voltage alter the timing period by less than 0.005 percent per volt. Therefore

poorly or nonregulated supplies with considerable ripple can be used. The chip typically consumes only 2.5 milliamperes in the quiescent state (i.e., no external load being driven).

Figure 2-22 is a straightforward on-after-delay timer given in National Semiconductor's data sheet on the LM3905. This circuit nicely simulates a thermal time delay relay of the kind often used to apply power to a circuit or device a fixed interval after a main power switch is closed.

Operation of the circuit in Fig. 2-22 is straightforward. When S1 is closed, the LM3905 enters a timing cycle with a duration of R1C1 seconds. After the cycle is completed, the relay is

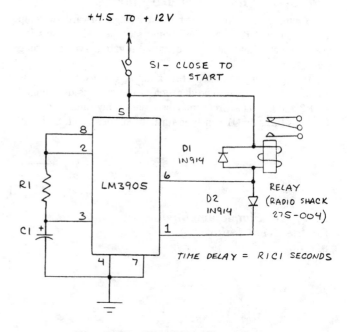

+4.5 TO +12V

S1 - CLOSE TO START

LM3905

D1
1N914

D2
1N914

RELAY
(RADIO SHACK
275-004)

R1

C1

TIME DELAY = R1C1 SECONDS

Fig. 2-22. On-after-delay timer.

actuated. D1 protects the LM3905 from back emf produced by the relay coil when S1 is opened.

For variable delays, use a 1-megohm potentiometer for R1. For very long delays, use large values of capacitance for C1.

Incidentally, the circuit can be easily modified as shown in Fig. 2-23 to operate in the converse manner. In this configuration, which is also given in the LM3905 data sheet, the relay is closed when S1 is closed. After a time interval of R1C1 seconds, the relay is opened.

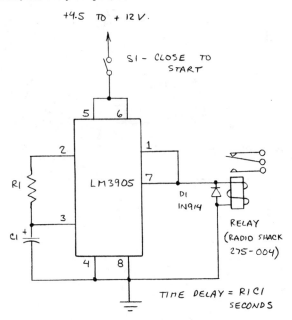

Fig. 2-23. Off-after-delay timer.

For those of you who are skeptical about circuits given in application notes, such as the two discussed here, you can be assured I've breadboarded both these timers. They work just as described.

A Tunable Notch Filter

Sometimes a narrow frequency signal greatly interferes with the operation of an electronic circuit. The best known interfering signal is 60 Hz line noise. Shortwave radio listeners often find a broadcast they are attempting to monitor is partially masked by an annoying tone.

Figure 2-24 shows a passive filter that can be tuned to reject or block a narrow band of audio frequencies. Such a circuit is called a *band-rejection filter*, *band-stop filter*, or *notch filter*. The particular filter in Fig. 2-24 is called a *bridged-differentiator* tunable-notch filter. It is easier to adjust than the better known twin-tee (or twin-T) notch filter.

The filter in Fig. 2-24 uniformly attenuates frequencies outside its band-stop region by about 4.1 dB (i.e., about 48 percent of a non-band-stop frequency is blocked). This loss can be virtually eliminated by following the filter with an op amp to form an *active* notch filter as shown in Fig. 2-25. This circuit has a uniform attenuation outside its stop band of only 0.4 dB (i.e., about a 5 percent attenuation).

I measured this attenuation with the help of a signal generator, oscilloscope, and a breadboard version of the active filter. When potentiometer R1,R2 was adjusted so that R1 = R2 = 50 kΩ, then the notch frequency was 1431 Hz. The signal from the filter at points outside the stop band had an amplitude of 2.75 volts. At the bottom of the notch the signal's amplitude was 0.15 volt. This represents an attenuation of 0.15/2.75 or 9.45 percent.

Many engineers prefer to express such relationships in terms of decibels. In this case, the attenuation in decibels is 20 log (V_{out}/V_{in}). Therefore, the attenuation is 20 log (0.055) or −25.27 dB.

Figure 2-26 is a plot of the frequency response of the circuit in Fig. 2-25 when the circuit has been adjusted for a stopband centered at 1431 Hz. Note that the slope of the notch is much sharper on the low frequency side. Also note that the amplitude

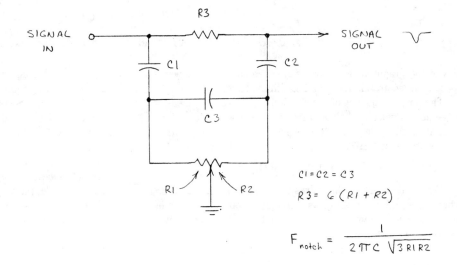

Fig. 2-24. A passive bridged differentiator tunable notch filter.

$$C1 = C2 = C3$$

$$R3 = 6 (R1 + R2)$$

$$F_{notch} = \frac{1}{2\pi C \sqrt{3 R1 R2}}$$

39

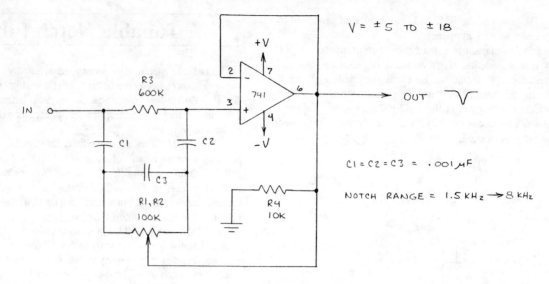

Fig. 2-25. An active tunable notch filter.

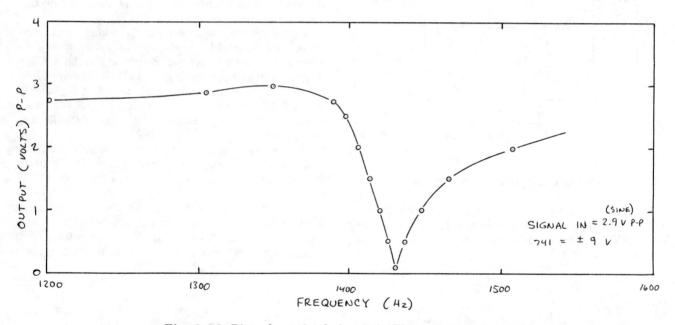

Fig. 2-26. Plot of notch of circuit in Fig. 2-25 (see text).

axis of the graph has a linear scale. Often such frequency response curves are plotted on a logarithmic scale.

The notch frequency of the circuit in Fig. 2-25 can be easily tuned across much of the audio spectrum. According to various texts about filters (e.g., H. Berlin's *The Design of Active Filters*, E&L Instruments, 1977), the notch frequency is given by

$$F_{notch} = 1/2\pi C \sqrt{3R1R2}$$

However, inserting the values for the circuit in Fig. 2-25 when R1 = R2 = 50 kΩ gives a predicted notch frequency of 1838 Hz. Recall from above that the actual notch frequency I measured was 1431 Hz.

Similarly, when R1 is 10 kΩ and R2 is 90 kΩ, the equation predicts a notch frequency of 3063 Hz. However, I measured an actual notch frequency of 2334 Hz.

These discrepancies are likely due to tolerance variations in all three capacitors and R3 in the test circuit. For optimum results, C1, C2, and C3 should have exactly equal values. Likewise, the resistance of R3 should be exactly six times the resistance of R1 + R2.

The attenuation at higher frequencies is not as sharp. At 10 kHz, for example, the amplitude at the notch frequency was 13.4 dB below the out-of-notch frequencies.

Incidentally, I connected the circuit in Fig. 2-25 between my shortwave receiver and an external power amplifier in an effort to tune out annoying interference tones. If the frequency of the interfering tone was steady, the filter did indeed greatly reduce its amplitude. I also found that the filter is ideal for attenuating the shrill tone from a piezoelectric alerter being operated in a room monitored by an intercom. This permits the intercom to be comfortably monitored without the annoyance of the alerter's tone.

Two 60-Hz Hum Filters

The most prevalent noise in electronic equipment is probably 60-Hz hum from the ac power line. The hum can be transmitted directly through the power supply. Or it can be inductively coupled into a circuit by nearby power lines. In this fashion even battery-powered equipment can be susceptible to 60-Hz hum.

Audio frequency amplifiers are particularly vulnerable to 60-Hz hum. Well designed audio frequency preamplifiers often include on-chip a power supply decoupler-regulator circuit to provide rejection of power supply ripple that might give rise to 60-Hz hum. For example, the LM381 provides 120-dB supply rejection. The LM 387 low noise preamplifier provides 110-dB supply rejection.

Typical op amps provide less supply rejection. The 741 and 308, for example, provide from about 80- to 96-dB supply voltage rejection.

A common 60-Hz hum entry point for a properly designed and shielded audio preamplifier is the input jack. Using an improperly or poorly shielded input cable or simply touching the input with a finger may couple more 60-Hz hum than signal into the preamplifier. In both cases, the cable and human body serve as antennas that pick up 60 Hz radiated by nearby power lines.

Several kinds of passive and active notch filters can be used to trap 60-Hz hum. Passive filters are disadvantaged by insertion loss, so let's examine a couple of representative active notch filters.

Wien-Bridge 60-Hz Notch Filter

Figure 2-27 shows an active Wien-bridge 60-Hz hum filter. This filter directs the input signal along two routes leading to the inputs of an op amp. The components are selected so that at a given frequency the gains of both the inverting and non-inverting inputs match and thus cancel one another. An advantage of the circuit is that the two capacitors and the five resistors have equal values.

The notch frequency of the circuit in Fig. 2-27 is the reciprocal of $2\pi RC$. The component values in Fig. 2-27 give a predicted notch minimum of 58.95 Hz. A breadboard version of the circuit I built gave a notch minimum of 61 Hz, certainly very close to the predicted result considering the large tolerance of the components I used. At the notch frequency, 93 percent of the signal was rejected. This corresponds to a rejection of -12.5 dB.

Twin-Tee 60-Hz Notch Filter

Figure 2-28 is an active twin-tee 60-Hz notch filter with better rejection than the Wien-bridge version. This circuit requires one less component than the Wien-bridge version, but it requires two resistor and two capacitor values.

The notch frequency of this circuit is, like the circuit in Fig. 2-27, the reciprocal of $2\pi RC$ where $R = R1 = R2 = 2R3$ and $C = C1 = C2 = C3/2$. The component values in Fig. 2-28 give a predicted notch minimum of 63.66 Hz. However, the breadboard version of the circuit I built gave a notch minimum of 53 Hz. The attenuation at this frequency was -32.47 dB.

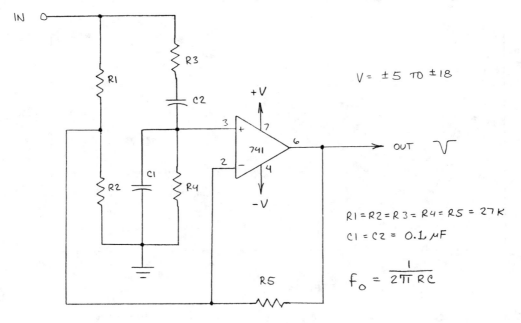

Fig. 2-27. Wien-bridge 60-Hz notch filter.

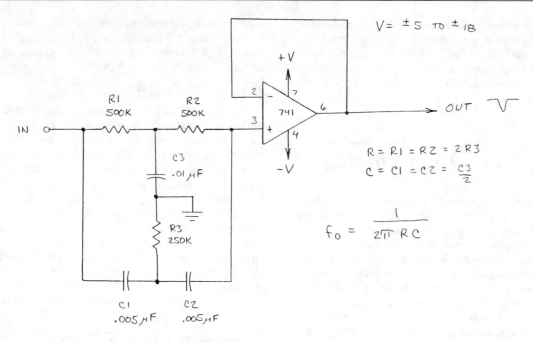

$$V = \pm 5 \text{ TO } \pm 18$$

$$R = R1 = R2 = 2R3$$
$$C = C1 = C2 = \frac{C3}{2}$$

$$f_0 = \frac{1}{2\pi RC}$$

Fig. 2-28. Twin-tee 60-Hz notch filter.

The difference between predicted and measured notch frequencies is due to the large tolerance of the components I used. Nevertheless, the filter provides an attenuation of -23.2 dB at 60 Hz.

Comparing the Two Filters

The actual frequency response of both 60-Hz notch filters described above is compared in Fig. 2-29. Note that although the twin-tee filter has a somewhat wider and deeper notch, the overall shape of both curves is very similar. Note also that the vertical axis in this plot is linear, not logarithmic.

Incidentally, some texts present an idealized, symmetrical notch to illustrate the performance of the various circuits and equations they present. The notches in Fig. 2-29 are certainly not symmetrical. Since they are derived from actual measurements made from breadboard versions of the circuits in Figs.

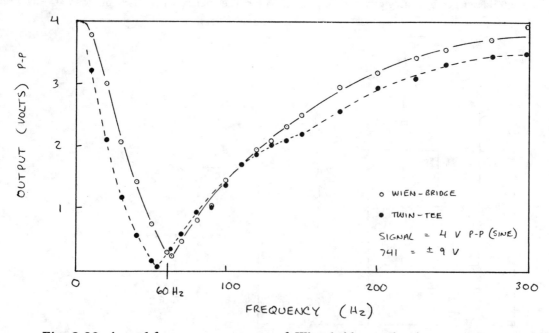

Fig. 2-29. Actual frequency response of Wien-bridge and twin-tee 60-Hz notch filters.

2-27 and 2-28, this illustrates the importance of plotting results from real test measurements rather than depending solely upon theoretical predictions.

An Easily Adjusted 60-Hz Hum Filter

Conventional active notch filters designed to block 60-Hz hum require a single operational amplifier and six or seven outboard resistors and capacitors. Newly developed switched capacitor filters such as National Semiconductor's MF-10 can provide improved 60-Hz hum rejection. A basic 60-Hz notch filter requires only three external resistors and *no* capacitors. A 300- or 600-Hz clock signal is required, however, so there is no component savings over conventional op-amp 60-Hz notch filters.

Though they are more expensive than conventional op-amp active filters, the center frequency (f_o), gain, and Q of a switched capacitor notch filter are all easily adjusted. This means virtually any degree of 60-Hz rejection can be achieved. Furthermore, a switched capacitor notch filter designed for 60-Hz rejection can be easily retuned to reject any frequency up to 30 kHz.

How It Works

Figure 2-30 is the circuit for a straightforward 60-Hz notch filter designed around the MF-10. The filter, which is operated in the inverting mode, provides a voltage gain of $-R2/R1$. The Q is f_o/BW (bandwidth at the -3-dB points).

The solid line in Fig. 2-31 is a logarithmic plot of the frequency response I measured for the circuit in Fig. 2-30. The

dashed line in Fig. 2-31 is a logarithmic plot of the filter's response when R2 is reduced to 1 kΩ. This simultaneously reduces the gain to -0.1 and increases the Q to 10.

Notice the excellent sharpness of the second notch. Note also that both notches are precisely centered at 60 Hz and that their stopband and notch maximums are separated by 20 dB. (A voltage gain of 1 is 0 dB, and a voltage gain of ± 0.1 is -20 dB.)

The filter is calibrated by setting the clock adjust potentiometer (R5) to 100 times the desired notch f_o or about 6000 Hz. For best results, the filter should be fine tuned with the help of a 60-Hz signal coupled into pin 4 of the MF-10 via R1. Adjust R5 while observing the 60-Hz output waveform at pin 3 of the MF-10. The filter is precisely tuned when the amplitude of the wave is at a minimum.

The clock frequency of the test circuit I built was 5964 Hz when the filter was fine tuned for a maximum notch at 60 Hz. This corresponds to a clock frequency-to-f_o ratio of 99.4:1. The MF-10BN chip I used is specified for a clock frequency-to-f_o ratio of 99.35:1, ± 0.6 percent (pin 12 low). The chip I used was within 0.1 percent of the specified ratio.

Going Further

A possible drawback of the circuit in Fig. 2-30 is the requirement of a dual polarity supply. Figure 2-32 shows a single supply version of the circuit which can be powered by a $+10$- to $+14$-volt supply.

Operation of the modified circuit is similar to that of the dual-supply version. You may, however, find it necessary to keep the resistance of R2 less than that of R1 and R3 when the supply voltage is 10 volts.

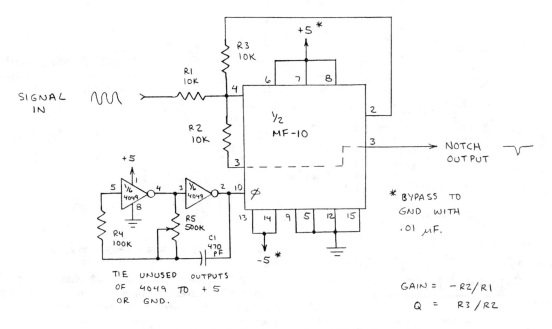

Fig. 2-30. Switched-capacitor 60-Hz hum filter.

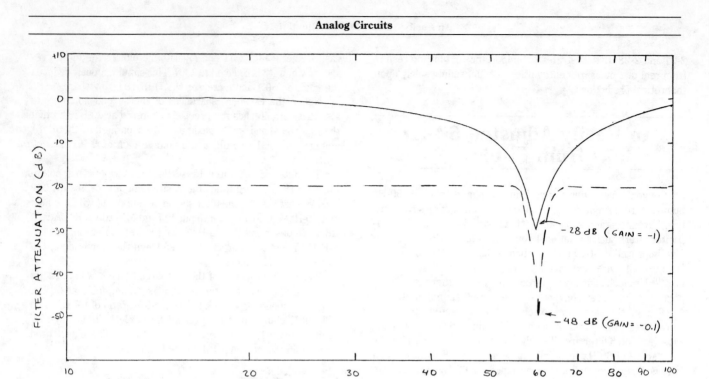

Fig. 2-31. Frequency response of 60-Hz notch-switched capacitor filter.

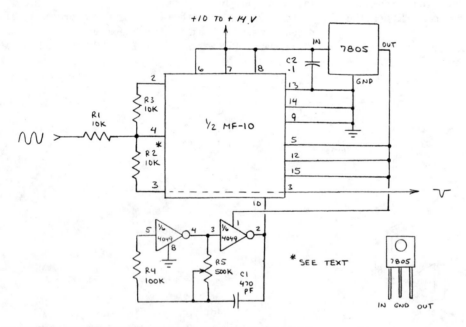

Fig. 2-32. Single-supply switched-capacitor 60-Hz notch filter.

An Easy-to-Use Universal Active Filter

Imagine a single-chip active filter which can be tuned with a single resistor, uses no external capacitors, and simultaneously functions as a lowpass, bandpass, and notch filter. Such a filter could save countless hours of design and breadboard evaluation time.

Happily a new generation of such filters is now available. They are collectively known as *switched capacitor filters*, and their operating parameters are easily tuned by a few external resistors. *No* external capacitors—the nemesis of conventional active filters—are required.

One such filter is National Semiconductor Corporation's MF-

10 universal switched capacitor filter. The MF-10 contains two independent filters in a 20-pin DIP. To say the MF-10 is easy to use is an understatement. In only a few short hours of bench time, I've used this new chip in as many roles as all the conventional active filter circuits I've built over the past ten years.

Of course conventional active filters have many important applications in both digital and analog circuits. For example, they are used to block undesirable signals and to detect touch tone and modem signals. In these and many other roles active filters are far superior to passive filters since they incorporate gain-restoring operational amplifiers.

But although conventional active filters are exceptionally useful, they can be very tricky to design. If you've never designed one, you might wish to take a quick tour through the pages of any of the many excellent books on the subject. You'll find numerous circuit arrangements, some fairly complex, and enough design equations to wear out your calculator finger.

Even if you successfully wade through the circuits and equations and design your own filter, you will almost certainly find that the first breadboard version you build has a resonant frequency that differs, perhaps substantially, from your design frequency. That's because the resonant frequency of all such filters is only as accurate as the tolerances of the circuit's resistors and capacitors. Precision resistors are relatively cheap and easy to find. But precision capacitors are much more expensive.

All these drawbacks are overcome by a switched capacitor filter like the MF-10. Indeed, the versatility and operating simplicity of this filter convince me that switched capacitor filters will soon obsolete shelves of conventional active filter books, design notes, and computer programs.

In this space we'll cover in some detail the operation and use of the MF-10. Then we'll experiment with a tunable MF-10 combination lowpass/notch/bandpass filter. Before finding out how the MF-10 works, however, let's review some active filter basics.

Active Filter Basics

Depending upon their function, most filters are designated *lowpass, highpass, bandpass,* or *notch* filters. The role of the filter is defined by its name. For example, a lowpass filter passes with little or no attenuation all frequencies below a specified cutoff point and blocking all other frequencies. A notch filter attenuates only a very narrow band of frequencies and passes with little or no attenuation those frequencies outside the notch.

Figure 2-33 illustrates in graphical form the operation of the four basic filter types. Other active filters can also be implemented. For instance, the *allpass filter* shifts the phase of a signal without causing attenuation.

The operation of an active filter is described by a number of terms you should understand. You can find a detailed explanation of these terms in any book about active filters. For now here is a brief explanation of some of the more important ones:

Passband refers to the band of frequencies passed with little or no attenuation through the filter.

Stopband refers to the band of frequencies attenuated by the filter.

Bandwidth (BW) of a bandpass filter is the difference between the upper and lower frequencies at points 0.707 times the filter's maximum response (−3-dB points).

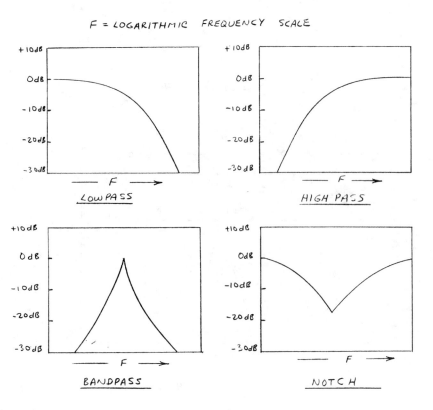

Fig. 2-33. Typical frequency response curves for various filter categories.

Center frequency (f_o) defines the pass and attenuation regions for, respectively, bandpass and notch filters.

Cutoff frequency (f_c) is the point at which the gain in the passband of a filter falls to 0.707 times the unattenuated gain (−3-dB point).

Q or *quality factor* of a bandpass or notch filter is f_o/BW. A high Q filter implies a very sharp or narrow frequency response.

Figures 2-34 and 2-35 show representative plots of the response of typical lowpass and bandpass active filters made in the conventional fashion with operational amplifiers. Note that the graphs are scaled logarithmically, the usual format for such frequency response curves.

Both filters show a 0-dB response (gain = 1) in their passbands. This coincides with a voltage gain of 1. The "ideal" cutoff (Fig. 2-34) and response (Fig. 2-35) are included to illustrate that no active filter has a perpendicular cutoff point.

The MF-10

The best conventional active filters typically require several op amps, at least two capacitors, and from three to seven resistors. The MF-10 contains *two* completely independent active filters which each perform five basic filter functions (lowpass, highpass, bandpass, notch, and allpass) with only a few external resistors and no external capacitors. Well, almost no capacitors. Unlike conventional active filters, the MF-10 requires an input clock and clock circuits usually incorporate a capacitor or two. The clock capacitor poses no tolerance problem, however, since the clock frequency can be easily tuned by changing the value of a single resistor.

The MF-10 is exceedingly easy to use since all five outputs

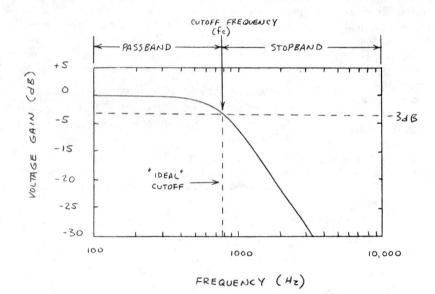

Fig. 2-34. Characteristics of a typical lowpass active filter.

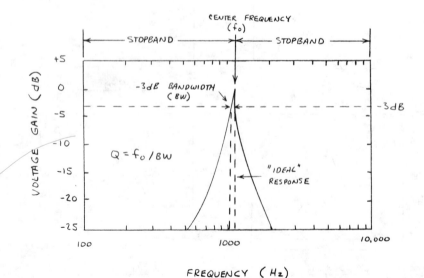

Fig. 2-35. Characteristics of a typical bandpass active filter.

are brought out of the chip. Furthermore, the two second-order filters in a single MF-10 can be cascaded to provide a fourth-order filter. Experienced active filter designers will be happy to know such classical filter responses as the Butterworth, Chebyshev, Bessel, and Cauer are easily implemented with the MF-10.

Another advantage of the MF-10 is its CMOS construction. This is why its typical power supply current demand is only 8 milliamperes. The supply voltage can range from ± 4 to ± 7 volts when operated from a dual polarity supply and 0 to $+10$ volts when operated from a single polarity supply.

Finally, as we'll observe firsthand in the test circuit that follows, the MF-10's response is easily tuned across a wide spectrum by simply varying the clock frequency. In applications requiring a precisely controlled notch or bandpass, the clock can be crystal controlled.

How the MF-10 Works

The key circuits in the MF-10 are a pair of noninverting integrators whose time constants are determined by their respective feedback capacitors and the clock frequency. A conventional integrator like the one in Fig. 2-36A has a time con-

stant of RC. The precision of the circuit is therefore only as good as the tolerances of the RC components.

The integrators in the MF-10 replace the resistor of the integrator in Fig. 2-36A with a capacitor that is alternately switched by the clock signal first to the input voltage and then to the feedback capacitor, V_{in} and C_F respectively. As shown in Fig. 2-36B, the switched capacitor performs the role of the resistor in Fig. 2-36A by transferring the voltage at the input to C_F.

The amount of charge transferred by C_{in} to C_F is determined by the time C_{in} is allowed to charge to V_{in} and the time C_F is allowed to charge to the voltage on C_{in}. In his paper "Introducing the MF-10," Tim Regan, a National Semiconductor applications engineer, shows how the current transferred through C_F to the output of the integrator is $V_{in}C_{in}$ divided by the clock period or $V_{in}C_{in}$ times the clock frequency (F_{clk}). Therefore, the effective resistance of C_{in} is the reciprocal of $C_{in}F_{clk}$. In other words, the effective resistance of C_{in} can be varied simply by altering the clock frequency. The time constant of the circuit is $C_F/C_{in}F_{clk}$.

Since the switched capacitor integrator employs on-chip capacitors, you may be wondering why its performance is better than a standard integrator with external RC components. The

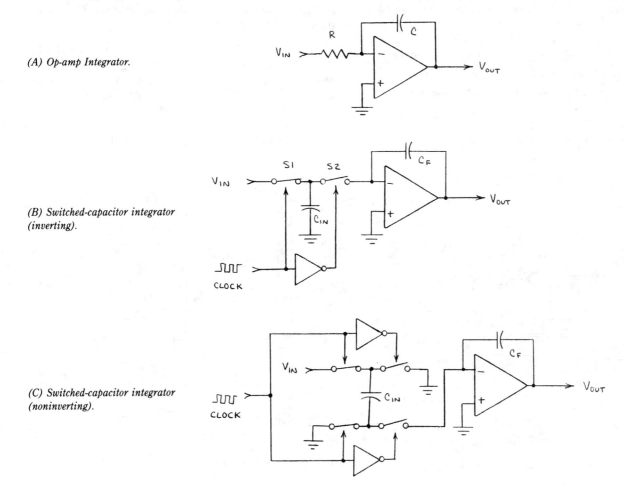

(A) Op-amp Integrator.

(B) Switched-capacitor integrator (inverting).

(C) Switched-capacitor integrator (noninverting).

Fig. 2-36. Operation of the switched capacitor integrator in the MF-10.

reason is that the time constant of the switched capacitor integrator is entirely dependent upon the *ratio* in C_{in} and C_F, both of which are fabricated on the same die and thus have precisely controlled dimensions and values.

Incidentally, Tim Regan notes in his paper that the integrators in the MF-10 are noninverting. This requires a somewhat more complex switching arrangement than that shown in Fig. 2-36B wherein *both* sides of C_{in} are alternately switched. As shown in Fig. 2-36C, first the upper and lower sides of C_{in} are switched, respectively, to V_{in} and ground. Then the upper and lower sides of C_{in} are switched, respectively, to ground and C_F. This reverses the polarity of the charge on C_{in} that's applied to C_F, thereby preserving the polarity of V_{in} at the output of the integrator.

For additional details about how the MF-10 works, see the application note for this chip and Tim Regan's paper. In the space that remains, we'll look at some practical aspects of using the MF-10 and experiment with an MF-10 in a typical circuit.

Using the MF-10

Figure 2-37 shows a simplified block diagram of one of the filters in an MF-10 and the pin outline for the actual chip. The MF-10 application note describes in detail the function of each pin. What follows is a brief summary.

LP, BP, and N/AP/HP are, respectively, the lowpass, bandpass, notch/allpass/highpass outputs of each filter.

INV is the inverting input of each filter.

S1 is used as a signal input pin when the filter is in the allpass mode.

$S_{A/B}$ activates the internal switch that connects the input of the filter's second summer amplifier (see Fig. 2-37) to

analog ground AGND ($S_{A/B}$ = low) or the lowpass output ($S_{A/B}$ = high).

V_A^+ and V_D^+ are the analog and digital positive supply pins. They can be interconnected if desired. They should be bypassed with a capacitor (two if separate supplies are used).

V_A^- and V_D^- are the analog and digital negative supply pins.

L_{SH} is the level shift pin. It allows the user to use the MF-10 with a variety of power supply and clock voltages. For example, when powered by a dual ±5 volt supply and driven by a CMOS clock (±5 volts), the L_{SH} pin should be tied to ground or V_A^-, V_D^-.

CLK is the clock input to each filter. The clock should have a duty cycle of close to 50 percent for optimum results.

50/100/CL determines the ratio of clock frequency to filter center frequency (50:1 when high and 100:1 when at analog ground).

AGND is the analog ground pin.

Again, be sure to refer to the MF-10 application note for additional information about these pin functions. The explanations given here are but brief summaries.

A Tunable Lowpass/Bandpass/Notch Filter

Figure 2-38 is the circuit for a straightforward LP/BP/N filter made from one-half an MF-10 and an external clock. The clock is made from two inverters in a 4049. Its frequency, thus the frequency response of the MF-10, can be altered by changing the setting of R5. Incidentally, although you may have often seen simple oscillators such as this made from two NAND gates

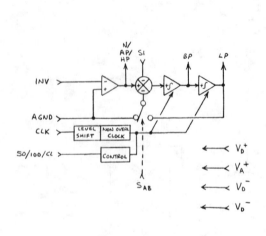 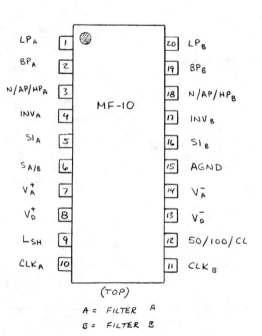

Fig. 2-37. MF-10 block diagram and pin outline.

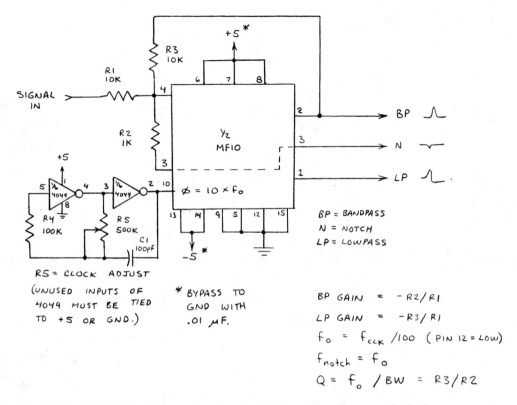

Fig. 2-38. MF-10 bandpass, lowpass, notch, active filter, and clock.

in a 4011, be sure to use the 4049 in this circuit. The 4011 may not oscillate at $V_{DD} = +5$ volts. (If you must use the 4011, try buffering its output with a third gate.)

The MF-10 is connected in what the application note calls the Mode 1 configuration. At least eight other modes are available. Some of the circuit's key parameters are defined in Fig. 2-38.

The circuit in Fig. 2-38 is extremely easy to use. The only significant constraints are to observe power supply limits and CMOS handling precautions. Otherwise, operating the circuit is a snap.

For example, with the values given in Fig. 2-38, the LP gain is −0.1 (near $f_o = 0$ Hz), the BP gain is −1 (at f_o) and the BP/notch Q is 10. The circuit can easily be tuned out to a maximum f_o of 30 kHz simply by altering the clock frequency (where the ratio of clock frequency-to-f_o is approximately 100:1). The gain and Q values can easily be altered by changing the outboard resistors.

When the clock frequency of the test circuit was 99.435 kHz, I measured an f_o for the bandpass and notch outputs of 1050 Hz. This is outside the clock-to-f_o ratio specified for the MF-10. The *rounded* clock-to-f_o ratios (pin 12) are 50:1 and 100:1. The *actual* ratios given in the data sheet are 49.94:1 and 99.35:1. In both cases and depending upon the grade of the chip, the actual ratios have a typical and maximum tolerance range of, respectively, ±0.2 and ±1.5 percent.

My tests, made on successive days with careful attention to

detail, gave a ratio of 94.7:1 (pin 12 low). I therefore tried two additional MF-10's and obtained almost exactly the same results. Although I cannot account for the apparent discrepancy in clock-to-f_o ratio, the three MF-10's exhibited virtually identical tolerances.

Figure 2-39 is a plot superimposing on a linear scale the LP, BP, and notch frequency response of the circuit in Fig. 2-38. The bandpass and notch curves appear quite normal, but the lowpass curve peaks sharply at f_o. This apparent anomaly results from the LP output being programmed by R3/R1 for a gain of −0.1 at $f = 0$ Hz. This means the LP output should approximate −0.26 volts peak-to-peak at $f = 0$ Hz when V_{in} is 2.6 volts peak-to-peak. Figure 2-39 confirms this is indeed the case.

Recall that frequency response curves for active filters are normally plotted on a logarithmic scale. Figure 2-40 is a log plot of the bandpass response in Fig. 2-39. The dashed plot represents the bandpass when the Q was increased from 10 to 100 by reducing R2 to 100 ohms. Both these plots as well as the linear scale plots in Fig. 2-39 are based upon actual measurements made with the test circuit in Fig. 2-38.

Going Further

Although exceptionally versatile and easily tuned, the multipurpose filter in Fig. 2-38 is only one of several ways you can use the MF-10. For example, it can be operated in a noninverting mode or it can be configured for simultaneous highpass, bandpass and lowpass filtering.

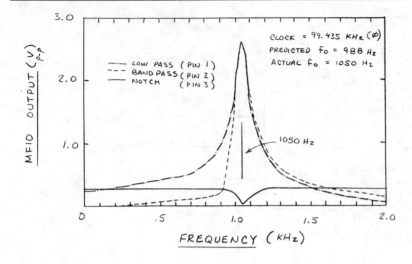

Fig. 2-39. Frequency response of MF-10 (Mode 1) at clock frequency of 99.435 kHz.

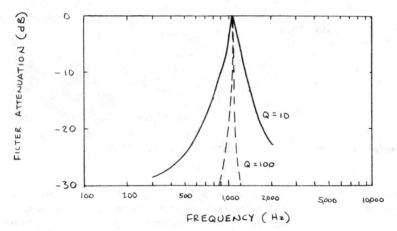

Fig. 2-40. Log plot of bandpass curve in Fig. 2-39.

A Fully Adjustable Pulse Generator

The 558 is a quad timer chip with many useful applications. Figure 2-41 shows one of the best, a highly versatile square pulse generator with fully adjustable pulse rate and pulse duration. The circuit is adapted from Exar's XR 558/559 data sheet.

In operation, the two timers in Fig. 2-41 are cross-connected so that the output of one timer is directly coupled to the trigger input of the second timer. When the timing cycle of the first timer is complete, the second timer is triggered. When its cycle is complete, the first timer is triggered. The result of this feedback cycle is an astable multivibrator.

The significance of this multivibrator circuit is that the RC time constant of each timer section is independently adjustable. This means both the pulse rate and duration are independently controllable. R1 controls the frequency of the pulses and R2 controls their duration. Notice that the circuit provides complementary outputs.

The pulse rate of the circuit is given by the reciprocal of (R1C1) + (R2C2). The minimum practical pulse duration is on the order of a microsecond, and the maximum pulse rate is about 100 kHz. The pulses remain very square until their duration falls to a few microseconds. Their amplitude can be altered by varying the power supply voltage.

The oscillation frequency of each timer in a 558 can be altered by changing the circuit's control voltage. The 558 has a single control-voltage pin which is common to all four timers. Therefore, the duty cycle of the circuit is unaffected by application of a control voltage. This means the circuit in Fig. 2-41 can be operated as a variable frequency oscillator having a fixed duty cycle.

You can add a fixed-duty-cycle frequency control by connecting the rotor of a 10-kΩ potentiometer to pin 4 of the 558. Connect the stator terminals of the pot to V_{cc} and ground, respectively. Adjust the pot to alter the frequency.

Driving External Circuits

The output stage of each timer in the 558 is a normally low, open-collector npn transistor that can sink up to 100 milliamperes. This means the circuit can directly drive an LED, a small

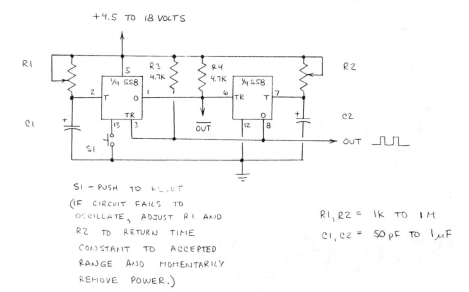

Fig. 2-41. Fully adjustable 558 pulse generator.

relay, or even a small speaker for audio effects. However, although each timer can sink up to 100 milliamperes, the power dissipation rating of the timer's package limits the maximum current per timer if all four are in use.

For example, the maximum dissipation for the plastic package 558 is 625 milliwatts. If the circuit is operated at 10 volts, 100 milliamperes per output would give a total dissipation of 4 watts (from Ohm's law, 10 volts × 0.1 ampere = 1 watt per timer). Therefore, the current delivered to each output would have to be reduced or external buffering would be required.

If each output is required to deliver only 10 milliamperes, then the total power dissipation would be 4 × 10 volts × 0.01 ampere = 0.4 watt, well within the 625-milliwatt maximum.

Incidentally, this circuit will cease oscillation if an attempt is made to operate it beyond its minimum pulse duration or maximum frequency. Should this occur, adjust the appropriate pot (or capacitor if you prefer) and momentarily disconnect the power to restart the circuit. If the circuit fails to operate when power is first applied, it is possible both RC time constants are out of range and require adjustment to bring the circuit to within its operating limits.

Power Pulse Generator

The LM150/LM250/LM350 is an adjustable voltage regulator capable of delivering from 1.2 to 33 volts at a maximum current of 3 amperes. Normally this chip is used in both fixed and adjustable power supplies, voltage regulators, and voltage references.

Always looking for a new way to obtain high current pulses for flashing lamps and driving motors, I have recently been experimenting with ways to switch the output of the LM350T on and off. Since the output voltage of this chip is fully adjustable,

a circuit which switches its output on and off provides a variable-amplitude power pulse generator.

Before looking at an LM350T power pulse generator, let's first examine the operating characteristics of this versatile chip. Figure 2-42 shows the pin diagrams for both the TO-220 packaged LM350T and the TO-3 versions of the LM150/LM250/LM350.

The chip requires a minimum of external components. For straightforward voltage regulation, only two essential resistors are required. As shown in Fig. 2-43, these form a voltage divider connected to the chip's voltage adjust terminal. Altering the setting of R2 varies the circuit's output voltage.

If the LM150/LM250/LM350 is located some distance from the power supply filter capacitor, the addition of a 0.1-microfarad bypass capacitor across the input is required. Transient response, as during momentary high current loads, is enhanced by connecting a 1-microfarad capacitor across the chip's output. The LM150/LM250/LM350 includes on-chip thermal shutdown circuitry to protect the chip from being damaged by overloads. This circuitry remains functional even if the voltage adjust terminal is disconnected.

Power Pulse Generator

Figure 2-44 shows one way to switch the LM350T between its minimum output of about 1.2 volts and the voltage set by R5. In operation, the 555 timer forms an oscillator which turns Q1 off and on. When Q1 is on, the output of the LM350T falls to its minimum 1.2 volt level as the voltage adjust terminal is shorted to ground through Q1's collector-emitter path. When Q1 is off, the LM350T functions normally.

R1 permits the pulse repetition rate to be altered, and R5 controls the amplitude of the output pulses. C1's value can be increased or decreased to, respectively, slow down or speed up the pulse rate. When C1 is 2.2 microfarads, the output pulses have a duration of 1.5 milliseconds. The pulse duration increases to 2.5 milliseconds when C1 is increased to 3.3 microfarads.

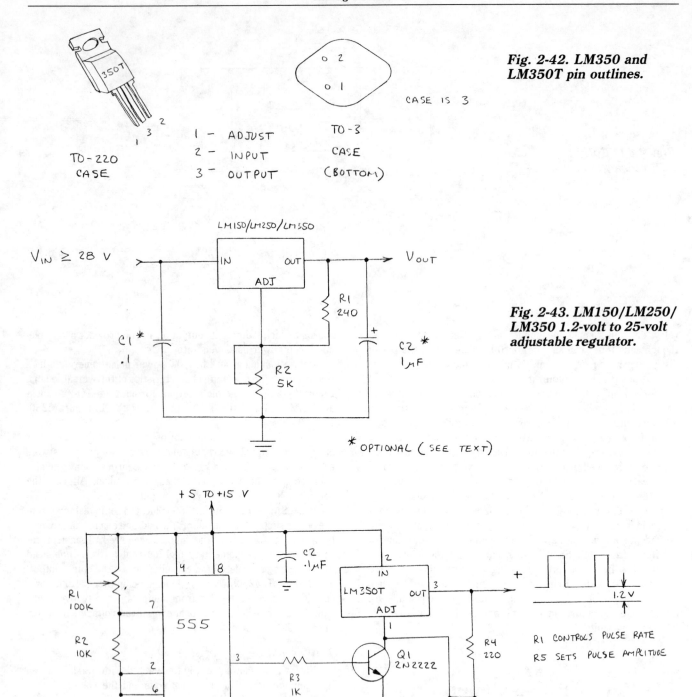

Fig. 2-42. LM350 and LM350T pin outlines.

Fig. 2-43. LM150/LM250/ LM350 1.2-volt to 25-volt adjustable regulator.

Fig. 2-44. Adjustable rate power pulse generator.

Though this circuit has many applications, I've found it to be particularly effective as a flasher for incandescent lamps. Between flashes, the voltage across the lamp's filament falls to about 1.2 volts instead of ground as in conventional lamp flashers. This minimal voltage can be considered a "keep alive" voltage which keeps the filament glowing dimly between flashes. When the lamp receives a current pulse, its filament reaches full brightness much faster than when a keep-alive current is not flowing.

You can operate an incandescent lamp at much more than its rated current level so long as the current is applied in brief pulses and the duty cycle is kept low. The resulting flashes are surprisingly sharp and brilliantly white. Application include emergency strobes, close-range flash photography, and novelty lighting.

Keep in mind other applications for the circuit. For example, it can deliver adjustable rate pulses to a dc motor, thus serving as a speed control. And it can be connected to a large speaker and used as an attention-getting warning tone generator.

A Programmable Function Generator

On a shelf above the busiest part of my workbench is a Heath SG-1271 Function Generator. For several years this piece of test equipment has injected square, sine, and triangle waves into hundreds of breadboard circuits with which I've experimented. Its variable frequency and amplitude controls enhance the utility of this function generator, and I'm sure it will provide additional years of useful service.

Sometimes, however, I require waveforms other than the simple square, sine, and triangle waves provided by most commercial function generators. One application for unusual or complex waveforms is electronic music. Another is sound effects generators. Unusual waveforms are also used to simulate mathematical functions and to imitate the unique signals or *signatures*

emitted by a wide range of natural and manmade objects. Signatures that can be electronically simulated include those of the human heart beat, nerve impulses, and earthquakes.

Another important application for specialized waveforms is the testing of electronic circuits. I've often used hastily breadboarded waveform generators to provide unusual transmitter signals in experimental fiber optic lightwave communication systems.

Figure 2-45 is a block diagram of an experimental programmable function generator which will allow you to produce customized, stepped waveforms. In operation, a variable frequency clock repetitively cycles a counter through its paces. The binary output from the counter is decoded into 1-of-n outputs by a decoder. In other words, for each state of the counter, one and only one output from the decoder is active.

The decoder outputs are connected to individual switches, each capable of applying a preselected voltage to a common OR-wired output. As the decoder sequentially actuates the switches, a stepped waveform appears at the output.

A Four-Step Programmable Waveform Generator

Figure 2-46 shows a practical four-step version of the block diagram in Fig. 2-45. The clock is designed around a 7555, the CMOS version of the 555 timer chip. The output from the clock is fed directly into the clock input of a CMOS 4017, a decade counter with a built-in 1-of-10 decoder. Nine of the ten outputs of the 4017 are normally low and the selected output is always high.

The four lowest order decoded outputs from the 4017 are connected to the control inputs of each of the four analog switches in a CMOS 4066. The analog inputs of each switch are connected to the wipers of miniature 10-kΩ trimmer resistors which serve as adjustable voltage dividers.

In operation, the first four decoded outputs from the 4017 sequentially actuate each of the analog switches. The voltages appearing at the inputs of each switch are then placed one at a time on the common, OR-wired bus which connects the outputs of the four switches. The output then assumes the high-imped-

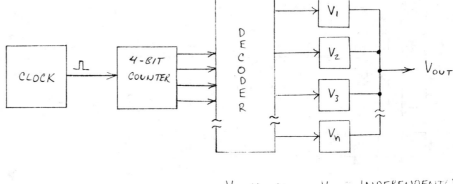

Fig. 2-45. Block diagram of a basic programmable function generator.

$V_1, V_2, V_3 \ldots V_n$ = INDEPENDENTLY PROGRAMMABLE VOLTAGES.

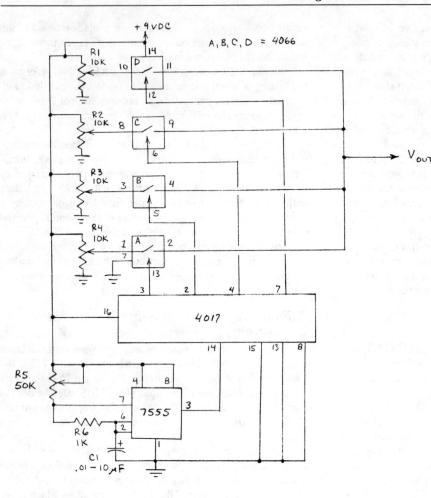

Fig. 2-46. Programmable four-step function generator.

ance (open) state for the next six clock cycles. The cycle then repeats, thus providing a repetitive waveform having a width of four clock cycles separated by intervals of six clock cycles.

Figure 2-47 is an oscillogram of a typical programmed stepped waveform produced by the circuit in Fig. 2-46. Simply by changing the adjustment of any or all the trimmer resistors (R1–R4), the waveform can be altered in any desired fashion. The period

of the waveform, hence the duration of each step, is controlled by the clock rate.

Since the 4017 incorporates a reset input (pin 15), the dead space between the stepped waveforms can be reduced in increments of one clock cycle or eliminated entirely. This is easily accomplished by connecting one of the six unused decoder outputs to the reset input.

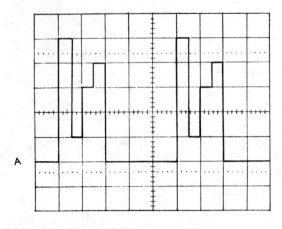

VOLTS/DIV:
A. ___1.0___
B. ___–___

TIME/DIV: ___100μsec___

Fig. 2-47. Typical programmed waveform.

If, for example, the fifth output (pin 10) is connected to the input, all the dead space will be eliminated and the stepped waveform will recycle immediately after the fourth step. A typical oscillogram of a waveform recycled in this fashion is shown in Fig. 2-48.

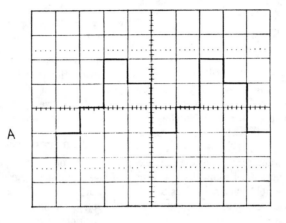

VOLTS/DIV:
A. ___0.5___
B. ___—___

TIME/DIV: ___100 μsec___

A

Fig. 2-48. A typical compressed four-step waveform.

An important operating feature of this circuit is that any desired stepped waveform can be preprogrammed *without* viewing the actual waveform on an oscilloscope screen. All that's necessary is to adjust each trimmer resistor while monitoring the resulting voltage at the trimmer's rotor.

A Programmable Tone Generator

Among the many applications for this function generator is the generation of repetitive sequences of programmable tones. This is readily accomplished by connecting a voltage controlled oscillator (VCO) such as the one shown in Fig. 2-49 to the generator's output.

The circuit in Fig. 2-49 is a straightforward astable multivibrator designed around a CMOS 7555 timer. Normally the 7555 oscillates at a fixed frequency determined by R1 and C1. Variations in the voltage applied to the control voltage input, however, alter the output frequency.

Incidentally, note that Fig. 2-49 specifies that either the 7555 or the standard 555 can be used in the VCO circuit. The 555 does produce slightly more volume from the small speaker. However, the 7555 has substantially reduced power consumption and a higher operating frequency.

A wide range of unique, attention-getting tone sequences can be programmed into the trimmer resistors. Simulated chirps, stepped tones, and twee-dell sirens are some of the sound sequences I've obtained while experimenting with a breadboard version of the circuit.

For best results, slow the function generator's clock rate to a few tens of hertz by increasing the value of C1 in Fig. 2-46

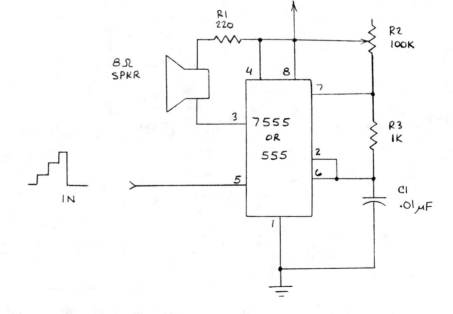

Fig. 2-49. A simple voltage-controlled oscillator.

to several microfarads. There's no need to remove the existing capacitor. Just connect the new, larger capacitor directly across the leads of the original capacitor.

If you're using an oscilloscope to program waveforms, you will need to keep the clock rate high to effectively monitor the waveform. After you program the desired waveform you can add the new capacitor to slow down the repetition rate. If you build a permanent version of the circuit, add a switch to allow you to increase or decrease C1 at will.

Expanding the Function Generator

This basic function generator of Fig. 2-46 is easily expanded to provide four or six additional output steps per waveform cycle by adding, respectively, one or two 4066 analog switches and their respective trimmer resistors. The switches are actuated by the unused decoded outputs of the 4017.

Figure 2-50 is the complete circuit diagram of the fully expanded circuit with ten stepped outputs. Despite its apparent complexity, this circuit can be assembled on a solderless breadboard in about fifteen minutes once you've assembled the necessary components and connection wires.

For best results, try to arrange the trimmer resistors in two rows of five each either on the left or right side of the board. Also, push the connection wires between the trimmers so they do not protrude above the board. These steps will simplify the programming procedure and encourage you to experiment with the circuit.

Adding Additional Steps

Since the 4017 has a carry output (pin 12), expanding the function generator to twenty or more stepped outputs is a straightforward procedure. All that's necessary is to connect the carry output of the first 4017 to the clock input of the second 4017. A twenty step waveform would require two 4017s, five 4066s and twenty trimmer resistors.

A Programmable Waveform Control Panel

If you build a permanent version of this circuit, consider installing the programming trimmer resistors on a control panel. For best results, use linear slide potentiometers instead of rotary action trimmers. By installing the slide pots side-by-side, the positions of their control handles will enable you to visualize the

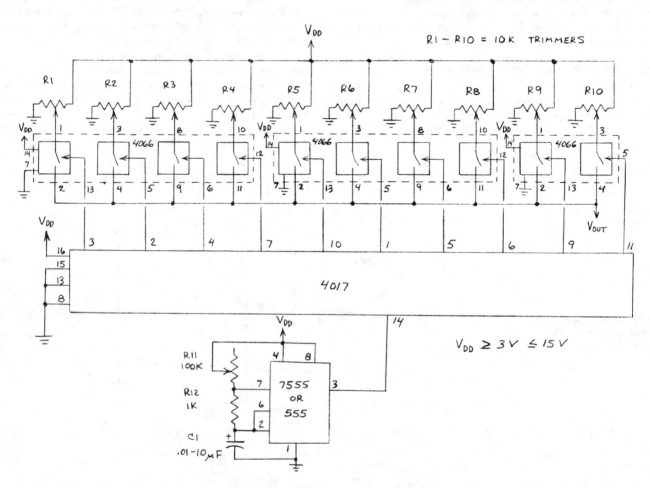

Fig. 2-50. Expanded ten-step programmable function generator.

approximate shape of the programmed waveform. And you will be able to make virtually instantaneous changes in even very complex waveforms.

A Sound Effects Generator

One way to produce attention-getting sound effects is to control the frequency of an oscillator by means of periodic impulses from a second oscillator. This is an ideal role for a pair of 555 timers operated in their astable (free-running) mode.

Figure 2-51 shows how two 555s are connected to provide sound effects. The first 555 (IC1) is connected as an oscillator with an adjustable period of a few tens of hertz or less. The second 555 (IC2) is connected as a voltage controlled oscillator (VCO) with an adjustable frequency of from a few hundred to a few thousand hertz.

For best results, use potentiometers with knobs for R1 and R5. This will enable you to quickly change the circuit's cycle rate (via R1) and its tone frequency (via R2). A faster cycle rate resembles the sound of a chirping bird. A slower cycle rate makes a good warning alarm.

Be sure to experiment with the values of R3 and C2. Increasing C2 stretches the time required for IC2's tone to fall from its highest to its lowest frequency. Increasing R3 has a similar effect. If the values for C2 or R3 are too high, C2 will not fully charge during each cycle, thus reducing the dynamic range of the circuit's tone frequency.

Incidentally, note that Fig. 2-51 specifies either a 7555 or 555 for IC1 and IC2. The 755 is the CMOS counterpart of the 555. It consumes much less power and can operate from a lower voltage (less than 3 volts) than the standard 555. It also has a higher oscillation frequency.

This project is an excellent way to become acquainted with either the 555 or the 7555.

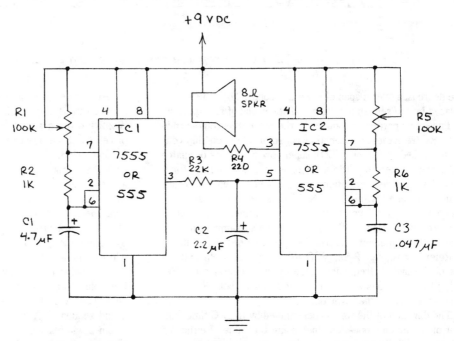

Fig. 2-51. Sound effects generator circuit.

C2 is the key to the unique sounds produced by the circuit, so let's assume for a moment C2 is not present. IC2 will then oscillate at a fixed frequency determined by the voltage at pin 3 of IC1. Negative going 10-millisecond pulses from IC1 will produce brief, click-like interruptions or changes in the oscillation frequency of IC2.

Now let's return C2 to the circuit. During intervals between negative going pulses from IC1, C2 charges through R3 to the voltage at pin 3 of IC1. The relatively slow charging rate of C2 produces a gradual *decrease* in IC2's oscillation frequency. When IC1 switches, C2 is immediately discharged and the frequency of IC2 is suddenly *increased*. C2 then begins to recharge, and the cycle repeats. The resulting sounds from the speaker are far more interesting than the rather boring interrupted tone sequence produced when C2 is not present.

Bomb Burst Synthesizer

Here's a circuit that realistically synthesizes the whistle and explosion of an aerial bomb or shell. This attention-getting sound sequence is generated by an SN76488, the Texas Instruments complex sound generator chip used in some video arcade games to synthesize the sounds of race cars, airplanes, gun shots, and explosions.

Figure 2-52 is the circuit diagram for the bomb burst synthesizer. The circuit is adapted from one suggested in a Texas Instruments application note. In operation, closing and then opening S1 starts the sound sequence. Initially a tone having a frequency of a few kilohertz is produced. The frequency of the

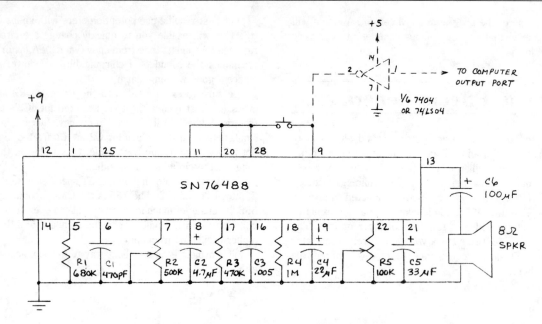

Fig. 2-52. Bomb drop and explosion sound effects generator.

tone decreases over a span of several seconds. The tone is then replaced by the sudden and very realistic sound of an explosion.

The sound patterns produced by the SN76488 are governed by the RC (resistor-capacitor) time constants of the on-chip voltage controlled oscillator (VCO), super low frequency oscillator (SLF), noise generator filter, and monostable multivibrator (one-shot). The detailed operation of these and the various other circuits contained within the SN76488 is beyond the scope of this book, and I refer you to the Texas Instruments' data sheet for more information. In the meantime, let's examine how the various RC components control the operation of the circuit.

Referring to Fig. 2-52, VCO RC components R3 and C3 control the overall frequency range of the preburst tone. Reducing the time constant by reducing either or both R3 or C3 produces a high pitched whistle.

The duration of the tone is determined by the RC time constant of the on-chip one-shot. Increasing the value of either or both R5 or C5 stretches the duration of the preburst tone (or whistle), thus simulating a high altitude bomb drop.

The sound of the explosion is governed by the noise generator's output filter, R1 and C1. Reducing R1, for example, to less than 100 kΩ suppresses the low frequency component of the noise and gives a sound which resembles a stretched gun shot. Increasing R1 to a megohm or more produces a deep, rumbling explosion.

The duration of the explosion is controlled by the RC time constant of an on-chip decay circuit. Reducing R2 or C2 gives a very brief explosion which closely simulates a gun shot if the low frequency component of the output from the noise generator is suppressed by reducing R1.

For best results, R2 and R5 should be trimmer resistors or potentiometers. For even more flexibility, R1 and R3 can also be pots.

An interesting variation of the tone-explosion sound sequence can be obtained by reducing R4. When, for example, R4 is 2.2 kΩ, the gradually falling pitch of the whistle will be replaced by three up-down cycles of a siren followed by the explosion sound.

Other variations are also possible. Disconnect pins 1 and 25 and the explosion sound will be eliminated. Disconnect pin 11 to produce a continuous, nondecaying explosion after a silent delay period. Disconnect pin 20 to produce a gun shot or explosion after a silent delay. The length of the delay is controlled by R2 and C2. Substitute pin 27 for pin 28 to give a descending whistle followed by a continuous explosion.

I've had so much fun with a breadboard version of this circuit I'm going to assemble a permanent version in a small plastic enclosure. A push button will permit manual activation of the sound sequence. A 7404 (or 74LS04) hex inverter connected to pin 9 will permit the sequence to be actuated by the output port on a computer. A gadget like this will make a great gift for a young electronics enthusiast.

An ICM7209 Ap Note

The ICM7209 is a CMOS clock generator made by Intersil. Packaged in an 8-pin miniDIP, the chip has a guaranteed maximum frequency of at least 10 MHz. With a maximum supply voltage of 6 volts, the ICM7209 is capable of driving many 5-volt logic systems. For example, it can drive up to five TTL loads.

Figure 2-53 shows both the internal logic diagram of the clock generator and the pin outline. An external crystal is con-

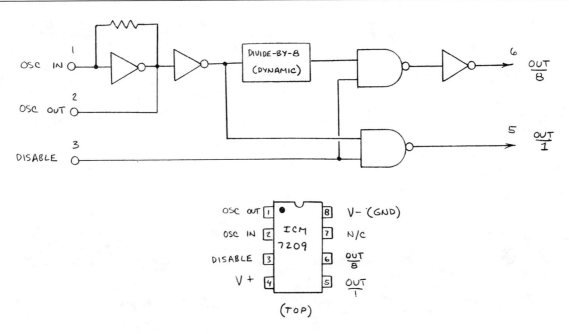

Fig. 2-53. Internal logic diagram of the ICM7209.

nected across pins 1 and 2 to provide the resonant feedback necessary for the oscillator. Two buffered outputs are available. One supplies the oscillator frequency while the other supplies the base frequency divided by 8.

The Disable input provides a convenient way to control the clock with an external logic signal. When this input is high, the generator oscillates. When low, it turns off the oscillator. The base frequency output (pin 5) then goes high and the divide- by-8 output goes low.

The ICM7209 has typical rise and fall times of 10 nanoseconds. At 10-MHz, the chip consumes only about 50 milliwatts.

Figure 2-54 shows just how simple the ICM7209 is to use. I built the circuit on a plastic solderless breadboard and, when used with a 10-MHz crystal, it worked exactly as specified. It also worked with a 3.579545-MHz television colorburst crystal. And it even worked with a 455-kHz ceramic resonator!

Well, it *mostly* worked with the resonator installed. The divide-by-8 output became a divide-by-1.5 output. This is because the divide-by-8 output works properly only when the oscillation frequency of the oscillator circuit exceeds 2 MHz.

Why? Intersil uses a *dynamic* design scheme in its divide-by-8 circuit wherein data is stored in very small integrated capacitors rather than flip-flops. Conventional or *static* dividers use cascaded flip-flop chains called registers. Intersil chose the

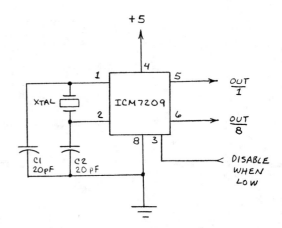

Fig. 2-54. A single chip 10-MHz clock generator.

dynamic approach to keep power consumption low and to improve high frequency response. The dynamic approach also requires less chip real estate. If all this seems vaguely familiar, it's because dynamic and static design methods are used to make most semiconductor RAMs.

59

Digital Circuits

Digital Circuits

A 000 to 999 Event Counter

The circuit in Fig. 3-1 functions as a simple 000 to 999 event counter. It can be used to count such events as rotations of a wheel or shaft, objects passing by a point and even flashes of lightning. It can also be configured to count events whose amplitude exceeds a programmable threshold.

12) of the 4553 advances the counter one count. The BCD status of the count is converted into a seven-segment format by an MC14511B (or 4511) BCD-to-seven segment latch/decoder/driver. The 4511 can directly drive the segments of a common cathode LED display at a maximum current of 25 milliamperes. Like the 4553, it is a CMOS device with a very low quiescent current (about 0.01 microampere at 25°C).

The digits of the three-digit LED readout are sequentially strobed by the three digit select outputs (pins 2, 1, and 15) and

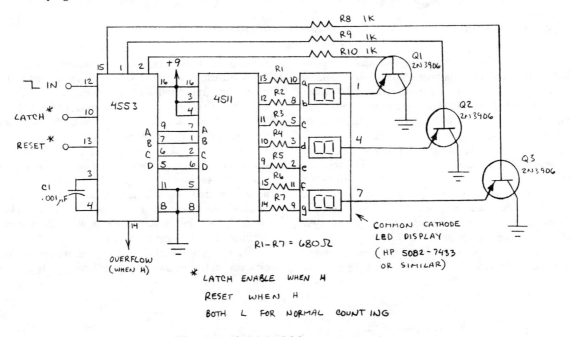

Fig. 3-1. A 000–999 event counter.

The key to the simplicity of this circuit is an MC14553B (or 4553) three-digit BCD counter, a CMOS LSI chip that typically consumes only about 0.02 microampere (25°C) between counts. The 4553 includes a quad latch at the output of each of its three counters to provide storage of the current BCD count status. The outputs from the quad latches are steered to the chip's four BCD output pins by a time division multiplexing arrangement.

Three digit select outputs indicate which of the outputs of the counter latches has been steered to the BCD outputs. This provides a means for driving a three-digit multiplexed readout.

In operation, a high to low transition at the clock input (pin

transistors Q1–Q3. The strobe rate is controlled by an internal scan oscillator whose frequency is determined by C1. For special purpose applications, the internal scan oscillator can be overridden by omitting C1 and applying an external clock signal directly to pin 4.

Triggering the Counter

The counter circuit can be triggered by external logic signals or by a relay or magnetic reed switch. Buffering is generally not required since the 4553 provides an internal pulse shaping stage at its clock input.

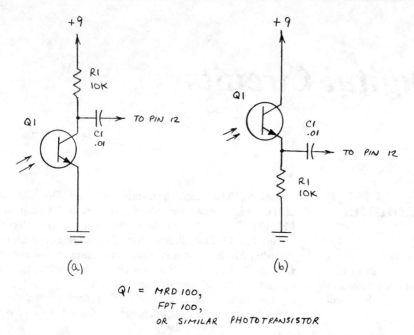

Fig. 3-2. Phototransistor inputs for 3-digit event counter.

Q1 = MRD 100,
FPT 100,
OR SIMILAR PHOTOTRANSISTOR

(A) Triggered by application of light pulse.   (B) Triggered by interruption of light beam.

Figure 3-2A shows a simple phototransistor input stage you can connect to the counter to provide an optically triggered input. A light pulse of sufficient magnitude turns on Q1 which forces the clock input of the 4553 low and advances the count. With this input, the counter can be triggered by a flashlight, an infrared LED, a xenon strobe, or lightning flash.

To trigger the circuit by interrupting a continuous light source, as when counting objects such as people or cars, you can use the phototransistor input stage in Fig. 3-2B. Here the phototransistor is normally illuminated, thus keeping the clock input to the 4553 high. When the illuminator is interrupted, the phototransistor is turned off and the input to the 4553 is brought low, thus causing a count to occur.

In both modes of operation, I've found that lasers and collimated infrared emitting diodes make excellent light sources.

A low power (e.g., 1 milliwatt) helium-neon laser, for example, easily actuates the circuit over a range of tens of feet. Since their near infrared emission more closely matches the peak spectral response of silicon phototransistors, collimated beams from GaAs, GaAs:Si, and AlGaAs LEDs and diode lasers can activate the circuit over longer ranges.

For more sensitivity, hence longer ranges, an op-amp gain stage can be inserted between the phototransistor and the 4553 clock input as shown in Fig. 3-3. In the count-when-interrupted mode, the circuit will register a count immediately when a continuous light falling on the phototransistor is blocked. In the count-when-flashed mode, the circuit will indicate a count at the *trailing* edge of a light flash.

Be sure direct sunlight does not strike the phototransistor or it may not work properly. If sunlight or bright ambient light

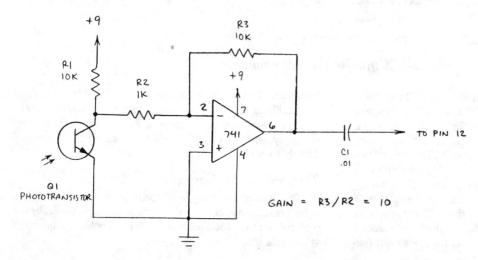

Fig. 3-3. Adding gain stage to phototransistor input to 3-event counter.

GAIN = R3 / R2 = 10

is a problem, try a filter and a light shield. Reverse biased PIN photodiodes such as the TIL413 are more immune to sunlight and can be used in place of the phototransistor.

A Truly Random "Coin Flipper" Circuit

Electronic "coin flipper" circuits have been popular with digital electronics hobbyists for more than a decade. Although there are various ways to design such circuits, most designs incorporate a clock coupled through a normally open push-button switch to a flip-flop or counter. The circuit is "flipped" by momentarily pressing and then releasing the push button.

This arrangement can provide a truly random 1-of-2 output since it's impossible for the operator to second guess the circuit when the clock is operating at a frequency of thousands of hertz. Ideally, of course, the clock should be operating as fast as the circuit will allow, consistent with allowable current drain.

Figure 3-4 shows a simple coin-flipper designed along these lines. In operation, two NAND gates in a 4011 are connected as a clock generator that oscillates at a frequency of about 15 kHz. The frequency can be increased by reducing the value of C1.

The signal from the clock is directly coupled to the count input of a 4017 decade counter. The 4017 is connected as a 1-of-2 flip-flop by tying its third lowest order output terminal (pin 4) to its reset input (pin 15).

The circuit is "flipped" by momentarily closing push-button switch S1. This grounds the clock enable input (pin 13) and permits the 4017 to respond to incoming pulses. Both LEDs, which are current limited by R2, appear to glow continuously since the clock rate is much faster than can be discerned by the human eye.

When S1 is opened, the 4017 is disabled and only one of the two LEDs remains illuminated. R3 pulls the clock disable input up to +9 volts to keep the 4017 from being inadvertently enabled by stray signals.

Does the circuit provide truly random flips? In an initial trial of 50 flips, 23 came up heads and 27 were tails. In a second trial of 50 flips, heads and tails came up 25 times each. That's random enough for me. I'll leave to you the task of determining if the circuit is perfectly random. It only takes a few minutes to build the circuit on a solderless breadboard, so have fun.

Incidentally, you might be wondering why anyone would bother building an electronic "coin-flipper" when it's so easy to write a program for a computerized version. Good question.

No doubt a creative programmer with nothing better to do could write a coin-flipper program that would graphically display in at least four colors a high resolution, simulated coin as it is flipped, spun through the air and dropped into an open hand! Of course the program could be a good deal simpler, perhaps a few lines using the machine's random number function.

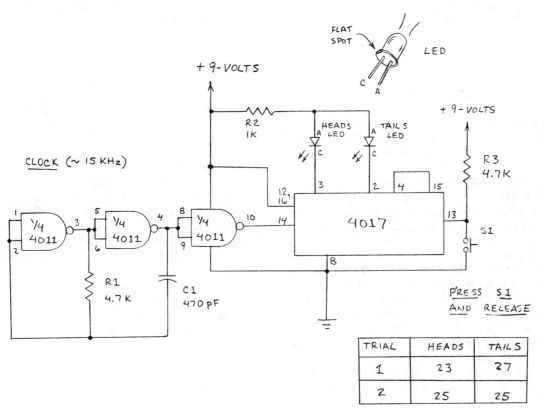

Fig. 3-4. Simple coin flipper circuit.

TRIAL	HEADS	TAILS
1	23	27
2	25	25

Still, there's something to be said about hard-wiring a dedicated circuit. It's always ready to use, it can be made small enough to fit in a pocket and anyone can use it.

Adding an Output Interface to a Digital Timer

One of the handiest gadgets in my office is a West Bend Electronic Timer. This compact, crystal-controlled device is a digital countdown timer that sounds an alarm at the conclusion of a preset, user-programmable interval. The interval can range from 1 second to 99 minutes and 99 seconds.

Figure 3-5 is a pictorial view of the timer's front panel. The unit is operated by keying in the desired interval and pressing START. When the count reaches 00M 00S, the timer emits a rapid series of attention-getting chirps. The chirps will sound for one minute or until the STOP/RESET key is pressed.

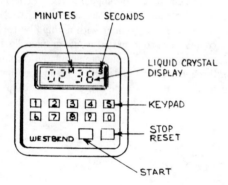

Fig. 3-5. Front panel of West Bend timer.

The countdown can be halted temporarily by pressing STOP/RESET. The countdown will resume when START is pressed. Pressing STOP/RESET twice clears the display to 00M 00S.

Opening the Timer and Adding Output Leads

On the back of the timer is a handy clip that allows the unit to be attached to a belt or pocket. It also functions as a desk stand. A magnet in the clip permits the timer to be temporarily attached to a metal surface.

Just below the bottom of the clip is a small Phillips screw. When this screw is removed, the back panel of the timer can be gently pried away from the front panel by inserting a flat screwdriver blade in the continuous, thin slot that separates the two panels. Be sure to pry along the bottom slot near the screw hole to avoid breaking the plastic snap retainers on the inside, upper edge of both panels.

After you lift the back panel away, temporarily set aside the power cell. Note the wires leading to a cylindrical component in the back panel. This is the piezoelectric alerter. The volume of sound that can be delivered by an alerter so small is truly remarkable.

Use a grounded or battery-powered soldering iron to remove both alerter leads from the circuit board. Wrap appropriately colored lengths of wrapping wire around the exposed wires of each lead, solder in place and then solder the two pairs of connection wires back to their original locations on the circuit board. Be sure to solder the leads to the correct locations, for if their polarity is reversed the alerter will not sound.

Drill a small hole in the timer's back panel. Then thread the two wrapping wire leads through the hole and replace the back panel. Be sure the battery contacts are properly aligned *before* replacing the screw. (You can see the contacts by removing the battery cover on the back panel.)

Replace the battery (plus side up) and check the timer for proper operation. Now you are ready to assemble the interface circuit.

An SCR Interface

The timer alarm signal consists of approximately eight bursts per second of 125-microsecond pulses having a 2.4-volt peak-to-peak amplitude. The pulses have a frequency of 4 kilohertz. Figure 3-6 shows a simple SCR interface that switches on upon the arrival of the first pulse and remains on until manually reset.

The SCR alone can control loads up to its rated capacity. I used a relay to provide isolation between the timer and the load and to increase the circuit's switching power. D1 reduces the possibility of 8-Hz chatter while the timer's alarm is sounding. D2 protects the SCR from reverse voltage developed across the relay coil when the reset switch is opened.

Not all SCRs will be reliably triggered by the timer's alarm signal. I had good results with Motorola's SCR1128 and Radio Shack's 276-1067.

Incidentally, you're on your own when you modify a manufactured electronic device such as this timer! Any warranty may be voided. And you must exercise caution to avoid damaging the timer's internal circuitry. You *must* also use caution and follow appropriate safety requirements should you use the modified timer to control line-powered devices.

Applications for the Interface

You can use the modified timer to control a darkroom enlarger, outdoor lighting, battery charger, appliance, radio, or television. For best results, install the interface and relay on a small board and mount it adjacent to a 9-volt battery holder in a suitable enclosure. Provide suitable plugs and jacks for connecting the timer to the interface and the interface to the device being controlled.

The timer can be clipped to a suitable shoulder or extension on the interface enclosure. Or you can use adhesive-backed squares of hook and loop fasteners. This will allow you to remove the timer when it's not being used to control the interface.

WARNING: You *must* follow safe wiring procedures if you use this circuit to control devices powered by the household line. Insulate *all* exposed connections. Do *not* exceed the contact ratings of the relay. The timer and interface should *not* be used for any application in which a circuit malfunction might cause injury to people or property.

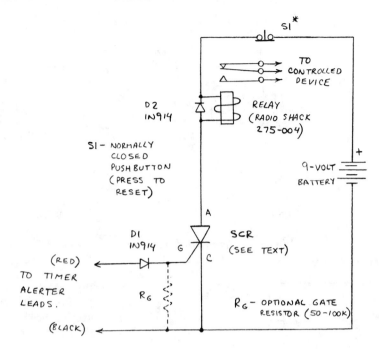

Fig. 3-6. Output interface for the timer.

Going Further

The timer I used is made by The West Bend Company. I purchased mine at a department store for about $12. Recently I've seen advertisements from three other companies showing timers having an almost identical appearance to the West Bend unit but selling for as much as $29.95! You may be able to modify these and other digital timers with the help of the circuit in Fig. 3-6. In any case, shop around for the best buy. You may save more than enough to pay for the interface circuit . . . and possibly enough to buy a second timer!

Adding an Output Interface to a Clock Module

The wide variety of inexpensive liquid crystal display watches and clocks now available is truly remarkable. Equally good news for the electronics experimenter is that watches and clocks with a built-in alarm function can be easily modified to control external devices such as radios and lights. All that's necessary is to connect a suitable interface circuit to the clock's alarm outputs.

In the space that follows we'll add an output interface to an LCD clock module designed to be installed in a user-supplied enclosure. Similar interfacing methods may work with preassembled digital alarm clocks and watches.

The Clock Module

The module which I've interfaced, the model PCIM-161A, is one of a family of miniature LCD clock modules made by PCI Displays. It can also be purchased from PCI representatives for $20 (or $15 in quantities of 10 to 99). This price is subject to change. If it seems too high, you might prefer to attempt adding an output interface to a standard, premanufactured digital alarm clock or watch. Keep in mind, however, that the circuit to be described has been tried only with the PCIM-161A. It may not work with other clocks.

Figure 3-7 shows both front and back views of the module. The crystal-controlled module has an accuracy of ± 2.5 minutes per year. It should be powered by a 1.3- to 1.6-volt supply. At 1.5 volts it typically consumes 6 milliamperes.

Figure 3-8 shows how the module can be connected to a piezoalerter (muRata PKM11-6A0, Radio Shack 273-064, or similar). When the alarm is activated, the alerter emits a tone interrupted at a 1-Hz rate with a 25 percent duty cycle.

Setting the calendar, actual time, and alarm time of this clock module requires a procedure similar to that for making the same settings for any nonkeyboard digital clock. The PCIM-161A specification sheet describes the setting procedure. Here is a brief summary:

1. When the module is first powered up, the display defaults to 1:00. Pressing SET will start the clock and cause the colon to flash once each second. The hours are set by first pressing MOD and then holding SET until the correct hour appears. Minutes are set by repeating this procedure. Press MOD and then SET to start the clock at the new setting.

2. The date is set by pressing MOD for 3 seconds or until the month and day appear. Press MOD and then hold SET until the correct month appears. Repeat this procedure for the day. Pressing MOD again will show a two letter mnemonic for the day of the week. Press SET until the correct day appears. Then press MOD to return to the normal timekeeping mode.

3. The alarm hour is set by pressing ALS twice within 3 seconds and then pressing SET until the correct time appears.

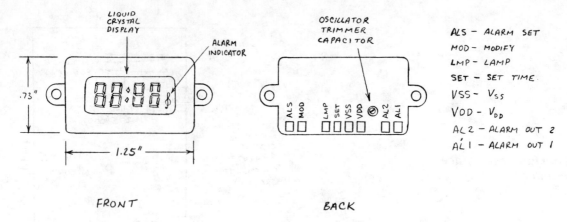

Fig. 3-7. Front and back views of PCIM-161A clock module.

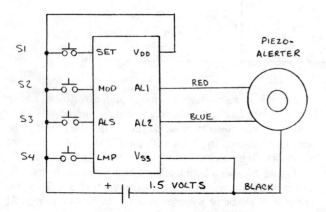

Fig. 3-8. Adding an alerter to the module output.

Press ALS and then SET to set the minutes. Pressing ALS a third time automatically returns the clock to the normal time-keeping mode. The display will now show the symbol indicating the alarm is set.

The alarm can be turned off at any time by pressing and holding ALS. Momentarily pressing ALS turns the alarm back on. In both cases the alarm time is displayed for a second before the time of day is restored to the readout.

An SCR Output Interface

Figure 3-9 shows a straightforward output interface circuit for the PCIM-161A. In operation, the alarm signal applied to the base of Q1 is amplified and applied through D1 to the gate of the SCR. This turns on the SCR and, in turn, pulls in the relay. The SCR stays on after the alarm signal ends unless the

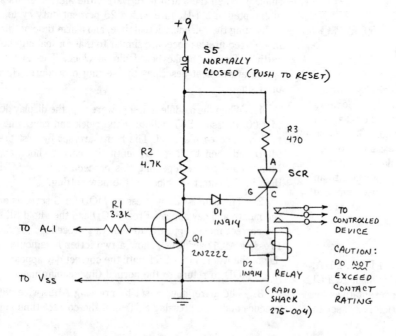

Fig. 3-9. An output interface for the clock module.

reset switch (S5) is momentarily opened. D2 absorbs any back emf that is generated by the relay coil when the circuit is reset.

Applications

The interfaced clock module is ideal for turning lights and appliances on or off at any desired time. For best results, install the module on the front panel of a small, plastic enclosure. Install push buttons S1–S5 on the front panel below the clock module. Install the interface circuit and holders for a 1.5- and 9-volt battery inside the enclosure. Include a suitable jack for connecting devices to be controlled by the relay contacts.

Before mounting the clock's control switches (S1–S4), test the clock for proper operation. It may be necessary to use double pole switches that hold the inputs low when they're not switched high.

WARNING: You *must* follow safe wiring procedures if you use this circuit to control devices powered by the household line! Insulate all exposed connections. Do not exceed the contact ratings of the relay. The clock and interface should not be used for any application in which a circuit malfunction might cause injury to personnel or property.

Going Further

Keep in mind you might be able to use the methods outlined here to interface most commercial digital clocks with an alarm output. For example, I plan to try the circuit with a keyboard programmable digital alarm clock.

The Electrostatic Discharge Problem

Everyone has experienced the static discharge that occurs when one touches a metal object after walking across a carpet on a dry winter day. But few people are aware that high voltages are accumulated by many common objects.

Things made from plastic are notorious generators and accumulators of very high static charges. Styrofoam cups, cigarette and candy wrappers, parts trays, and some kinds of solder removal tools are all potential high voltage generators. These and many other plastic objects are commonly found on or near electronic work benches.

It's surprisingly easy to demonstrate the accumulation of a static charge by plastic objects. For example, rub a piece of plastic packing snow between two sheets of dry paper, and the plastic will adhere to a surface having an opposite charge. Or rub a balloon on a flannel shirt and it will stick to a ceiling.

A neon glow lamp makes a handy visible indicator of static electricity. Walk across a rug while wearing leather soled shoes to accumulate a charge and touch one lead of a neon lamp to a metal object while holding the other lead between a thumb and forefinger. The lamp will flash when the discharge occurs.

It's very important to isolate MOS, CMOS, and other components which are vulnerable to electrostatic discharge (ESD) from objects which can generate a static charge. Ideally, all static generating objects should be removed from the vicinity of vulnerable components. Soldering irons should be grounded (or battery powered) as should workers who handle components.

Manufacturers often ship components and circuit boards that are vulnerable to ESD in antistatic polyethylene bags known as "pink poly." These special purpose bags do not develop the high potential which ordinary polyethylene bags easily generate when rubbed or flexed.

3M Static Control Systems makes a different antistatic bag. The 3M bag, which is more expensive than pink poly, consists of an inner layer of antistatic polyethylene and a polyester strength layer coated with a 10-micron thick film of nickel.

When this subject was discussed in one of my columns, Dan C. Anderson of the Richmond Division of Dixico, Inc. responded to this item with a thick package of literature about his firm's antistatic products. He also sent along some samples of Richmond's pink poly as well as some special purpose RCAS 3600 antistatic bags which provide both rf and EMI shielding.

Being a long time static electricity experimenter, I was particularly attracted to Dan's method of demonstrating the static electricity produced when transparent adhesive tape is unrolled. He says to place a neon lamp, whose leads have been spread apart, near a spool of tape. The lamp will glow as the tape is unrolled. I tried this demonstration and it worked even on a very rainy day. (For best results, dim the lights and pull the tape rapidly.)

The primary purpose of Dan's package, however, was to explain the merits of pink poly. According to their literature, Richmond RCAS 1200 was the first pink poly. Prior to its development, the chief antistatic wrap was Velostat, a product of Custom Materials, a company since acquired by 3M. Velostat is made by mixing finely ground carbon particles with polyethylene or a similar resin. It is used to protect electronic components, printed circuit boards, and explosives from ESD. Unlike Velostat, pink poly is transparent. The pink hue is added to distinguish the material from ordinary plastics.

According to Richmond, the development of its pink poly was stimulated by a 1964 tragedy at Cape Canaveral in which three men were killed by the accidental ignition of a solid propellant rocket motor inside a hangar. The rocket was apparently ignited when a static discharge generated by a polyethylene dust cover over the rocket caused a spark to jump across the ignition squib.

Pink poly is made by impregnating ordinary polyethylene resin with an antistatic liquid. According to Richmond, the antistatic liquid ". . . forms a self-renewing, noncorrosive 'sweat layer' on all its exposed surfaces by combining with the moisture found in normal air." If removed by a solvent or abrasion, a new layer of antistatic compound is eventually formed.

Apparently there is a good deal of healthy competition between 3M, Richmond, and other companies over the relative merits of their respective antistatic products. Richmond, for instance, is quick to point out that categorical criticism of pink poly is unfair since the product is "widely and poorly imitated." They also note that their RCAS 1200 meets the requirements of military standard MIL-B-81705, Type II, "and is still the only material meeting this as determined by the government's Qualified Products List."

On the other hand, 3M observes "No one product . . . no one technology . . . can offer full protection from static." They then boast that "Only 3M has the products and the trained static analysts to give you total control of the static in your business."

If reports in various technical journals and trade magazines are a reliable indicator, protection against component damage due to ESD is becoming a matter of major concern and importance. For example, at a forum on ESD sponsored by *Electronic Products* magazine, several conferees noted that though ESD damage to components and assembled circuit boards is a serious problem, many companies don't have the technical expertise necessary to trace their rejects and failures to ESD. Some are unwilling to invest the funds necessary to equip and maintain a static-free work environment.

You can learn more about the *Electronic Products* forum in that magazine's June 1980 issue (pp.31–38). If you're involved in the manufacture of circuit boards or systems which use components vulnerable to ESD damage, the Department of Defense has published a detailed standard on the subject. It's designated 1686 and is entitled "Electro Static Discharge Control Program for Protection of Electrical and Electronic Parts, Assemblies and Equipment." You can request a copy of the standard by writing the Navy Publications and Forms Center, 5801 Tabor Avenue, Philadelphia, PA 19120.

In the meantime, pay particular attention to antistatic procedures to protect vulnerable components, especially MOS and CMOS chips, from ESD. Richmond has formulated a set of antistatic rules you may wish to follow. They're called "The S-I-G-H of Relief from ESD" and here they are:

1. *Surround* . . . the device or assembly with antistatic materials (bag, lidded box, or other shaped container) except when it is being actively worked on.

2. *Impound* . . . all plain plastics and textiles, foams and cushionings from near approach to the items. Replace with approved antistatic types or treat with topical antistatics.

3. *Ground* . . . the skin of all item-handling personnel with safely resistive wrist straps. Where this is not possible, use conductive floor mats and appropriate footwear.

4. *Hound* . . . personnel and management to see that the above rules are observed, for without breaking one of them it is virtually impossible to cause electrostatic damage.

Richmond's Dan Anderson acknowledges Fred Mykkanen of Honeywell Defense Systems for originating the "S-I-G-H of Relief" idea. Mr. Mykkanen is an authority in the field of ESD control.

I hope this discussion on the importance of protecting sensitive components from ESD damage doesn't cause a single reader to avoid using MOS and CMOS chips and transistors! CMOS is the best way to go, in my opinion. It's very flexible, simple to use, and consumes little power.

My CMOS chips are inserted in aluminum foil covered styrofoam salvaged from the grocery store's meat counter. The foam plastic is cut to fit inside ordinary plastic parts trays. Although the contact between the foil and the IC leads may cause some reaction to occur, thus far none of my CMOS chips have been damaged by ESD . . . to the best of my knowledge. I have, however, zapped a few chips or individual gates by foolish or accidental circuit errors. I always touch a grounded object before handling CMOS chips and I use a battery powered soldering iron. Finally, loose chips are laid on a sheet of aluminum foil until they are inserted into a circuit or placed back in their foil-covered carrier.

The Varistor

The varistor, a nonlinear resistor that does not obey Ohm's law, is one of those wonderfully simple components whose operating principles are not yet fully understood. What is understood is that the varistor is well suited for protecting electronic circuits from transient voltage surges.

The current through a standard resistor is directly proportional to the voltage across the resistor. The current through a varistor, however, varies according to a power n of the voltage where n is from 2 to 25. Figure 3-10 is a curve of the voltage versus current characteristics of a typical varistor.

Varistors are made from silicon carbide or metallic oxides, and their construction more closely resembles that of a resistor than a semiconductor device. Metal oxide varistors, for example, are made from powdery grains of zinc oxides and small amounts of other metal oxides. The zinc oxide powder is pressed and sintered to form ceramic discs. The discs are coated on opposite sides with silver to provide solderable contact regions.

The varistor's voltage and current handling capacity are determined, respectively, by the thickness and width of the disc. The size of the zinc oxide grains also determines the varistor's voltage rating.

Varistors are primarily used to protect electronic circuits from transient voltage surges. They are particularly important when it is necessary to protect a computer from power surges that cause memory errors or even loss of memory contents.

As you can see by referring back to Fig. 3-10, the operation of a varistor resembles that of two back-to-back zener diodes. When the voltage across the varistor is below its rated continuous voltage, the varistor is in a high impedance state. When the voltage exceeds the rated level, the varistor suddenly begins to conduct and its formerly very high impedance falls to a few ohms.

Voltage transients can be caused by power line surges, lightning, switch arcing, component failures, and the sudden removal of a voltage across a transformer, inductor, or relay coil. Figure 3-11, for example, shows a relay coil connected through a switch to a power supply. When the switch is opened, the field in the coil collapses and induces a series of high voltage spikes across the coil terminals. The varistor serves to short to ground the spikes having an amplitude above the varistor's voltage rating.

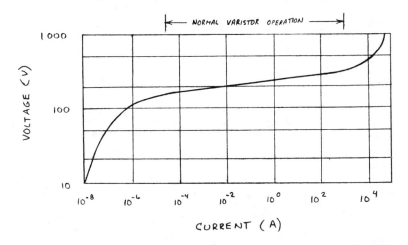

Fig. 3-10. Typical voltage vs current characteristics of a varistor.

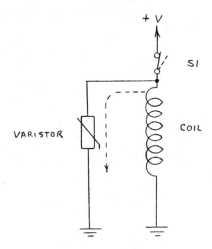

Fig. 3-11. Using a varistor to clamp inductive spikes across a relay coil.

A Pair of Low Voltage Varistors

General Electric includes in its line of GE-MOV II metal oxide varistors two low voltage devices designed specifically to protect TTL integrated circuits from transient voltage spikes. Designated V8ZA1 and V8ZA2, the two varistors are rated for a continuous voltage of 5.5 volts. The V8ZA1 is rated for a maximum clamping voltage of 22 volts at 5 amperes and a peak transient current of 100 amperes. The V8ZA2 is rated for a maximum clamping voltage of 20 volts at 5 amperes and a peak transient current of 250 amperes.

General Electric claims these varistors perform better in overstress situations than competing silicon devices. Another advantage is low cost. In large quantities, the General Electric devices sell for around 35 cents each.

Incidentally, in addition to its low voltage varistors, General Electric also manufactures an entire line of high-voltage, high-current devices. For example, B-series devices can protect industrial electrical systems from transients with peak currents of 70,000 amperes and energy levels up to 10,000 joules. At 200 amperes, the maximum clamping voltage of one such device is 7800 volts.

For more information about metal oxide varistors, contact a General Electric distributor or write the Semiconductor Products Department of General Electric. General Electric has published an excellent applications manual on varistors, *Transient Voltage Suppression*. The manual is GE publication number 400.3, and it sells for $5.00. The company offers for an unspecified price, design kits which include a low voltage GE-MOV II varistor.

Varistor Operating Precautions

When exposed to voltage or current transients above its peak ratings, a varistor may fail. General Electric's varistor literature describes a possible consequence being " . . . mechanical rupture of the package accompanied by expulsion of package material in both solid and gaseous forms." An explosion by any other name is still an explosion. For this reason, if overstress conditions are expected, it is wise to fuse the varistor or to mechanically shield it from other components and people.

FOUR

LEDs, Laser Diodes, and Optoelectronics

- The Rainbow LED
- Super Bright LEDs
- Visible Light Laser Diodes
- Experimenting with CW Laser Diodes
  Part 1. How-To Basics and a Laser Pulse Transmitter
  Part 2. Two CW Laser Diode Sources
- Experimenting with Electronic Flash Circuits
- A Photonic Door and Window Intrusion Alarm
- Optoisolators: The Photon Connection
- A Two-Way Optoisolator

LEDs, Laser Diodes, and Optoelectronics

The Rainbow LED

A light emitting diode capable of emitting each of the three primary colors (red, yellow and blue) in any combination would be an optical display designer's dream. Such an LED would be far more than a tricolor emitter for it could generate any color or hue. With all three colors simultaneously activated, it would appear to be a white light source.

So far the rainbow LED remains an elusive goal. The major obstacle is the design of an effective blue emitter. Experimental gallium nitride and silicon carbide blue emitting diodes have been demonstrated, but they are very difficult to make. A mass produced blue LED has yet to appear.

I'm convinced a practical rainbow LED will someday be developed. In the meantime, the next best substitute may be a gallium phosphide (GaP) multicolor LED developed by Takao Yamaguchi and Tatsuhiko Niina of the Sanyo Electric Co., Ltd. This LED consists of a stacked pair of pn junctions arranged as shown in Fig. 4-1.

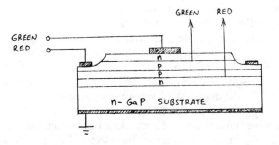

Fig. 4-1. Cross section of the new monolithic dual color LED.

The LED is constructed by first depositing n- and p-type layers of crystalline GaP on an n-type GaP substrate. The resulting pn junction emits *red* light when forward biased.

Next, a second pair of n- and p-type layers is deposited over the first two layers. This junction emits *green* light when forward biased. And it permits the passage of red light emitted by the underlying junction.

The green emitting GaP layers block electrical access to the p-side of the red emitting junction. Therefore, it is necessary to etch through the uppermost n-type layer to provide electrical access to the top side of the red junction. This explains the mesa structure shown in Fig. 4-1.

A nonmesa or *planar* diode may be formed by etching channels through the uppermost n-type layer. Contacts are more easily applied to a planar version of the diode.

You may be wondering why the first GaP pn junction emits red light and the second junction emits green. The key is the doping of the GaP crystalline layers which are epitaxially deposited (grown) over the GaP substrate. When oxygen is added to p-type GaP, a red emitting junction is formed. A green emitting junction is formed when nitrogen is added to both p- and n-type GaP.

Either junction of this new LED may be biased independently of the other. This permits the diode to function as an emitter of either red or green light. When *both* junctions are simultaneously forward biased at various current levels, the eye perceives the resulting two-color combinations as red, orange, yellow, or green. The perceived hue is determined by the relative brightness of the two junctions.

An identical effect can be obtained by switching each junction on and off at a rate greater than about 20 hertz. Even though only one junction is biased at any given instant, the human eye's persistence of vision causes the red and green to merge into intermediate hues. The color perceived is governed by the duty cycle and current level through each junction.

Applications for Sanyo's LED include multicolor indicators and displays. The most interesting applications incorporate arrays of the diodes to form multicolor characters, digits, and even images.

Takao Yamaguchi and Tatsuhiko have published an interesting paper on this LED which describes in detail its construction and operation. It's entitled "A High Brightness GaP Multicolor LED" and it appeared in the *IEEE Transactions on Electron Devices* (Vol. ED-28, No. 5, May 1981, pp. 588–592). You can find this journal at most technical libraries.

Conventional Multicolor LEDs

Until single-chip multicolor LEDs like the one in Fig. 4-1 become available commercially, you can experiment with two-chip versions which are available *now*. These devices are made by installing red and green emitting chips side by side on a single LED header.

Dual-chip LEDs made in this manner are called bicolor or tristate LEDs. Tristate refers to the possibility of obtaining a third color, yellow, by biasing both diodes simultaneously or in rapid sequence.

You can obtain bicolor LEDs from Radio Shack, Opcoa, AEG-Telefunken, and possibly other sources. The Radio Shack and Opcoa diodes are two-lead devices in which the two chips are internally connected in reverse-parallel. This means the direction of current flow must be reversed to switch colors. For the intermediate color of yellow, an ac bias must be provided.

Figure 4-2 shows a simple circuit to apply an ac bias to two-chip diodes like these. The two TTL gates form an astable multivibrator whose switching rate is controlled by potentiometer R1. At slow switching rates the diode alternates between red and green. The effect is quite striking. At faster rates, the two colors merge into a washed-out orange or yellow hue.

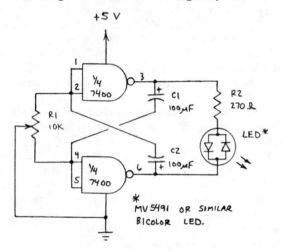

Fig. 4-2. Simple bicolor LED driver.

AEG-Telefunken's dual-chip LED, designated the V 518 P, incorporates separate anode connections to each of the two chips. The header to which the chips are soldered serves as a common cathode. Though I've not yet experimented with this LED, I suspect it is better suited for the production of yellow than the two-lead, tristate LEDs. A simple battery and potentiometer arrangement should be all that is necessary to produce red, yellow, green, and intermediate hues.

Super Bright LEDs

In this era of increasingly complex microprocessors and very-large scale memory arrays, it's hard to get excited about a simple component like a light-emitting diode. That's why a few days passed before I got around to evaluating three ordinary-looking LEDs that were sent by Lance Kempler, a long-time reader of my columns and books.

Lance was then with A.C. Interface, Inc., the U.S. Company that represents Stanley Electric Co., Ltd., one of Japan's major manufacturers of LEDs and photodetectors. Before sending the

sample diodes, Lance had called to say they were on the way and that I'd be surprised at their brilliance. That prediction turned out to be an understatement. The LEDs are the most amazing I've seen since first viewing the soft red glow of an early gallium arsenide phosphide (GaAsP) red-emitting diode during a visit to Texas Instruments back in 1966.

So bright are the LEDs that when I first tested one of them there appeared to be a problem with the digital multimeter used to monitor the current through the diode. When the power supply was cranked up so that the LED was emitting a bright red glow, the multimeter indicated a current of only 3 milliamperes. A normal LED would have required considerably more current than that to achieve a similar level of brightness, so I substituted another meter. But the result was the same. Still not trusting the separate but equal current readings, I determined the current by measuring the voltage across the 100-ohm current limiting resistor in series with the diode and then used Ohm's law to calculate the current (I = E/R). The result was still 3 milliamperes.

After making that third measurement, I suddenly remembered Lance's remarks about being surprised and decided to apply a full 50 milliamperes to the diode. The LED emitted such an astonishingly bright beam it was impossible to stare directly into the end of its epoxy lens.

These super-bright LEDs are designated by Stanley as part number H2K. According to the specification sheet Lance sent with the diodes, the H2K is a gallium aluminum arsenide (GaAlAs) device that has a brightness of 2,000 mcd (millicandelas) when biased with the industry-standard forward current of 20 milliamperes. At 50 milliamperes, the brightness increases linearly to an incredible 4500 mcd.

How bright is 2000 or more millicandelas? Prior to the arrival of the H2Ks, the brightest red LEDs in my benchstock were some Hewlett-Packard "ultrabright" HLMP-3750s. According to the HP data sheet, these diodes are made from GaAsP on GaP and have a typical brightness of 140 mcd at 20 milliamperes. In other words, the brightness of the H2K is more than 14 times greater for the same forward current. In fact, at 2-mA forward current, the H2K achieves a brightness of 200 mcd, higher than that of the HP diodes at 10 times the current level.

As might be expected, the new LEDs have an exceptionally high power conversion efficiency. An incandescent lamp transforms only about 5 percent of the electrical current flowing through a hot filament into visible light. The new super bright LEDs are three times as efficient as incandescent lamps. If not for losses of light caused by internal reflection and reabsorption inside the LED chip and additional losses contributed by the LED package and chip header, the new diodes would possess considerably higher efficiency.

Applications for Super Bright LEDs

Stanley's state-of-the-art LEDs are so much brighter than previous devices that several important new applications are made possible. One of the most obvious uses is to form arrays of the new diodes into burnout-proof lights for automobile tail-lights and traffic signals. When placed behind a plastic diffuser, a single super bright LED can function as a bicycle taillight. Linear arrays of the new diodes can function as high-brightness light sources for xerographic-style copy machines that also function as computer printers. A warning flasher could be made simply by connecting one of the new LEDs to a simple pulse generator circuit.

A more sophisticated role for the new LEDs is as a light source for both optical fiber and free-space lightwave communication systems. Low-cost plastic fibers transmit well at the 660-nanometer wavelength emitted by the new LEDs. The optical diodes used in free-space systems are almost always the infrared-emitting variety since they are so much more powerful than conventional red LEDs. However, and this is truly impressive, the new LEDs are just as powerful as most of their infrared-emitting counterparts. Therefore, in applications where a visible beam is acceptable, the new red LEDs will make an excellent alternative. If you've ever attempted to point the invisible beam from an infrared communicator directly at a distant receiver, you can better appreciate the value of an easily modulated source like an LED that happens to be emitting a very powerful yet visible beam of light.

In preliminary tests, the new LEDs perform well as light sensors. Therefore, they can be used as dual-function source-sensors in two-way lightwave links. I have used this method extensively in the design of lightwave communication systems that send and receive information over a single fiber.

The features that make the new super LEDs well-suited for lightwave communications are equally desirable for such applications as reflection-mode and break-beam detection systems. The new diodes could also be used in visible light versions of infrared remote control units for home appliances and toys.

For applications in which the LED is paired with a sensor, it's important to consider the spectral sensitivity of the sensor. Figure 4-3, for example, compares the spectral emission of the H2K with the spectral sensitivity of two photodetectors. Some photodiodes have better sensitivity to the 660-nm wavelength of the H2K than the one shown in the illustration. Note how the 880-nm near-infrared emitter more closely matches the peak sensitivities of the two photodetectors.

Finally, a number of scientific applications for super-bright LEDs come to mind. For example, since LEDs can be pulsed on and off in tens of nanoseconds, an array of super LEDs could be used to illuminate ultrafast moving objects in high-speed photography. Another scientific application would be to use a 660-nm super LED in an array with a conventional AlGaAs 880-nm diode and a GaAs:Si 950-nm emitter to quickly measure the reflectance of objects at three distinct wavelengths.

Measuring the Power of Super LEDs

Prior to trying the sample super bright H2K LEDs in some simple circuits, I first used a Centralab CSC-12 calibrated silicon solar cell to measure their power output. Figure 4-4 shows both the basic drive circuit for these tests and the formula for calculating the value of the required current limiting resistor. The actual tests were made with power supplied by an adjustable supply.

According to a technical paper prepared by scientists from Stanley Electric Co., Ltd., at a forward current of 20 milliamperes the typical H2K emits 6 milliwatts. At 50 mA, the typical power output rises to 14 mW. At the 660-nm wavelength emitted by the H2K, the CSC-12 detector used to measure the power in the beam from the diodes has a responsivity of 0.42 mA/mW.

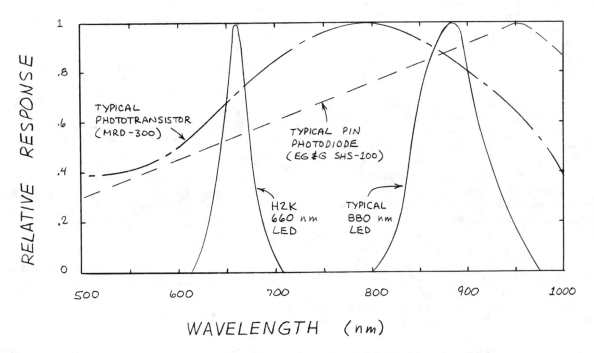

Fig. 4-3. Comparing the spectrum of the H2K LED with other devices.

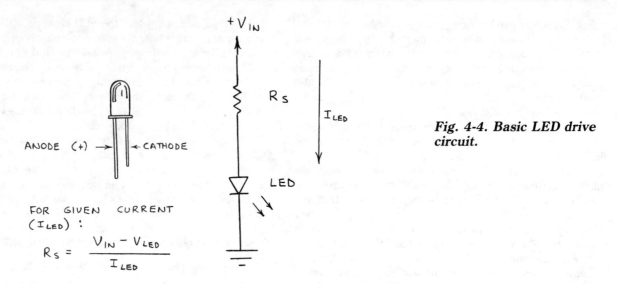

$+V_{IN}$

R_S

I_{LED}

LED

ANODE (+) → ← CATHODE

FOR GIVEN CURRENT (I_{LED}):

$$R_S = \frac{V_{IN} - V_{LED}}{I_{LED}}$$

Fig. 4-4. Basic LED drive circuit.

In other words, for every milliwatt of 660-nm radiation that strikes the detector, an output current of 0.42 milliampere is produced.

The three sample diodes I tested emitted less power than specified in the Stanley paper. At 50-mA forward current, the measured output power for each diode was:

LED 1: 5.55 mW

LED 2: 7.10 mW

LED 3: 5.05 mW

Each LED was measured within one second after power was applied since the power output falls somewhat as the chip becomes warm. At 50-mA forward current, the minimum specified output for these LEDs is 70 percent of the typical value of 14 mW or 9.8 mW, still significantly more than the measured results. Why?

My tests measured only the power contained within the focused beam emerging from the lensed end of the LEDs. However, some light is emitted from the sides and even the base of epoxy-encapsulated LEDs. Manufacturers usually measure the *total* power emitted by an LED by collecting virtually all the radiation with an *integrating sphere*, a hollow sphere coated on its inside with highly reflective white paint. The LED being measured is inserted in one aperture and a detector is inserted in another. Though the calibrated detector I used collected all the focused beam, it collected none of the off-axis light spilling out the sides of the epoxy package.

In other words, the power within the central beam of the H2K is only about half the total power emitted by the LED. Nevertheless, the power is still substantial. Only a few years ago, the best near-infrared emitting diodes made from GaAs:Si and encapsulated in epoxy much like the H2K were considered good quality devices. However, at 50-mA forward current their central beam contained 3 mW, only about half the power of the H2Ks I tested. AlGaAs near-infrared 880-nm diodes emit about the same power as the red H2K LEDs.

Figure 4-5 shows the beam pattern emitted by an actual H2K. Like other visible and infrared optical diodes in which the chip is installed within a miniature reflector, the far-field beam structure is a bright central spot surrounded by a somewhat dimmer halo. The central spot is the imaged surface of the LED chip. The dark spot within the central spot is the point of attachment for the chip's upper lead.

The halo effect is caused by the tiny reflector in which the LED is installed. The reflector captures light emitted from the sides of the chip and reflects it toward the lens formed by the curved end of the epoxy package. Since the reflector is larger than the chip, it is imaged as a halo around the chip. The halo has a broader beam spread or divergence (about 20 degrees) than the chip (only about 7 degrees).

Note that considerable light emitted by the LED is *not* within the central spot and halo. Some is contained within a very broad, off-axis halo that surrounds the central halo and spot. The rest emerges from all sides of the diode's clear package.

Figure 4-6 shows the power output of the lowest powered H2K I measured. For these measurements, continuous power was supplied to the LED. The power output would have been somewhat higher if the measurements were made within a second or two after power was applied.

A Super Bright LED Flasher

Figure 4-7 shows an ultrasimple but effective flasher circuit for a super bright LED such as the H2K or similar devices. This circuit flashes approximately 1.5 times per second using the values shown. The flash rate is directly proportional to the size of capacitor C1.

For applications in which maximum brilliance is not required, R3 can have a very high value. Since the H2K has a brightness of hundreds of millicandelas at only a few milliamperes forward current, the current drain of the flasher circuit will be exceedingly low and battery life will be quite long.

Going Further

If you haven't previously worked with LEDs, you can find some basic information about these and other solid-state optoe-

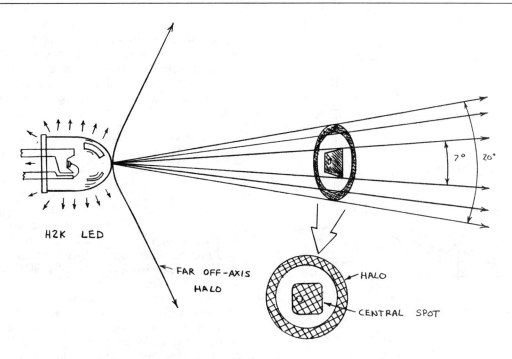

Fig. 4-5. Beam profile of the H2K LED as measured by the author.

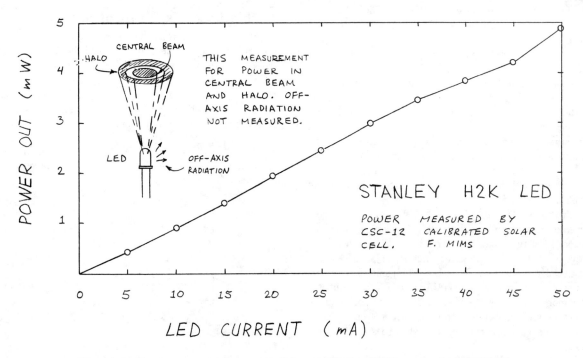

Fig. 4-6. Output power in central beam of the H2K super bright LED.

lectronic devices in a book I've written for Radio Shack called *Getting Started in Electronics*. The book also includes some circuits you may wish to try. For additional information about the H2K and similar super bright red LEDs, write A.C. Interface, Inc. (address in the Appendix). Since the yield of the super-

bright H2K LED is low, the diode's price is well above standard LEDs.

Incidentally, at a time when high-power near-infrared emitting diodes can be purchased for a few dollars or less, the price of the H2K probably seems out of line. On the other hand, old

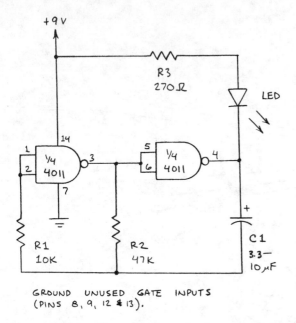

Fig. 4-7. Super bright LED flasher.

timers can readily remember when very dim red LEDs cost about the same as the H2K. As for near-infrared emitters, the first ones I experimented with back in 1966 cost $365 each! No doubt the price of the H2K device will decrease over time as the yield is improved.

If you want to explore fiber optic communication applications for super bright LEDs, Stanley makes the FH511 emitter and FS511 detector. These devices are encapsulated in special flat-ended packages in which the sensing and emitting chips are very close to the surface. Stanley also makes plastic fiber connectors for these devices.

Many other companies also make inexpensive LEDs and detectors designed to be installed within plastic connectors. At the present time, the chief advantage of the Stanley diodes is that their peak wavelength of emission is transmitted well by inexpensive plastic optical fiber.

Though the H2K is one of the most exciting LED developments to occur in some time, other big developments are on the horizon. LED technology was first invented in and extensively developed in the United States long before Japanese companies entered the business. Since U.S. companies have been making various kinds of powerful GaAlAs near-infrared emitting diodes for more than five years, there should be no fundamental reason why they can't also develop super-bright visible LEDs. Should this development occur, super-bright LEDs will become very cheap.

After several false starts, watch for the eventual availability of the elusive blue LED. Finally, considerable work is underway in several countries in an effort to solve the very difficult problem of reducing to below 700 nanometers the wavelength of continuously operating laser diodes. The objective of making laser diodes that emit at visible red wavelengths is to make them better suited as a readout device for audio and video laser disc players. Considering the enormous market for laser disc players, visible light laser diodes that emit about the same power as the H2K LED should one day be very affordable. The advantage of the laser over the LED is that the light emitted from a laser can be collimated, by means of a simple convex lens, into a much tighter beam since it originates from a much smaller point.

Visible Light Laser Diodes

How would you like to have a compact laser no bigger than a pocket penlight? Moreover, what if this midget laser could emit a bright red beam just as narrow, intense, and powerful as that emitted by a much bulkier helium-neon (HeNe) laser? And would you be impressed if this laser included a built-in modulator circuit for transmission of both analog and digital signals from a battery powered circuit?

Though you cannot yet purchase a laser with these remarkable properties, I'm happy to report the technology for this amazing new laser has already arrived in the form of a new generation of *visible light emitting laser diodes*. We'll discuss this new technology shortly. But first let's briefly review the background of this new development so you can better appreciate its significance.

Visible Light Vs. Near-Infrared

Figure 4-8 is a graph that shows the light sensitivity of the typical human eye. The graph is called the *photopic luminosity curve*. The curve shows that visible light ends and near-infrared begins at or near a wavelength of about 720 nanometers (nm). However, the curve is asymptotic at both extremes since the actual point at which the eye no longer responds is determined by the intensity of the radiation.

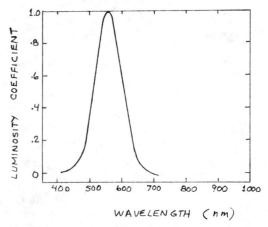

Fig. 4-8. The photopic luminosity curve.

For example, I know a laser technician who several years ago accidentally looked into the beam of a powerful neodymium doped YAG laser and reported seeing a bright orange glow. According to a quick glance at the photopic luminosity curve this is impossible since the YAG laser emits at a wavelength of 1060 nm, well beyond the range of human vision. Perhaps the technician, who was fortunate his eyes were not permanently injured by this experience, actually observed some of the visible light used to "pump" the YAG laser crystal. Or perhaps he saw the 1060 nm.

In any event, I've often clearly observed the "invisible" 880-nm radiation emitted by aluminum-gallium arsenide (AlGaAs) light emitting diodes (LEDs). Of course, in the interest of accuracy, I should note that these LEDs emit a spectrum of wavelengths *centered* at 880 nm. In other words, some of the radiation is closer to the visible wavelengths than the 880-nm figure suggests.

I've related these two examples to illustrate the controversy that surrounds the definition of where visible light ends and near-infrared radiation begins. Suffice it to say that the reports of researchers who claim to have developed "visible light" laser diodes are generally viewed with considerable skepticism. Until recently, that is. With these thoughts in mind, let's look at the evolution of the visible light laser diode.

Visible Light Emitting Laser Diodes

The first laser diodes were made in 1962 from gallium arsenide (GaAs), a semiconductor alloy still used to make many such lasers. Figure 4-9 compares the simplest possible light emitting diode chip with an early GaAs laser diode. The laser is a pn diode alike in nearly all respects to the LED. Unlike ordinary LEDs, however, the laser version has a very smooth and flat junction and two opposing ends of the chip have been cleaved to form perfectly flat and parallel reflective facets that function as end mirrors. The mirrors provide the optical feedback necessary to establish and sustain laser operation. Radiation is emitted from the chip at the junction region of both facets.

Today's sophisticated laser diodes have structures much more complex than the simple pn junction version shown in Fig. 4-9. Nearly all are made by forming multiple layers of semiconductor alloys into a sandwich-like configuration that confines to the junction region radiation that would otherwise be absorbed in the crystal or escape out the sides of the chip. I described several of these structures in an earlier article you may want to review ("The Laser at Twenty," *Popular Electronics*, December, 1980).

At room temperature, GaAs lasers emit invisible near-infrared radiation at a wavelength centered at about 904 nanometers. The wavelength range of these lasers is much narrower than similar diodes operated as nonlaser infrared emitting diodes.

However, even the very first laser diodes emitted red light because they could only be operated when cooled to the temperature of liquid nitrogen (−196°C). The light is shifted downward in wavelength from the near-infrared to the visible red because the emission wavelength of *any* light emitting diode, laser or otherwise, decreases with the temperature of the diode's junction. A liquid nitrogen cooled GaAs laser, for example, emits a wavelength of about 845 nm. When cooled to this temperature, the laser will operate continuously.

When I was a development engineer at an Air Force laser laboratory, I once designed and worked with such a cryogenically cooled laser diode. I well remember the brilliant, cherry red beam the laser emitted. Of course, operating a laser diode, or any other device for that matter, at the temperature of liquid nitrogen is a major hassle. That's why scientists have long sought to develop a laser diode that will emit visible light *and* do so continuously at room temperature.

Several kinds of room temperature laser diodes that emitted radiation at the far end of the visible red portion of the spectrum were developed in the United States during the 1960s and early 1970s. One was made from gallium arsenide phosphide (GaAsP),

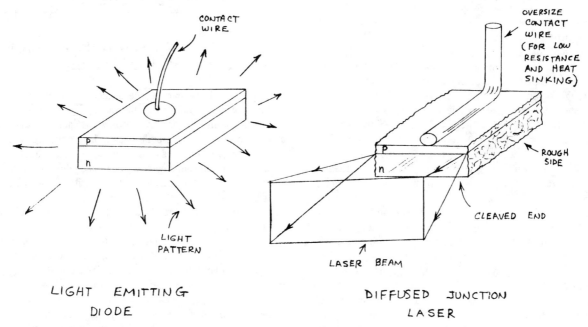

Fig. 4-9. The physical structure of a simple LED and injection laser.

an alloy that emits at about 860 nm. A second was made from aluminum gallium arsenide (AlGaAs), an alloy that then emitted wavelengths as low as 805 nm.

These early visible light lasers could not be operated continuously at room temperature. Instead, they were driven by brief (200 nanosecond) current pulses.

I've used both these kinds of lasers with various drive circuits. Although they do produce a visible red light, the intensity is very low. The light from the GaAsP lasers I've used could only be seen when the room lights were dimmed. The emission from the AlGaAs lasers was slightly brighter. When viewed straight on, the chip emitted a bright but tiny sparkle of red. For practical purpose, neither kind of laser can be considered useful as a visible light emitter.

During the late 1970s, RCA, a leader in laser diode research and development, produced laser diodes of AlGaAs that operated continuously at room temperature and emitted at wavelengths as low as 740 nm. As early as 1970, RCA scientists had produced lasers that emitted at 690 nm in a pulse mode, but they were unreliable due to the high current density required to attain laser operation.

In recent years the race to develop practical, visible light laser diodes has reintensified. The main incentives for the new interest are laser scanned audio and video discs and laser printers. In both cases, laser diodes could be better utilized as compact, low-power-consuming radiation sources if only their wavelength of emission could be reduced. This would permit the laser beam to be focused to a smaller spot than that available from a conventional near-infrared emitting laser diode.

The small spot size of focused light from visible laser diodes would mean audio and video discs could carry as much information as competing discs designed for use with the 632.8-nm radiation emitted by the HeNe laser. And the laser nonimpact printer could achieve better resolution. Furthermore, the laser printer would require less power since the same energy could now be focused into a smaller space.

Although virtually all important advances in laser diode technology over the past twenty years took place in the United States, several Japanese companies have recently made major breakthroughs in the development of visible light laser diodes. Until recently, the very best visible laser diodes made in this country and Japan emitted in the visible red area above about 700 nm. But recently scientists at Japan's Sharp Corporation announced the first laser diode to emit continuously at wavelengths *below* 700 nm. Various versions of their new AlGaAs laser deliver from 5 to 10 milliwatts (about the same power as a small penlight) at 683 nm.

Shortly after Sharp announced its new laser, an even bigger surprise came from Omron Tateisi Electronics, another Japanese company. Omron announced the development of indium gallium arsenide phosphide (InGaAsP) laser diodes that emit at about 621 nm at room temperature! This wavelength is lower than that emitted by the HeNe laser.

Will laser diodes that emit visible light replace HeNe lasers? In some applications they might. Although laser diodes produce a much more broadly diverging beam than helium-neon lasers, a single convex lens can focus the light from a laser diode into

a beam as narrow as that from a HeNe laser. Of even more significance, a laser diode can be operated from a compact battery powered supply, and its light output can be pulse or intensity modulated by a relatively straightforward circuit.

The laser diode's biggest drawback (yes, there's a catch) is its temperature sensitivity. The current supplied to the laser must exceed a certain threshold point before the laser will begin to function as a laser and not merely as an LED. The threshold point, however, *decreases* with temperature. To avoid destroying the diode by pumping too much current through it, the laser chip must be cooled to a constant temperature (perhaps by a thermoelectric module). Or it must be powered by a temperature regulated supply. Alternatively, the supply can be regulated by a photodiode circuit that monitors the radiation from one end of the laser. This setup automatically compensates for temperature changes since the radiation from a laser diode biased by a constant current increases in power as temperature decreases. And it guarantees that the laser will not be operated above the limit where excessive optical radiation causes damage to the chip's facets.

Its temperature sensitivity notwithstanding, the era of the visible light emitting laser diode has finally arrived. Hopefully, the demand created by the video disc and laser nonimpact printer industries will warrant mass production of these important new lasers. When that occurs, prices will fall. You can then expect to see various kinds of penlight-size laser diode assemblies, complete with battery powered, regulated supplies and, perhaps, self-contained modulator circuits.

Experimenting with CW Laser Diodes

Part 1. How-To Basics and a Laser Pulse Transmitter

Thanks in large part to the success of compact disc audio technology, laser diodes capable of continuous wave (cw) operation at room temperature are now available at very reasonable prices. When I first built circuits incorporating such lasers in 1975, individual lasers cost several hundred dollars or more. Now several different low-power cw laser diodes are available for as little as $26 each in single quantities from electronics distributors that represent Sharp Electronics Corporation.

These new low-power laser diodes are designed to emit up to about 3 milliwatts at a wavelength of 780 nanometers near the extreme end of the visible spectrum. Newly developed, highly visible red "super" LEDs made from AlGaAs can emit twice this power level when driven at similar current levels (50–60 milliamperes). But the laser diode is the preferred choice when a very narrow optical beam is required for applications such as free-space communications and intrusion alarms.

The emission from each end facet of a laser diode chip generally forms a fan-shaped beam having a divergence of about 15

× 30 degrees. Virtually all the radiation emitted from the front facet of a laser diode can be collected and collimated into a pencil-thin beam by means of a simple lens. Since LED chips are much larger than laser diode chips, it's not possible to focus their emission into as tight a beam. Moreover, since LEDs emit radiation in all directions, only a small fraction can be collected by a lens.

An important advantage of laser diodes over LEDs in some applications is the highly coherent nature of their emission. The new generation of cw laser diodes emit beams having a coherence that rivals or even exceeds that of the popular helium-neon laser.

Though cw laser diodes offer several important advantages over noncoherent LEDs, they are easily damaged if handled or operated improperly. For instance, most LEDs can withstand momentary surges of drive current. The drive current applied to a cw laser diode must never exceed the rated level or the device will be permanently damaged. Though the chip may continue to function as a low-power LED, the portions of the facets that provide the end mirrors for the laser will be destroyed by the high optical power density (several hundred thousand watts per square centimeter) produced by a current surge.

Most cw laser diodes can be damaged or destroyed by electrostatic charges or even the voltage spike that often occurs when a switch is flipped on or off. For these reasons, I have devoted a substantial portion of what follows to safe operating procedures for cw laser diodes. For best results it is essential that you read and heed these procedures. Having zapped my fair share of cw laser diodes over the years, I can assure you it pays to be very careful when installing and working with these devices.

Low-Cost Sharp Laser Diodes

While preparing this material, I contacted most manufacturers of laser diodes and managed to receive price information from several. Of these, the cw lasers made by Sharp Electronics Corporation are by far the least expensive and the easiest to buy. Single-mode cw laser diodes having lower thresholds and operating currents are available from Ortel, M/A-COM Laser Diode, Mitsubishi, and others, but their cost is from four to fifty times higher than the $26 Sharp lasers used in the circuits to be described in this discussion.

Figure 4-10 shows a pair of Sharp LT026MD cw laser diodes housed in packages similar to those used by other makers of cw laser diodes. The LT026MD and other members of the Sharp laser diode family include a photodiode to monitor the power emitted from the rear facet of the laser chip. The current from this photodiode is proportional to the light emitted from both facets of the laser. Therefore, the photodiode provides an effective sensor for a closed-loop regulator circuit designed to apply sufficient current to the laser so that its output remains stable as the ambient temperature changes.

Figure 4-11 shows the details of Sharp's laser diode package. Since I have used the LT020MC single-mode laser more than any other, its beam dimensions are illustrated in Fig. 4-11. The package dimensions shown in the figure apply to many cw laser diodes.

Note the internal connection of the laser diode and photodiode. This common-cathode configuration is by no means universal. In some Mitsubishi lasers, for example, the anode of the laser is internally connected to the cathode of the photodiode. In other devices the laser diode and photodiode are not internally connected, and all four connections are available externally.

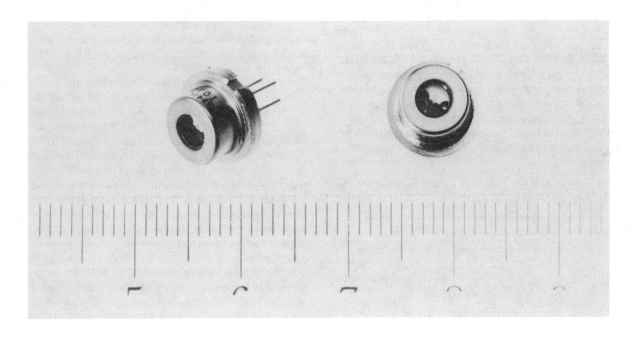

Fig. 4-10. Sharp laser diodes. (*Courtesy Sharp Electronics Corp.*)

Handling Precautions

There are many different kinds of laser diodes, and most are among the most delicate of semiconductor devices. Because of the very narrow width of their stripe-geometry junction, cw laser diodes are as susceptible to damage from destructive electrostatic breakdown as MOS semiconductors components and integrated circuits. For this reason cw laser diodes must be shipped and stored in packaging materials that do not generate or store electrostatic charges. Similarly, you must remove any charge on your body before handling a cw laser diode. This can be done by touching a grounded object immediately before handling the laser. A ground strap affixed to a wrist is even better. When soldering it is important to make sure the iron is either battery-powered or free of voltage leaks.

The laser diode's glass window should be considered a precision optical component. Dust and scratches will reduce the output power level and disturb the beam pattern. Sharp recommends that the windows of its laser diodes be cleaned with cotton soaked in ethanol.

The design of the laser diode package plays a vital role in removing heat from the delicate laser chip. Therefore, never cut, drill, or machine the package. Since the laser chip is affixed to its heatsink with a low-melting point solder, never attempt to solder a heatsink or wire to the package.

Operating Precautions

Laser diodes can be instantaneously and irreversibly damaged or destroyed by current surges that exceed the maximum allowable. Consider a cw laser diode that emits 5 milliwatts. For a typical stripe-geometry device, the light is emitted from regions of the front and rear facets measuring about 0.6×3 micrometers each. This corresponds to a power density of about 280,000 watts per square centimeter! From studies conducted at RCA and elsewhere, the facets of cw laser diodes can be irreversibly damaged when the power density falls in the range of 200,000 to 400,000 watts per square centimeter. Therefore, a current only slightly higher than the rated maximum can cause a cw laser to emit enough power to damage its reflecting facets.

Because of their exceptional vulnerability to damage, it is essential that you exercise great care when using cw laser diodes in working circuits. Here are some precautions which, if followed, can save your lasers from the unfortunate fate several of mine have met:

1. Never connect the probes of a multimeter across the leads of a laser diode.

2. Always make sure the leads of a laser diode are installed correctly.

3. Never connect a cw laser diode to a battery through a series resistor. For cw operation, always drive the laser with a closed-loop current regulator that derives its feedback signal from the photodiode inside the laser package.

4. Never connect a laser diode directly to a line-powered power supply. Voltage spikes generated when the supply is turned on can destroy the laser.

5. When testing or troubleshooting a drive circuit, use great care to avoid shorting the leads of the laser diode to other circuit leads.

6. If your workbench is metal, it should be at the same potential as the ground line of the laser's power supply.

7. Use care when operating laser diodes near equipment that generates high-frequency surges. The leads of the laser may couple such surges into the chip and destroy it.

8. The threshold current of a laser diode increases with temperature. For this reason alone, it is important to drive cw laser diodes with a photodiode-coupled current regulator (see 5 above).

9. Follow the manufacturer's recommendations about proper heatsinking. Sharp recommends a copper or aluminum heatsink measuring about $20 \times 30 \times 2$ millimeters. Some lasers include heatsink attachment holes. Use a spring-loaded or push-on heatsink for those that do not. A solderless RG59/U cable connector (Radio Shack 278-215) will snap snugly over laser diodes made by Sharp.

Laser Safety

It's important to be aware of safety precautions that apply to the use of laser diodes. The Bureau of Radiological Health, now the National Center for Devices and Radiological Health (NCDRH), has formulated extensive regulations governing the safe operation of all types of lasers. Under these regulations, Class 1 lasers are exempt from regulation. Most laser diodes are Class 3B devices and are required to bear a warning label that reads in part:

"DANGER—INVISIBLE LASER RADIATION. AVOID DIRECT EXPOSURE TO BEAM."

Since laser diodes are much too small for such a verbose warning label, manufacturers usually include a replica of the label in the laser's specification sheet or attach a label to the box in which the laser is shipped.

There is an ongoing controversy about the government's efforts to regulate laser diodes. Since the beam from such lasers is much more divergent than that from most other kinds of lasers, the safety hazard may be more imagined than real. Indeed, some light-emitting diodes can produce a higher power density at the eye than low-power (3–5 milliwatt) cw laser diodes when both devices are viewed at the same distance. (But the laser radiation will be focused to a smaller spot on the retina.)

A recent research project funded by Bell Labs and the U.S. Army explored the effect on the eyes of monkeys which were exposed to cw and pulsed beams from various kinds of laser diodes. One conclusion of this project is worth including here: "It required from 6 to 8.4 mW of GaAs radiation entering the eye for periods ranging from 400 to 3000 sec to produce a detectable lesion (on the retina). Since the spot size on the retina was >50 micrometers in diameter, it is difficult if not impossible to imagine how the human eye could remain focused on such a source for an appreciable time, even if 8 mW were entering the pupil." (William T. Ham, et al, *Applied Optics*, July 1, 1984, pp. 2181–2186.)

Though this study seems to indicate that low-power cw laser diodes are relatively safe, the authors recommend that government safety standards be followed until more data becomes available. Therefore, you should follow these safety precautions when working with cw laser diodes:

1. Avoid staring at the raw beam from a laser diode closer than arms length.

2. Never stare at the beam of a laser diode whose emission has been focused into a narrow beam by a lens.

3. Never point the beam from a collimated laser diode toward the eyes of onlookers or toward specular surfaces that might reflect the beam toward you or onlookers.

4. Observe the beam from laser diodes with an infrared image converter or infrared phosphor screen. If the wavelength of the beam falls in the visible spectrum, you can safely observe the beam by directing it toward a nearby white card (matte, not glossy, surface) in a dark room.

Should you wish to use laser diodes in a product to be sold, you may be considered a "laser manufacturer" by the NCDRH and the governments of some eleven states. For additional information, obtain a copy of "Performance Standards for Laser Products" (21CFR 1040) from the NCDRH and "ANSI Standard for the Safe Use of Lasers" from the American National Standards Institute.

A Miniature Laser Diode Transmitter

Figure 4-12 shows a simple circuit designed to deliver current pulses to low-power cw laser diodes such as Sharp's LT020MC or LT022MC. In operation, the 555 is connected as a pulse generator whose oscillation frequency is given by $1.44/[(R1 + 2R2) \times C1]$. With the values in Fig. 4-12, the frequency is about 320 Hz, and the duration of each pulse is 60 microseconds.

R4 permits the current applied to the laser to be adjusted to a safe operating level, a procedure which requires an oscilloscope. The scope's probe is connected across R3, a 10-ohm resistor that serves as a current monitor. From Ohm's law, the current in amperes passing through R3 is the voltage across R3 divided by 10. Since R3 is in series with the laser, the current flowing through R3 also flows through the laser.

Incidentally, reducing R1 will increase the pulse repetition rate of the circuit. This may also affect the current delivered to the laser diode. Therefore, always monitor R3 when the circuit's pulse repetition rate is altered.

IMPORTANT: Before applying power to the circuit for the first time, it is absolutely essential that R4 be adjusted to provide its maximum resistance. If you have any doubts, practice performing the adjustment that follows with a red LED installed in place of the laser diode. Before removing the LED, adjust R4 for its highest resistance and switch off power to the circuit.

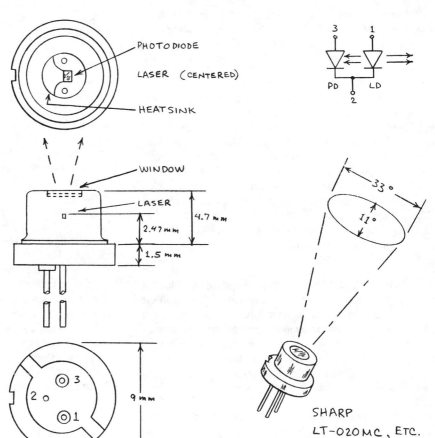

Fig. 4-11. Package design of Sharp LT-020MC laser diode.

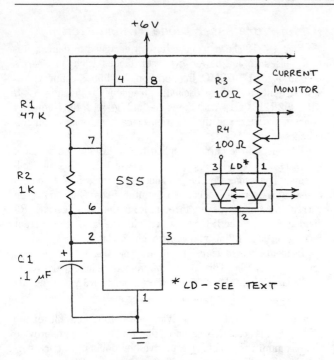

Fig. 4-12. Laser diode pulse generator.

Adjust R4 to give a current midway between the laser diode's threshold (usually about 50 milliamperes) and the maximum allowable operating current (usually about 60 milliamperes). Though this means the laser will not emit the full 3 milliwatts of which it is capable, the likelihood of irreversible damage will be greatly lessened.

IMPORTANT: If you did not receive a data sheet giving threshold and operating current levels for the laser(s) you purchased, check with your supplier. Each individual laser has unique operating specifications. Never attempt to guess the specifications of a laser or operate a laser without knowing its specifications! Some companies (Mitsubishi, M/A-COM Laser Diode Labs, etc.) provide threshold and operating current values on the packages or boxes in which lasers are shipped. In the case of Sharp laser diodes, these numbers are given on a batch printout that lists the specifications of 50 serially numbered lasers. The serial numbers are marked on small adhesive labels affixed to each laser.

The circuit in Fig. 4-12 does *not* take advantage of the laser's monitoring photodiode. Therefore, it is very important to adjust R4 when the transmitter is at the temperature at which you plan to operate it. Should the temperature later rise, the threshold of the laser will rise, and the laser's output power will then be reduced. Should the temperature later fall, the threshold of the laser will also fall. The output of the laser will then increase, perhaps to a point at which the laser may be irreversibly damaged.

Figure 4-13 shows how the laser transmitter, complete with battery and lens, can be installed in a small plastic box. I used a box measuring ¾ × 1 × 2 inches which was purchased from

an arts and crafts store. Note how the laser diode is installed in a fuse clip mounted inside the box. The clip secures the laser in place and doubles as a heatsink.

The collimating lens should be a convex lens with a small f/ number. I used a 10-mm diameter lens having a focal length of 12 mm (f.1.2). The lens is installed in an aperture carefully reamed into the end of the box and secured in place with silicone cement. An excellent variety of lenses is available from Edmund Scientific.

The mounting hole for the fuse clip should be made larger than necessary so the laser can be properly focused. Then the clip can be secured in place. Since the beam from the laser is just barely visible when focused onto a white card in a dark room, this task is difficult and will take time.

I have adjusted the laser in the prototype transmitter to give a perfectly circular pattern 6 inches in diameter at a distance of 285 feet. This corresponds to a divergence of 1.75 milliradians, about the same as that of many helium-neon gas lasers.

Figure 4-14 shows the beam pattern produced by the prototype transmitter. The concentric rings of light that form the beam are caused by spherical aberration of the lens. This effect can be eliminated and the beam made narrower by using a more expensive achromatic lens. A 35-mm camera lens can also be used with excellent results.

A Miniature Laser Receiver

Figure 4-15 shows a simple lightwave receiver that will transform the pulsed emission from the laser transmitter to an audible tone. If the divergence of the laser beam is adjusted to be as narrow as possible, this receiver will detect the laser at ranges in excess of 1000 feet. The detection range can be extended by placing a lens in front of the phototransistor.

The receiver allows the laser to be used as a remote signalling device or for demonstrations. If a missing pulse detector is added to the receiver, an excellent break-beam intrusion alarm can be built. A 567 tone decoder circuit can be added to the basic receiver to form a long-range remote control system. Suitable circuits are given in *The Forrest Mims Circuit Scrapbook* (McGraw-Hill, 1983) and various books I have written for Radio Shack.

It's important to realize that receiving the laser beam can be very difficult if you don't have an infrared image converter. One possibility is to place a large bicycle reflector at the receiver location. Then place the transmitter on a camera tripod and, while looking along the top of the transmitter, carefully adjust the device until you observe a red reflection from the reflector. The laser beam will now be centered on the reflector where it can be easily detected with the receiver.

This method works only when it is dark. Of course this makes it difficult to know where to point the laser. In other words, be prepared to spend lots of time learning how to align the laser. I have problems pointing the laser in Fig. 4-13 across a room. At 20 feet the beam is not much larger than the end of a thumbtack.

CAUTION: Do not stare at a nearby reflector illuminated by the laser. The reflector should be placed sufficiently far away

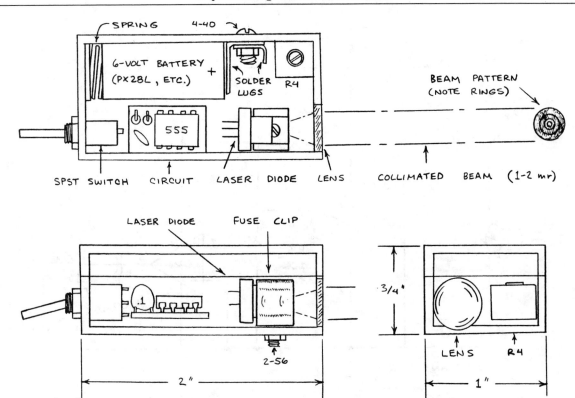

Fig. 4-13. Miniature laser diode pulse transmitter (see text).

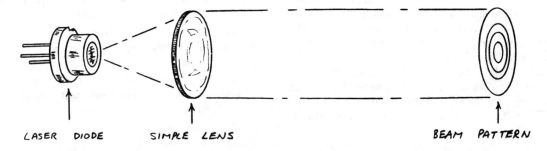

Fig. 4-14. Using a lens to collimate emission from a laser diode.

so that the laser beam has spread to at least several inches in diameter where it strikes the reflector.

Part 2. Two CW Laser Diode Sources

The new generation of highly coherent cw (continuous wave) laser diodes designed primarily for use in audio disc players opens up many fascinating applications. Only a few years ago laser diodes having the coherence properties of these new devices cost many hundreds or even thousands of dollars. Now such lasers can be purchased in single quantities for as little as $26 each, a price that is likely to fall even farther in coming years.

In Part 1, I described the operation of these new highly coherent cw laser diodes. I also discussed in some detail the handling, operation, and safety precautions associated with their use. And I presented the construction details for a miniature laser diode pulse transmitter.

In this part I'll describe how to operate in a continuous mode cw laser diodes that are equipped with a monitoring photodiode. I'll also describe in detail the construction of two different cw laser diode illuminators and a portable battery pack suitable for operating them. Finally, I'll discuss some applications for these amazing new lasers.

Laser Diode Operating Precautions

Being a frequent user and sometimes zapper of laser diodes since 1967, I've learned firsthand that it is absolutely essential

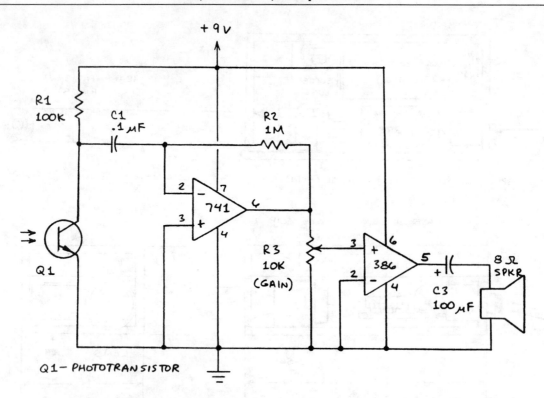

Fig. 4-15. Simple receiver for pulsed laser diode transmitter.

to be aware of the idiosyncrasies of these fascinating devices before attempting to use them in actual circuits. You must also be aware of the possible safety hazards associated with their use. Both these topics were covered in detail in Part 1, and it is important that you review that information before you attempt to work with actual laser diodes.

A summary of the various precautions associated with the use of cw laser diodes is included here. This summary, however, is *not* a substitute for the detailed precautions given in Part 1 or those given in the descriptive literature for a particular laser diode.

Handling Precautions

1. Cw laser diodes are susceptible to damage from electrostatic discharge and must be stored and handled like MOS semiconductor devices.

2. Avoid touching or scratching a laser diode's glass window. Should it become dusty, clean it with a cotton swab soaked in ethanol.

3. Never solder, cut, drill, or machine a laser diode package.

Operating Precautions

1. Never connect the probes of a multimeter across the leads of a laser diode.

2. Always observe proper polarity when connecting the leads of a laser diode.

3. Never connect a cw laser diode to a battery through a series resistor. Instead, use a current driver circuit or IC designed specifically for laser diode operation.

4. Never connect a cw laser diode directly to the leads of a line-operated power supply. Voltage spikes generated when the supply is switched on can destroy the laser.

5. When testing or troubleshooting a laser diode drive circuit, use great care to avoid shorting the leads of the laser to other circuit leads.

6. Always use the minimum heatsinking recommended by the laser's manufacturer.

Safety Precautions

1. Avoid staring at the raw beam from a low-power (>5 milliwatts) laser diode closer than arms length.

2. Never stare at the beam of a laser diode whose emission has been focused into a narrow beam by a lens.

3. Never point the beam from a collimated laser diode toward the eyes of onlookers or toward specular surfaces that might reflect the beam toward you or others.

4. Observe the beam from a laser diode with an infrared image converter or infrared phosphor screen. You can safely observe the visible red beam from cw laser diodes that have a wavelength of 780 nanometers by projecting their beam toward a white card (matte, not glossy, surface) in a dark room.

These safety rules must be viewed with common sense in

mind. For example, the collimated beam from a 780-nanometer laser diode can be safely viewed from a distance, but only if the beam has expanded so that only a small fraction of the light can enter the pupils of your eye.

For additional laser safety information, obtain a copy of "Performance Standards for Laser Products" (21CFR 1040) from the National Center for Devices and Radiological Health and "ANSI Standard for the Safe Use of Lasers" from the American National Standards Institute.

Constant-Output CW Laser Drivers

The optical output of a laser diode increases as temperature decreases. Many different circuits have been developed that monitor the optical output from a laser diode and then regulate the drive current so that the output remains constant. Among the simplest such circuits are a pair of laser diode driver chips made by the Sharp Corporation and available from Sharp distributors.

Figure 4-16 shows how one of the Sharp chips, the IR3C01, is used in a practical circuit. This chip is supplied as an 8-pin miniDIP and costs only $1.18 in single quantities.

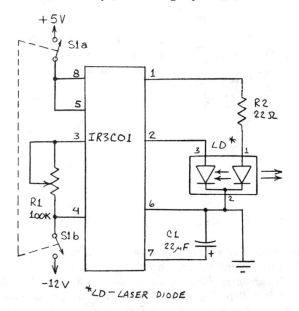

Fig. 4-16. Constant output cw laser diode driver circuit (see text).

Referring to Fig. 4-16, note how the internal monitor photodiode installed in the laser package is connected to the IR3C01 to form a closed-loop feedback system. The current output terminal of the IR3C01 (pin 1) is connected to the anode of a cw laser diode through series resistor R2. Though the chief function of R2 is to limit the current through the laser diode, it can also be used to monitor the current flowing through the laser.

R1 controls the current delivered to the laser at pin 1. When the circuit is operated with a new laser for the first time, R1 should be adjusted for its highest resistance. As the resistance of R1 is decreased, the current from pin 1 is increased.

Pin 5 permits the laser to be gated on or off. When pin 5 is connected to +5 volts, the laser is on. Connecting pin 5 to ground switches the laser off.

The chief drawback of the circuit in Fig. 4-16 is that it requires supplies of +5 volts and −12 volts. Sharp has recently announced a new laser diode driver chip, the IR3C02, that operates from supplies of +5 and −5 volts. Figure 4-17 shows a working circuit for this new chip. Other than the change in the negative supply voltage, the circuit is very similar to the one in Fig. 4-16.

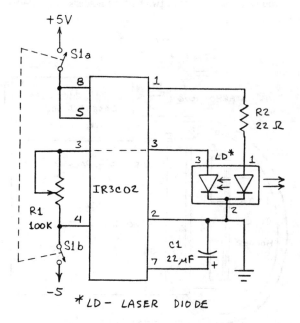

Fig. 4-17. Improved cw laser diode drive circuit (see text).

The IR3C01 is available both in 8-pin miniDIP and miniature surface-mount packages. Thus far I have not assembled a working circuit using the IR3C02. However, I have assembled several different cw laser diode circuits using the IR3C01 driver, two of which are described next. These circuits both require +5 and −12 volts, you may prefer to use the IR3C02 since it requires +5 and −5 volts.

Suitable CW Laser Diodes

Both the IR3C01 and IR3C02 will drive any of the laser diodes currently available from Sharp Corporation. The least expensive such lasers are the LT020MC and the LT022MC. Both these devices cost only $26 in single quantities. These lasers have a typical output power of 3 milliwatts and emit radiation having a wavelength of 780 nanometers at the far end of the visible spectrum.

The LT020MC is a single-mode, highly coherent laser designed for general use. Applications include measuring instruments, communications, readout devices, and so forth. The LT022MC is a low-noise device specifically designed for use with compact disc players. The output from cw laser diodes can fluctuate (become noisy) when some of the radiation is reflected back into the laser by the highly reflective surface of a compact disc. One way to reduce the effect of external reflections is to slightly increase the thickness of the pn junction region so that several longitudinal modes, each having a slightly different wavelength, can propagate within the laser.

A Laser Diode Battery Pack

Both the laser illuminators described below are best powered by batteries. Figure 4-18 is an outline view of a compact battery pack you can assemble from two battery holders available from Radio Shack.

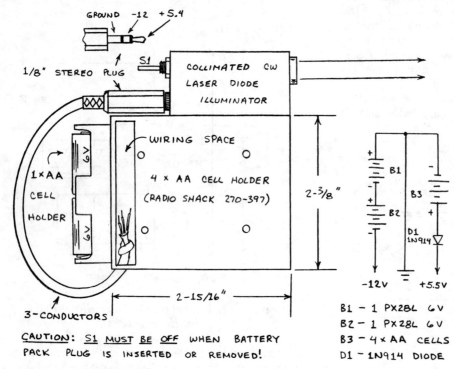

Fig. 4-18. Laser diode illuminator installed on battery pack.

The battery pack shown in Fig. 4-18 provides +5.4 and −12 volts and is intended for use with the IR3C01 driver chip. The −12-volt supply is obtained from two 6-volt lithium or silver oxide batteries (PX28 or similar) installed in series in a single AA penlight cell holder. The 5.5-volt supply is provided by four AA penlight cells. A 1N914 diode drops the voltage from 6 volts to about 5.4 volts. Penlight cells are used for this supply since it must provide a current of from 50 to 65 milliamperes.

The two battery holders are fastened together with 2–56 hardware or plastic cement and then wired as shown in Fig. 4-18. A stereo phone plug serves as the power output connector.

CAUTION: When the batteries are installed, slip the end of the phone plug into a piece of plastic tubing to prevent an inadvertent short.

A Miniature Collimated CW Laser System

Figure 4-19 is a drawing of a miniature cw laser diode il-

luminator I have assembled. Though I used an IR3C01 driver chip, an IR3C02 can be used instead.

The entire system is installed in a clear plastic box measuring $2 \times 1 \times \frac{3}{4}$ inches. Suitable boxes are available from arts and craft stores and specialty shops. The circuit is assembled on a perforated board measuring $\frac{1}{2} \times 1\frac{3}{16}$ inches.

R1 is a miniature 100-kΩ cermet trimmer potentiometer (No. Q0G15) available from Digi-Key. Other trimmers can also be used (see the next transmitter), but this particular trimmer is very compact.

Figure 4-20 shows how the circuit board is connected to the dpdt power switch and a miniature stereo jack that serves as a power jack. Use wrapping wire to make the connections between the circuit board, switch, and jack.

The prototype illuminator I assembled uses a Sharp LT020MC laser diode. Referring to Fig. 4-19, note how the laser fits snugly in the end of a solderless Push-F-Type RG59/U television coaxial cable connector such as Radio Shack's 278-215. At least two different versions of this connector are available. For best results use the longer of the two and saw off the knurled end just beyond the internal collar.

The sawed-off RG59/U connector will fit nicely within a ¾-inch length of ⁷⁄₁₆-inch (inside diameter) brass tubing available from a hobby shop. A 10-mm diameter lens with a focal length of from 10 to 15 mm will fit inside this tube. The lens should be held in place by a pair of rings cut from ⁷⁄₁₆-inch (outside diameter) brass tubing slipped in the end of the larger tube. Figure 4-19 shows how the lens fits between these two rings. Suitable lenses are available from Edmund Scientific.

The laser is carefully inserted into the expansion end of the RG59/U connector. The opposite end of the connector is then

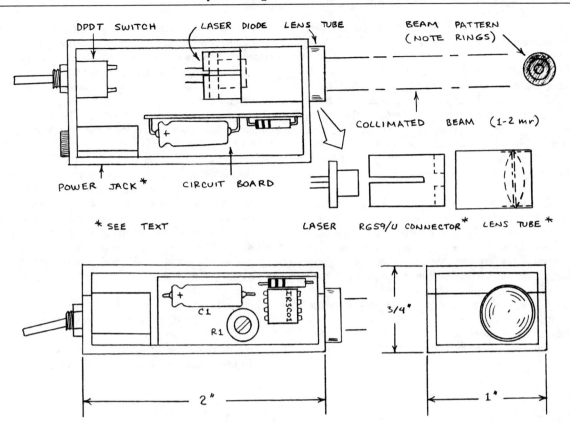

DPDT SWITCH   LASER DIODE LENS TUBE   BEAM PATTERN (NOTE RINGS)

COLLIMATED BEAM (1-2 mr)

POWER JACK *   CIRCUIT BOARD

* SEE TEXT

LASER   RG59/U CONNECTOR *   LENS TUBE *

3/4"

2"   1"

Fig. 4-19. Miniature collimated cw laser diode illuminator.

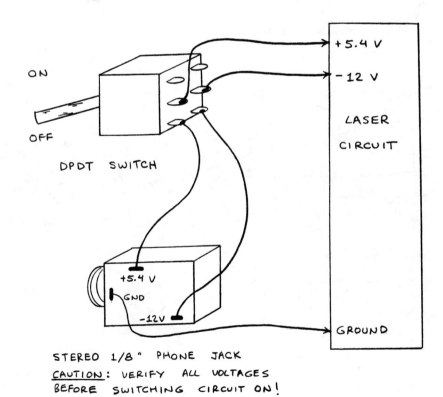

Fig. 4-20. Power switch and power jack wiring details.

ON

OFF

DPDT SWITCH

+5.4 V

-12 V

LASER CIRCUIT

GROUND

+5.4 V

GND

-12V

STEREO 1/8" PHONE JACK
CAUTION: VERIFY ALL VOLTAGES BEFORE SWITCHING CIRCUIT ON!

slipped into the brass lens tube. The connector and lens tube double as a heatsink for the laser.

CAUTION: The laser diode may be destroyed by electrostatic discharge. Ground your body to remove any residual charge before touching the laser. (Touch a cold water pipe or the metal case of a grounded electrical appliance.) Before installing the laser diode and switching the circuit on for the first time, use a multimeter to determine which direction to rotate R1's rotor to give the highest possible resistance. R1 *must* be set for its highest resistance before the laser is installed.

Figure 4-21 shows the pin outline of the laser diodes made by Sharp. Being sure to follow the precautions given previously, attach lengths of color-coded wrapping wire to each of the three leads of the laser diode. If you use a wrapping tool, be sure none of the free ends of the wrapped connections can short against one another! If you use a soldering iron, make sure it doesn't expose the delicate laser to line voltage.

Double check all the wiring connections for possible errors. Also, again make sure R1 is set to give a resistance of 100 kΩ. Finally, make sure the power switch is in the off position. You are now ready to connect the battery pack to the laser unit for the initial current adjustment.

Spread the various portions of the circuit slightly outward on a nonconductive surface. Plug the power pack plug into the circuit's power jack. Then connect the probes of a multimeter across R2, using great care to avoid touching any other leads. (I used small, insulated clip leads connected from R2 to the multimeter probes.)

Refer to the data supplied with the laser connected in the circuit to determine its threshold and operating currents. Use Ohm's law to calculate the voltage across R2 that will coincide with the desired operating current.

For instance, one of my lasers has a threshold current of 48 milliamperes and an operating current (for 3 milliwatts out) of 60 milliamperes. To be on the safe side, I operate this laser at a forward current of 55 milliamperes. According to Ohm's law, the voltage across R2 equals the current through R2 times its resistance. Therefore, for a current of 55 milliamperes, the voltage across R2 should equal 0.055 ampere times 22 ohms, or 1.21 volts.

After you calculate the voltage across R2 for your laser, switch on the power to the circuit. Then gradually lower the resistance of R1 until the voltage across R2 begins to approach the value you calculated. Very carefully continue the adjustment process until the desired voltage is reached.

To verify that your laser is lasing, place a white card near

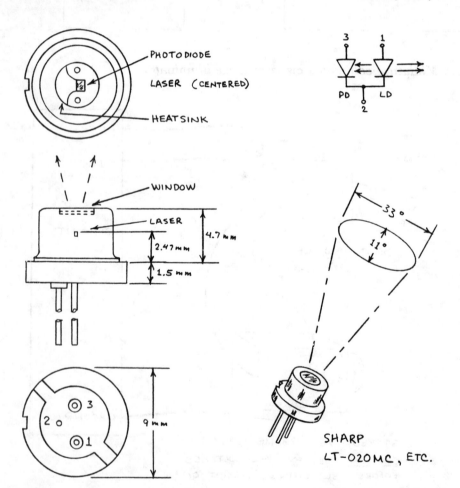

Fig. 4-21. Package design of Sharp LT-020MC laser diode.

the lens and switch off the room lights. If the room is sufficiently dark, you should see a dim but distinct red spot on the card. Switch off the laser and remove the battery pack plug. Then form holes in the plastic box for the lens tube, switch, and power jack and carefully install the circuit inside the box. Before again operating the system, inspect the circuit for possible shorts or broken wires.

Adjusting the lens to provide the tightest possible beam is difficult if you don't have access to an infrared image converter or a closed-circuit TV camera that is sensitive to near-infrared. In either case, point the lens tube toward a matte, white surface and move the RG59/U connector back and forth slightly until a spot about the diameter of the lens is formed. Then move the card farther away and repeat the adjustment until the spot is again about the size of the lens.

If the RG59/U connector is difficult to move, try moving the lens itself by changing the position of the rings that hold it in place. After you have adjusted the lens for the tightest possible beam, you may want to secure the lens tube and its lens with a small drop of cement.

Be sure to allow plenty of time for the focusing procedure. The beam from the prototype circuit is only about 10 inches across at a distance of 436 feet. From this distance on a dark night, the laser appears as a very bright red light in the distance. It is particularly spectacular when viewed with an image converter.

CAUTION: Never view the collimated beam from the laser unless it has spread to a safe viewing size.

A Miniature Uncollimated CW Laser System

Some applications for cw laser diodes require that the uncollimated beam be available. The laser system in Fig. 4-22 accomplishes this purpose. The circuitry, wiring details, and adjustments are identical to that of the previous system.

The absence of a lens system is the only significant difference between the system in Fig. 4-22 and the one in Fig. 4-19. Therefore, it is essential to provide a heatsink for the laser diode when the unit is in operation. Sharp recommends a piece of aluminum or copper measuring 20 × 30 × 2 millimeters for its higher power lasers that operate at a current of about 100 milliamperes. The LT020MC and LT022MC operate at a little over half that current, so less heatsinking is necessary.

Several heatsink options are available. Finned heatsinks designed to fit on a TO-5 transistor case can be used. So can an RG59/U connector, though long versions of this connector may have to be sawed off to give full access to the laser beam.

CW Laser Applications

The collimated laser in Fig. 4-19 can be used for many of the experiments and demonstrations for which helium-neon lasers are ordinarily used if you have access to an image converter. Since the light is highly coherent, you can use the laser in con-

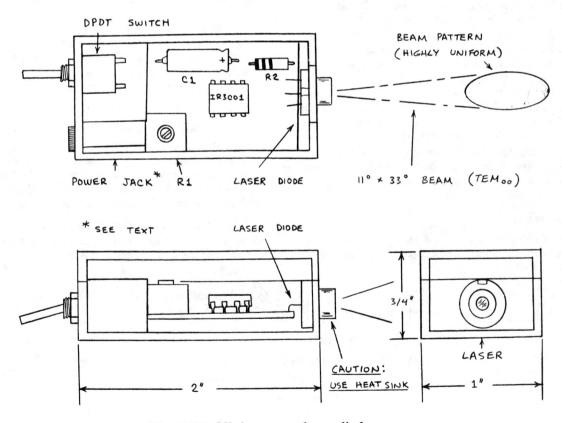

Fig. 4-22. Miniature cw laser diode source.

junction with a simple interferometer to create interference patterns. With such a device you can detect movements of less than half the 780 nanometer wavelength of the laser.

The uncollimated laser in Fig. 4-22 can also be used for many experiments that usually specify a HeNe laser. Since the fan-shaped beam is very uniform, it should be possible to make holograms using this laser. Another possibility is fiber optic sensors that require a coherent light source.

Figure 4-23 shows a very simple way to connect a plastic multimode plastic fiber to the laser in Fig. 4-22. For best results, use the RG59/U solderless connector like the one illustrated. (The long version of this connector is used for the laser in Fig. 4-19.) Cut the end of the fiber with a hobby knife and polish it with ultrafine sanding paper. If the fiber is jacketed, remove a portion of the jacket(s) as shown in Fig. 4-23. Otherwise, wrap an inch or so of plastic tape around the fiber a short distance from its polished end.

Insert the end of the fiber into the large end of a plastic wall anchor previously trimmed to fit inside the threaded end of the coax connector. Twist the anchor firmly to secure the fiber in place. Ideally, the fiber should be perfectly centered in the hole inside the connector. When the laser is inserted into the open end of the connector, its window should just touch the end of the fiber.

Coherent light passing through a multimode fiber undergoes considerable interference. The result is a microscopic pattern of speckle at the output end of the fiber. If the end of a fiber connected to the laser in Fig. 4-22 is caused to illuminate the lens of a phototransistor connected to the input of a high-gain audio amplifier, a speaker connected to the amplifier's output will emit a low hiss if the fiber is perfectly motionless. If, however, the fiber is barely moved, the interference pattern will move also and the speaker will emit a sound.

Depending upon the degree of movement, the sound may range from a quiet "pock" to crashing twangs and drum-like sounds. The fiber is so sensitive this system will respond to a puff of air! A level-detector connected to the amplifier will allow an alarm to be triggered when the signal from the fiber exceeds a preset level. Used in this mode, the fiber could be concealed under a carpet or even buried in a driveway and used to detect the presence of visitors or intruders.

Going Further

The applications for the cw laser diode circuits presented in this discussion are limited only by your application. Pulsed lasers are ideal for communications and intrusion alarms and cw lasers are well-suited for interference experiments, holography, and fiber sensing. If you wish to know more about lasers and laser diodes in particular, visit any good technical library and begin exploring the many books and technical papers on this subject.

More About Laser Safety

A few months after this discussion was published in *Modern Electronics*, I attended the International Congress on Applications of Lasers and Electro-Optics in San Francisco to give a paper on the surreptitious interception of conversations using lasers. While there I renewed acquaintance with R. James Rockwell, Jr., one of the foremost experts on laser safety. I first met Mr. Rockwell in 1968 when he was involved in some pioneering work in the medical applications of lasers.

Mr. Rockwell is now president of Rockwell Associates, Inc., a company that specializes in laser safety products (signs and protective eyewear) and training courses. Recently he sent me a thick package of literature related to laser safety along with a cover letter that included the following observations:

"The hazards to the eye associated with near infrared laser diodes are generally considered 'less' than those associated with visible laser wavelengths, but hazards are possible—especially if one views the diode directly and captures the beam with a collecting optic (such as a pair of jewelers loupes) so as to put ALL of the output into the eye.

"For example, for a cw diode operating at 850 nm, in the condition of optically unaided viewing at a distance, the 'allowed' irradiance incident on the eye is 0.64 mw/cm. Therefore the 'worst case' power limit into the eye is 0.26 mW (7-mm pupil diameter)."

Mr. Rockwell then observed that

". . . viewing an emitting diode under magnification may be the more hazardous viewing condition, even though the retinal image of the source is larger. Obviously, one recommends caution when working with any laser source to never look directly into the beam and, with diode lasers, NEVER observe the emission using magnifying optics."

This is good advice. For additional information about laser safety, contact Rockwell Associates at the address given in the Appendix and the Laser Institute of America.

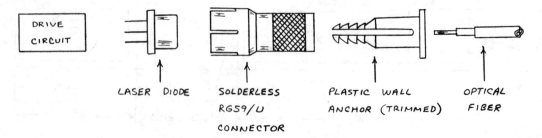

Fig. 4-23. Simple laser optical fiber coupler.

Experimenting with Electronic Flash Circuits

The brilliant white light emitted by the xenon flash tube has many practical applications in photography, solar simulation and optical pumping of various laser materials. Xenon flash lamps are also used in many kinds of safety, rescue and warning lights for emergency vehicles, aircraft, bicyclists, and hikers.

Though specially designed high-pressure xenon lamps can be operated continuously if suitable cooling is provided, most xenon-filled lamps are operated in a flash mode. Such lamps can have many different configurations. A few have envelopes of metal and glass, but most are glass cylinders or bulbs containing discharge and trigger electrodes. The cylinder configuration, the most common of all, is merely a hollow glass or quartz tube with an electrode at each end.

Cylindrical xenon flash tubes can be as small as a matchstick or, in the case of very high power glass lasers used in fusion research, as big as a fence post. The glass tubes can be straight or shaped in the form of an L, U, or spiral.

Figure 4-24 shows in block diagram form the essential ingredients of a circuit for operating a xenon lamp in a flash mode. A high voltage supply simultaneously charges a large energy storage capacitor and a much smaller trigger capacitor. When the main capacitor is charged, the lamp is flashed by dumping through the primary of the trigger transformer the charge on the trigger capacitor. The high voltage spike appearing at the secondary of the transformer is coupled to a small metal strap or wire wrapped around the flash tube.

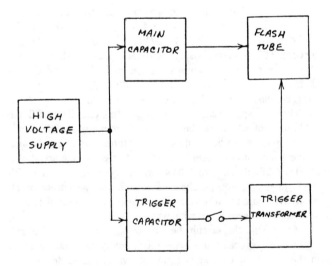

Fig. 4-24. Block diagram of a flash tube trigger circuit.

The very high voltage spike coupled into the tube ionizes some of the gas within the tube and provides an electrically conductive path for the charge in the main capacitor. The main capacitor then discharges through the tube and excites the xenon atoms into emitting an intense white flash.

Figure 4-25 is a circuit that implements the operation of the block diagram in Fig. 4-24. Note that C1, the main energy storage capacitor, is connected directly across the flash tube. No leakage of charge occurs since the xenon does not conduct unless it is first ionized by a voltage higher than that across C1.

After both C1 and C2 are charged, the flash tube is triggered by pressing S1. This dumps the charge on C2 through the primary of trigger transformer T1. Since T1 has a very high turns ratio, several thousand volts appear across the secondary. This voltage is coupled to the flash tube's trigger electrode where it ionizes some of the xenon gas and provides a low resistance path for the energy stored in C1.

The discharge of C1 through the xenon is accompanied by a brilliant flash of light. The xenon resumes its nonconductive state immediately after C1 has discharged. As soon as C1 and C2 are recharged, S1 can again be closed to obtain another flash.

The energy in joules stored in the main flash capacitor is half the product of the capacitance and the square of the voltage. For example, a 400-microfarad photoflash capacitor charged to 350 volts has a stored energy of 24.5 joules.

The duration of the flash is determined by the RC time constant of the flash capacitor and the discharge path through the flash tube, Sometimes intervening networks are included to shape the discharge event into a square pulse with fast rise and fall times.

Since a very large capacitance implies a long discharge time, hence a long flash, for short flashes it is necessary to use small values of capacitance. To obtain equal illumination, the voltage must then be increased. For example, to match the 24.5 joules in the preceding example, a 10-microfarad capacitor would have to be charged to 2214 volts.

Alternatively, the flash can be electronically ended at almost any point in time by an appropriate solid-state switch. This is the method utilized by "computer" strobe flashes so popular with photographers. Fast pulses can be obtained, but energy levels are low. A plus is that energy remaining in the capacitor can be used for one or more subsequent flashes, thereby extending battery life and reducing recycle time.

You may be wondering why R1, which should have a resistance of a megohm or more, is included in the circuit in Fig. 4-25 since it plays no role in the charge-discharge cycle. Its only role is to bleed the charge from C1 should the high voltage supply be turned off. Even if the flash tube is triggered *after* the high voltage supply is turned off, any residual charge remaining in C1 will be discharged through R1.

The circuit will operate without R1, but including R1 is a very important safety precaution. The hazards of the high voltage required to operate xenon flash circuits are so profound that it's important to discuss them in detail before looking at some working circuits you may wish to assemble.

IMPORTANT FLASH TUBE PRECAUTIONS: Though the power supply potential of some specialized xenon flash units may exceed several thousand volts, most flash tube power supplies produce from 150 to 500 volts. You should, therefore, exercise

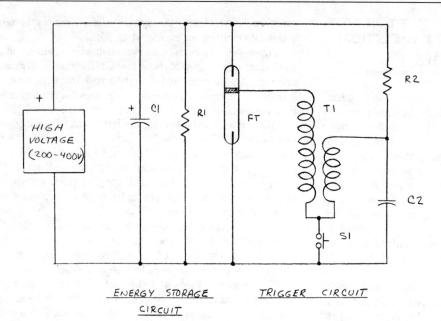

Fig. 4-25. Basic flash tube circuit.

ENERGY STORAGE CIRCUIT

TRIGGER CIRCUIT

terminals and be sure the capacitor is *fully* discharged before handling it.

The hazards of charged photoflash capacitors are so real I want to take space to relate a personal experience. Ten years ago I was experimenting with a voltage divider I had assembled from a string of 10-microfarad, 450-volt capacitors. The input to the multiplier was connected to a 350-volt miniature power supply powered by a 3-volt battery. The output from the multiplier was about 1000 volts.

Because of the hazard, I followed the traditional safety practice of keeping one hand in a pocket to avoid a dangerous through-the-body shock. All went well until the springy ladder of capacitors and diodes suddenly slid from the workbench and into my lap. I grabbed one end of the multiplier with my free hand just as its high voltage output lead touched my pants leg. Suddenly there was a loud pop, a puff of pungent smoke and a terrific jolt that threw me from my chair and onto the floor. The discharge burned a hole in my trousers and formed a small cavity in my left thigh. Suffice it to say that this unsettling experience left with me a lasting impression about the hazards of charged capacitors.

Incidentally, those rather foolish individuals who like to brag they have no fear of high voltage should realize that incidental consequences of a shock can cause more personal injury and property damage than the shock itself. For instance, in "Electronic Hobbyists' Handbook," (Tab Books, 1958), Rufus Turner wrote of a technician who was rendered unconscious by a high voltage shock. Upon recovering hours later, he found that his soldering iron had burned a hole through his workbench. The technician could have lost his life *and* his home.

A Single-Shot Flash Circuit

Figure 4-26 shows a single-shot flash circuit designed around commonly available components. T1 is a standard filament transformer. Q1 and Q2 form a simple oscillator. In operation, the

fast risetime pulses from the oscillator are directed through the 6.3-volt winding of T1. When powered by a 1.5-volt dry cell, the initial portion of the output pulse appearing at T1's 120-volt winding has a peak potential of about 170 volts and a duration of about 40 milliseconds. The pulse amplitude then falls to about 100 volts for the remainder of the 110 microsecond pulse.

The high voltage from T1 is stored in C2 and C3. D1 prevents these capacitors from discharging through T1. R3 is a bleeder resistor that discharges C1 and C2 should the power supply be turned off.

The very high voltage required to ionize the xenon in flash tube FT1 is provided by R2 and C4. C4 is charged through R2 to the power supply voltage. When S2 is closed, C4 is discharged through T2's primary. A spike of several kilovolts then appears at T2's secondary and ionizes the gas in FT1. C2 and C3 are then discharged through FT1. After C2, C3, and C4 are recharged, closing S2 will initiate a second flash.

The only specialized components in this circuit are T2 and FT1, both of which can be purchased from various electronic parts suppliers. Various kinds of flash tubes and trigger transformers may have different pin orientations, so none are shown in Fig. 4-26. Be sure to follow any pin outlines provided with the components you purchase. The high voltage output of T2 is often indicated by a red dot. The primary winding is of heavier wire than the secondary winding.

If you build this circuit, be sure to observe carefully all relevant safety precautions. *Never* touch any connections or leads in the boxed high voltage section shown in Fig. 4-26. Ideally, all such leads and connections should be well insulated.

An Automatic Flashing Strobe

The circuit in Figure 4-27 automatically discharges a pair of capacitors through a flash tube every 1 to 2 seconds. In operation, the output from a 555 oscillator is directed through the 6.3-volt winding of T1. When powered by a 9-volt battery,

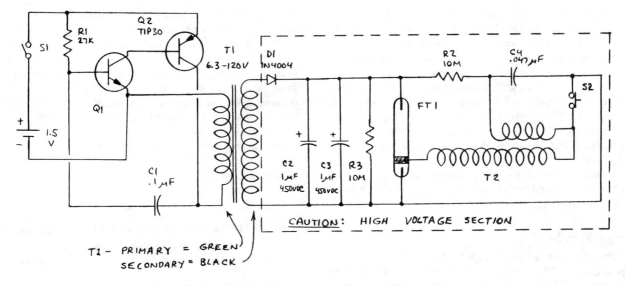

Fig. 4-26. Single flash xenon strobe circuit.

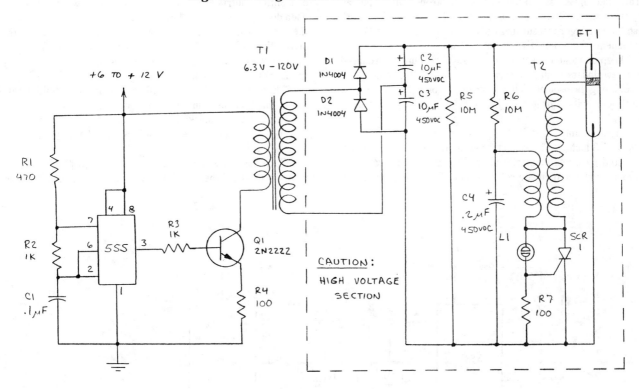

Fig. 4-27. Xenon strobe flasher circuit.

a 200-volt square wave appears at T1's 120-volt winding. D1, D2, C2, and C3 form a voltage doubler that rectifies, increases, and stores this voltage.

R6 charges C4 to the power supply voltage. When the voltage reaches neon lamp L1's turn-on point (80 to 100 volts), C4 begins to discharge through the primary of T2 and L1. Simultaneously, SCR1 is turned on by the voltage appearing across the lamp. SCR1, which should be rated for 400 or more volts,

provides a very low-impedance path between T2 and C4. The resulting high voltage spike across T2's secondary ionizes the gas in FT1, thus providing a low impedance path for discharging C2 and C3. After the resulting flash, C2, C3, and C4 begin recharging until the trigger cycle is automatically repeated.

The flashes from this simple circuit are sufficiently bright to enable the circuit to be used as a warning light. For brighter flashes, C2 and C3 can both be increased. The flash rate, however, will be reduced as C2 and C3 are increased in value.

If the circuit fails to flash, L1 may be switching on at a voltage below that required to provide sufficient ionization potential across FT1. Try another neon lamp. You may also try connecting two or more neon lamps in series to increase the ionization potential. Another cause for circuit malfunctions is low-impedance leakage paths between ground and the high voltage output from T2. These paths may be directly between exposed or poorly insulated wire leads. Or they may even take the form of moisture and contamination on circuit boards or even the glass surface of the flashtube itself.

Be sure to follow the safety precautions given for the previous circuit. Remember that the boxed portion of the circuit is potentially hazardous.

Going Further

After you have experimented with the basic flash circuits in Figs. 4-26 and 4-27, you may wish to replace the bulky filament transformer (T1) with a more compact dc-dc converter transformer like those used in photographer's strobe units. You can buy such transformers, but you can also salvage them along with flash tubes, trigger transformers, photoflash capacitors, and other components from defective or surplus flash units.

I keep on hand a stock of a dozen such flash units purchased for a few dollars each at the camera department of a discount store. Using the oscillator circuit and other components salvaged from these units, I have built half a dozen miniature automatic flashers.

A typical unit is installed in a plastic case measuring about 1 × 2 × 3 inches. The reflector assembly and xenon lamp are protected by a yellow plastic filter. The unit's AA nicads are recharged by a homemade solar battery. This unit clips on my bicycle shorts or bike packs and has accompanied me on many long distance cycling trips over the past five years.

Should you attempt to build such a flasher or work with salvaged flash units, be especially careful of the high voltage generated by these units. It is *essential* that you discharge the main capacitor *and* turn off the power before attempting to disassemble or modify any such device. Make sure the batteries are disconnected or removed before beginning work. For automatic flasher units, you will want to replace the large photoflash capacitor with no more than 10 to 20 microfarads. Make sure the replacement capacitor is rated for the proper voltage (usually 450 volts).

A Photonic Door and Window Intrusion Alarm

The circuit in Figure 4-28 is the photonic equivalent of a conventional intruder alarm that uses magnet switches to detect open doors and windows. The circuit consists of two 555 timers, the first of which is connected as a free-running oscillator that drives the LED in a slotted optoisolator (or optocoupler) at a frequency given by $1.44/(R1 + R2)\,C1$. R3 limits current through the LED to a safe value.

The second 555 is connected as a monostable multivibrator which functions as a missing pulse detector. When the slot in the optoisolator is blocked, pulses from the LED do *not* reach the phototransistor in the optoisolator and the output from the one-shot is *low*.

When the slot is opened, the phototransistor receives a pulse from the LED and the one-shot begins its timing cycle. Its output

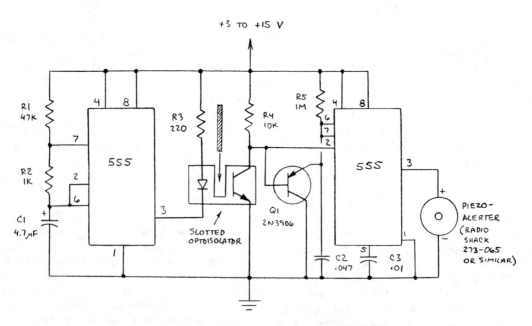

Fig. 4-28. A photonic door and window intrusion alarm.

then goes *high* until the timing cycle is completed. When the one-shot's output is high, the piezoelectric alerter is activated.

The time constant of the one-shot is R5C2. If the time constant is adjusted so the timing cycle is *longer* than the interval between incoming pulses from the LED, the output from the one-shot will stay high until a new pulse arrives. Therefore, the alerter will emit a continuous alarm tone.

If, on the other hand, the timing cycle is *briefer* than the interval between pulses, the one-shot will complete its timing cycle before the next pulse from the LED arrives. This will cause the alerter to emit a pulsating warning tone.

The component values in Figure 4-29 have been selected to provide the aforementioned pulsating output tone because of its attention-getting impact. An added benefit is that in this mode the circuit consumes less current when the alarm is sounding.

Although I used a General Electric H20A1 slotted optoisolator, any LED-phototransistor optoisolator will work. Some of these devices, such as G.E.'s H13B1, have mounting holes. If you cannot locate one of these devices, you can make your own with a phototransistor and an infrared LED. Install the two components facing one another on a small phenolic board. Leave a gap of about 0.25 inch between them.

To operate this circuit as an intruder alarm, the optoisolator should be installed on an immovable portion of the frame of the door or window to be protected. Attach an opaque projection flag such as a small aluminum L-bracket to the door or window so the flag resides in the slot of the optoisolator when the door or window is closed. When the flag is moved, the alarm will sound.

The circuit can be powered by a supply providing from 3 to 15 volts. When powered by a 9-volt transistor radio battery, my prototype version consumed about 8.5 milliamperes in its standby mode.

Going Further

An obvious simplification of the basic circuit in Fig. 4-28 is to replace the pair of 555 timers with a 556 dual timer. The circuit in Fig. 4-29 is the result. Though it is functionally identical to the circuit in Fig. 4-28, I have included it to preclude the possibility of pin errors should you wish to try it.

Other variations are also possible. For example, the piezoelectric alerter can be replaced by a relay (Radio Shack 275-004, or similar) which, in turn, can switch a siren or other powerful alarm signal. For silent alarms, substitute a 270-ohm resistor in series with a red LED for the alerter. The LED will flash at a rate of a few hertz when the alarm is triggered.

Still another variation is to replace the slotted optoisolator with a reflective sensing transducer. Consisting of an LED and phototransistor facing in the same direction, the sensor can detect the presence of an object a few millimeters away. If the object is sufficiently reflective (add white tape if it is not), the alarm will sound. When the object is moved away from the sensor, the alarm will cease sounding.

Finally, be sure to assemble and install your alarm circuit with care for the quality of your work will determine the reliability of the alarm. Be sure to avoid exposed wires leading to the optoisolator and replace the battery when necessary.

Fig. 4-29. Single IC photonic door and window intrusion alarm.

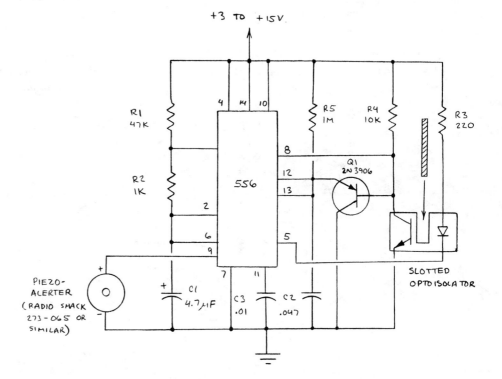

99

Optoisolators—The Photon Connection

The *optoisolator* is a solid-state component which permits one circuit to influence or control a second circuit while maintaining virtually total electrical isolation between the two circuits. As its name implies, the optoisolator achieves this goal with the help of optoelectronics.

There are many kinds of optoisolators and they are described by various names, the most common alternative being *optocoupler*. Nearly all are made by combining a semiconductor light emitter and a detector in a single package. The light source and detector are electrically isolated from one another within the package, which is usually made from an opaque plastic. Optical access between the two is provided by an air space or a transparent glass or plastic channel.

Figure 4-30 is a pictorial view of a typical LED-phototransistor optoisolator. The LED chip is almost always made from GaAs or GaAs:Si. A transparent glass window called the *dielectric channel* insulates and isolates the LED chip from the phototransistor.

The phototransistor in optoisolators installed in 6-pin miniDIPs like those of Fig. 4-31 is generally mounted directly on the package's frame. This provides electrical contact to the collector. Flying bonding wires connect the base and emitter to their respective pins. Electrical access to the LED is also provided by flying bonding wires. The electrical isolation between the two chips is enhanced by bringing their respective bonding leads to opposite sides of the package. An isolation of 2500 volts or more is readily achieved.

Various source-detector combinations can be used to make optoisolators. Here are some of the most common:

Source	Detector
Incandescent Lamp	Cadmium Sulfide Photoresistor
Neon Glow Lamp	Cadmium Sulfide Photoresistor
GaAsP Red LED	Cadmium Sulfide Photoresistor
GaAs IR LED	Phototransistor
GaAs IR LED	Photodarlington
GaAs IR LED	Photodiode
GaAs IR LED	SCR
GaAs IR LED	Triac
GaAs IR LED	PhotoFET

The devices which use cadmium sulfide detectors are much slower than those using silicon detectors. But they are very simple to manufacture, often consisting of a plastic tube with a light source installed in one end and the detector in the other.

Silicon detectors can be assembled in this way also. Indeed, you can quickly make a working optoisolator by installing a phototransistor and infrared LED in opposite ends of a short length of heat shrinkable tubing. This arrangement may provide excellent high voltage isolation, but the physical distance between the active elements is such that the LED will require more bias current to provide the same degree of coupling as an optoisolator in which the LED and phototransistor are separated only by a thin glass window.

This brings us to the optoisolator parameter known as the *current transfer ratio* (CTR). Simply put, CTR is the ratio of current through the isolator's LED to the output current from the isolator's detector.

Commercial optoisolators have CTRs ranging from less than 20 percent to more than 500 percent. Since higher isolation voltages are achieved by moving the LED away from the phototransistor, the CTR is usually inversely proportional to the CTR.

Incidentally, a CTR higher than 100 percent doesn't imply the creation of something from nothing. High CTRs result from the inherent gain of phototransistor and, especially, photodar-

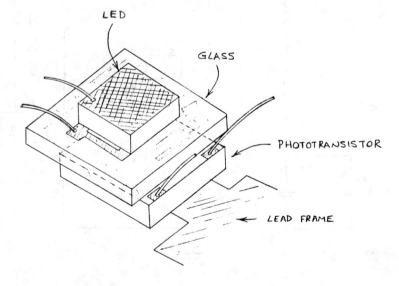

Fig. 4-30. Typical construction of optoisolator.

lington detectors. Optoisolators using these detectors can be considered a unique kind of optoelectronic amplifier or gain block with applications which go beyond voltage isolation and signal coupling.

Applications for Optoisolators

One of the most important applications for optoisolators is high voltage isolation. There have been instances, for example, of electrocution of hospital patients by the very electronic monitoring equipment intended to help save their lives. Figure 4-

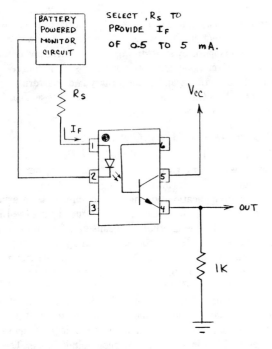

Fig. 4-31. Patient monitoring circuit with 2500-volts isolation.

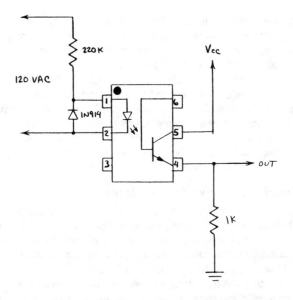

Fig. 4-32. Line voltage monitor.

31 shows how a low cost photodarlington optoisolator can provide up to 2500 volts isolation between battery powered monitoring apparatus attached to a patient and a bedside monitoring system.

This same idea is used to isolate low voltage circuits from high voltages. In addition to reducing shock hazards, the low voltage circuits are protected from the possibility of catastrophic damage.

Figure 4-32 shows how an optoisolator can monitor a high voltage line and emit an output signal when the high voltage is present. This idea can be adapted to monitor a telephone line to detect a ring signal.

Figure 4-33, for example, shows a telephone ring detector published by Hewlett-Packard (Application Note 951-1). The circuit provides 2500-volts isolation between the telephone line

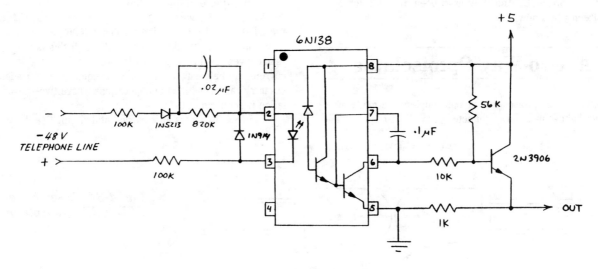

Fig. 4-33. Telephone ring detector.

and any external circuit. Although it extracts power from the ring voltage to bias the LED in the optoisolator, the loading of the telephone line is minimal. The circuit, for example, borrows only about 100 microamperes from a ring signal.

It's important to realize that the voltage isolation capability of the optoisolator in these and other applications is not greater than the isolation provided by the circuit board upon which the unit is mounted. Solder bridges, bits of wire, nearby metal objects or wire leads can greatly reduce the isolation voltage. The isolation can be intermittently reduced by contaminants ranging from solder flux to moisture.

Special Purpose Optoisolators

Thanks to optical fibers it's possible to make optoisolators which can provide complete isolation of potential differences of many millions of volts. One application of such sophisticated optoisolation is the monitoring of high voltage transmission lines and the control of high voltage switching circuits.

Another use for optical fiber isolation is monitoring the performance of electronic equipment on an aircraft subjected to massive electromagnetic pulses (EMP). The Air Force subjects various of its aircraft to such pulses at several specially designed facilities near Albuquerque, New Mexico. The pulses simulate the EMP which accompanies a nuclear blast. Conventional shielded cables do not provide sufficient shielding to eliminate noise spikes from the EMP. Optical fibers intercept absolutely none of the EMP.

Optical Integrated Circuits

Several research laboratories in the United States and abroad are developing what may one day become the *optical* integrated circuit. In these circuits, photons will replace electrons, thus making possible faster switching speeds and direct interfacing with optical fibers.

The news releases from some of these laboratories would have us believe that the optical IC is a radically new development. Actually, of course, ordinary optoisolators are a form of hybridized optical IC. And they have been available for more that fifteen years now.

A Two-Way Optoisolator

Conventional optoisolators or optocouplers are made by installing a light source and a light detector in a light-tight package.

The source is electrically separated from the detector by an optically clear dielectric such as epoxy, glass or air. This arrangement allows two separate circuits or two portions of the same circuit to interact without any intervening electrical connections.

Some optoisolators use a neon glow lamp or an incandescent lamp as a light source. Most, however, use a visible or near-infrared light emitting diode. Detectors include cadmium sulfide photoresistors, phototransistors, photodarlington transistors, photodiodes, light activated SCRs, light activated triacs, and other light sensitive semiconductor devices.

Conventional optoisolators are *unidirectional*. In other words, they transfer an incoming signal from the source to the detector in only one direction. It's possible to make a *bidirectional* or two-way optoisolator by substituting for the source and detector semiconductor components that function as both sources *and* detectors. Many LEDs and some ternary and quarternary photodiodes (such as GaAsP photodiodes) can be used as dual-function emitter-detectors.

A Practical Two-Way Optoisolator

Figure 4-34 shows how to make a functional two-way optoisolator by installing two LEDs face to face in either end of a short length of heat shrinkable tubing. For very high voltage isolation or for applications in which two circuits to be interfaced by the two-way optoisolator are some distance apart, the LEDs can be coupled to one another by means of a fiber optic cable.

Though many different commercial LEDs can function as both sources and detectors, GaAs, GaAs:Si and AlGaAs:Si near infrared emitters work better as detectors than do most visible light emitting diodes. Figure 4-35 is a plot that shows the current transfer of a pair of TRW Optron OP-195 GaAs:Si near infrared emitters arranged in the configuration shown in Fig. 4-34. The data from which this plot was made was obtained by operating the detector LED in an unbiased, photovoltaic mode. The detector LED can also be operated in a reverse biased, photoconductive mode. The resulting current transfer curves are almost identical to those in Fig. 4-35.

As Fig. 4-35 confirms, the current transfer ratio (I_{out}/I_{in}) for the OP-195 LEDs I used is only about 0.06 percent when the input current is 20 milliamperes. Although this is much lower than conventional LED-phototransistor optoisolators, the output current can be easily amplified. The requirement for additional circuitry can be justified in applications where two-way optoisolation is required.

Figure 4-36 shows an experimental circuit I've designed to

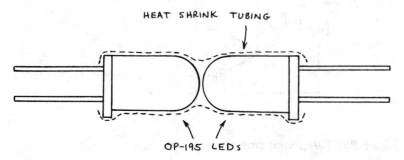

HEAT SHRINK TUBING

OP-195 LEDs

Fig. 4-34. Assembly of LED/LED optoisolator.

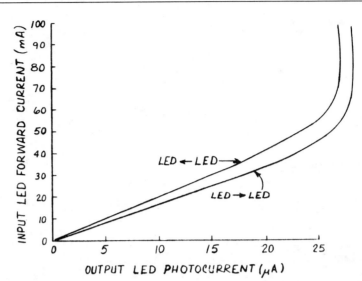

Fig. 4-35. LED/LED optoisolator current transfer.

implement under digital control two-way optoisolation. The circuit preserves input-output isolation by employing conventional LED phototransistor optoisolators. In operation, a low or high bit at the control input forward biases the LED in *one* of the two conventional optoisolators. For example, assume the control bit is *high*. This causes the LED in optoisolator 1 to be forward biased which, in turn, turns on its phototransistor. Any signal present at the phototransistor's collector can now forward bias LED 1 in the LED-LED (two-way) optoisolator. LED 2 functions as a detector. It cannot receive any signal present at the collector of the phototransistor in optoisolator 2 since that phototransistor is turned off. When the control bit is changed from high to low, this operating mode is reversed and the LED-LED optoisolator transmits in the opposite direction.

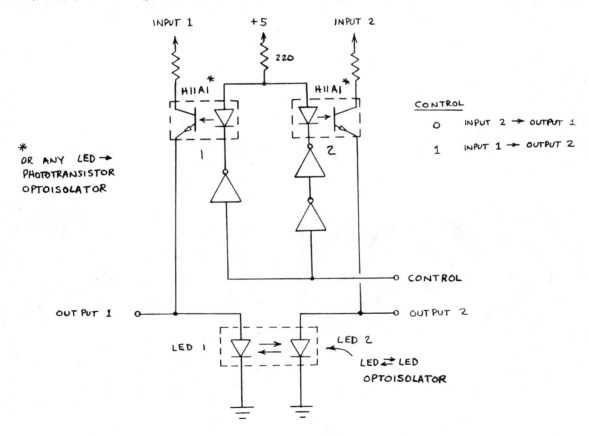

Fig. 4-36. One method for implementing two-way optoisolator.

103

FIVE

Lightwave Communications, Control, and Telemetry

- Getting Started in Lightwave Communications
- Optical Data Communications
- Experimenting with Low-Cost Fiber Optic Links
- A Pulse-Frequency Modulated Infrared Communicator
- An Infrared Temperature Transmitter
- An Experimental Infrared Joystick Interface
- A Single-Channel Infrared Remote Control System

Lightwave Communications, Control, and Telemetry

Getting Started in Lightwave Communications

The wires and cables that connect electronic devices with one another and their sources of power have always been considered a necessary nuisance. But are they?

Simultaneous advances in low power CMOS circuitry and high capacity disposable *and* rechargeable batteries have eliminated power cords from many electronic devices. Now the capability exists to replace the wires that connect many devices to one another with radio waves or beams of light.

Radio Vs. Light

Radio links have been used for decades to open garage doors and to control model airplanes, boats, and cars. They are easy to use and omnidirectional, but they require antennas and may be subject to government regulation. For these reasons, photonic systems that transmit information or control signals by means of near-infrared radiation and visible light often offer a viable alternative for short range wireless links.

Strictly speaking, the term *light* refers only to the range of wavelengths in the electromagnetic spectrum that are visible to the human eye. It's common practice, however, to classify both systems that use visible light *and* those that employ near-infrared as *lightwave links*. I'll use this terminology in this discussion.

Often lightwave signals can be transmitted directly to a suitable receiver in what is often called a *free-space link*. When total electronic security is important or when distance or obstacles preclude a direct optical link, information carrying beams of light can be injected into highly transparent fibers of plastic, glass, or silica.

Whether radio or infrared is the best choice for a particular wireless application depends upon the circumstances. My personal preference is to use a lightwave link when possible. But I don't hesitate to use radio when transmitting data from model rockets or triggering a camera suspended from a kite or balloon hundreds of feet in the air. In short, both radio *and* lightwave links each possess relative advantages and disadvantages.

Lightwave Links Today

Communications over beams of light was first pioneered in the United States by Alexander Graham Bell and Sumner Tainter in 1880. Prior to the 1950s, most lightwave communications research was conducted by the military and individual experimenters. During World War Two, Italy, Germany, and Japan developed advanced lightwave voice communications gear.

Solid-state light emitters and detectors as well as the laser were developed in the United States during the 1960s. Japan and West Germany were among the first to apply these components in lightwave links for consumer products.

For example, German companies were among the first to develop infrared remote control transmitters for television sets, toys controlled by infrared signals, and wireless, infrared-linked stereo headsets. And Japan's Canon makes a midget infrared RS-232 free space data link that allows a handheld computer to communicate with a nearby printer.

Canon's infrared computer-printer link is merely the first of what may become many such wireless links between computers and their peripherals. The keyboard of IBM's PC*jr* personal computer, for example, transmits keycodes to the system unit over beams of near-infrared emitted by a pair of light-emitting diodes.

Telecommunications is by far the biggest application for fiber-coupled lightwave links. If you make a phone call to or from a major metropolitan area in the United States, Japan, Brazil, England, Italy, Canada, and many other countries, chances are your voice travels at least part of the way as pulsations of near-infrared through silica fibers.

A Simple Free-Space Lightwave Link

A free-space lightwave link capable of transmitting your voice several hundred feet can be assembled from surprisingly low-cost components. I've been building such communicators since 1965 when, as a student at Texas A&M University, I used flashlight bulbs and newly developed GaAs near-infrared emitting diodes to send voice to receivers made by connecting a silicon solar cell to an audio amplifier.

Since then I've designed and built dozens of lightwave communicators. Because of its simplicity and low cost, the transmitter circuit in Fig. 5-1 is one of my favorites.

In operation, the small voltage generated when voice pressure waves are intercepted by a crystal microphone is coupled through C1 into a 741 operational amplifier. The signal is then amplified with a gain determined by the setting of R1.

The amplified signal is coupled through C2 to the base of driver transistor Q1. R5, R6, and R7 form an adjustable voltage

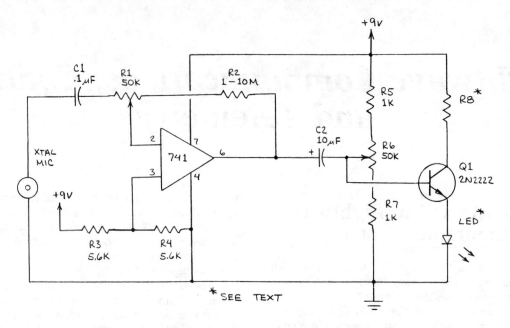

Fig. 5-1. Simple free-space lightwave transmitter.

divider that permits the bias on Q1's base to be adjusted for an optimum, distortion-free output from the LED.

The collector-emitter junction of Q1 acts like a variable current switch that controls the current flow through the LED. Maximum current flow is limited by R8.

Many different LEDs will work with this circuit. For best results, the LED should be an AlGaAs unit emitting at 880 nanometers (nm). A GaAs:Si unit that emits at 950 nm will also work, but at less than half the power efficiency of an AlGaAs unit.

In any event, it's necessary to limit the quiescent forward current through the LED to from about 10 to 40 milliamperes. Brief high-level audio surges will substantially raise this level.

The easiest way to establish the LED quiescent current is to temporarily replace R8 with a 1000-ohm potentiometer. Then, while monitoring a milliammeter inserted between the LED's anode and Q1's emitter, adjust the pot until the current flow is from 30 to 50 percent of the maximum allowable for the LED you're using. Remove the meter and pot, measure the pot's resistance and substitute a fixed resistor having a similar value.

The receiver, which is shown in Fig. 5-2, detects the voice-modulated beam from the transmitter by means of phototransistor which, together with load resistor R1, generates a voltage proportional to the amplitude of the signal. The signal voltage is amplified by a 741 op amp.

An LM386 audio power amplifier provides sufficient boost to drive a small speaker. R3 controls the signal level that reaches the LM386.

This transmitter and receiver pair will operate over a range of a few feet without external lenses. If you collimate the beam from the transmitter with a lens and place a second lens over the receiver's phototransistor, the range can be *greatly* increased.

Phototransistors saturate (i.e., turn fully on) in the presence of sunlight. Therefore, the receiver will work much better at night or if you block extraneous light with an infrared filter. Ironically, in total darkness a small amount of dc light falling upon the phototransistor will *improve* its sensitivity. This occurs because the light biases the transistor into conduction.

A Low-Cost Fiber Optic Communicator

The transmitter and receiver described previously can be easily linked to one another by means of an optical fiber. In fact, I've done so many times.

However, many experimenters who lack experience in building lightwave links have long hoped for an economical kit that will allow them to assemble, with minimum difficulty, a working fiber link.

One answer to their plight is the EDU-LINK Fiber Optic Kit. Available from Advanced Fiberoptics Corporation, this kit consists of a transmitter, receiver, and a 1-meter length of sheathed plastic fiber. Both the transmitter and receiver can be assembled in under an hour.

The EDU-LINK transmitter is TTL compatible and will convert incoming logic signals to optical pulses. The transmitter also includes a self-contained oscillator that provides a 1-kHz signal for test and demonstration purposes.

Figure 5-3 is the transmitter circuit diagram. IC1 (CD4093) is a quad of two-input NAND gates, each of which exhibits Schmitt-trigger operation on both inputs. Two of the gates (*a* and *b*) are connected as an astable oscillator that's enabled when the OSE (*OS*cillator *E*nable) and TXD (*T*ransmit *D*ata) inputs are both *high* (logic level 1).

A third gate (*c*) steers the signal from the built-in oscillator *or* an external source to driver transistor Q1. R2 and R3 supply

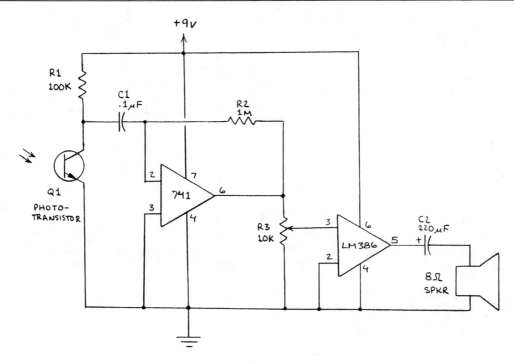

Fig. 5-2. Simple free-space lightwave receiver.

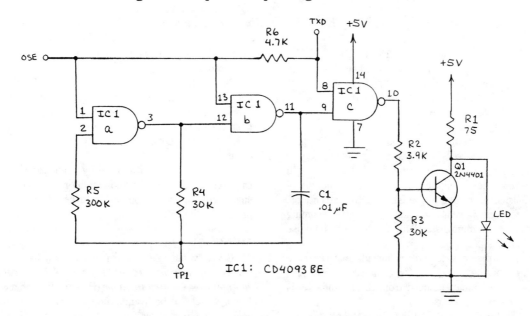

Fig. 5-3. EDU-LINK fiber optic transmitter.

base bias to Q1 and R1 limits the current through the LED to about 40 mA.

The LED is a Siemens GaAsP visible red (665 nm) emitter. Since the receiver's silicon photodiode is much more sensitive to near infrared at about 800–950 nm, the wavelength range of the powerful near-infrared LEDs I specified for the free-space transmitter in Fig. 5-1, you may be wondering why the EDU-LINK transmitter uses a less powerful, red emitting LED.

The principal reason is that plastic fibers transmit near-infrared radiation very poorly. On the other hand, they transmit visible red wavelengths quite well.

The simple circuit in Fig. 5-2 will detect signals from the EDU-LINK's transmitter. The EDU-LINK receiver, however, employs a clever design optimized specifically for the amplification and accurate reconstruction of incoming *pulses*.

The circuit of the EDU-LINK receiver is shown in Fig. 5-

109

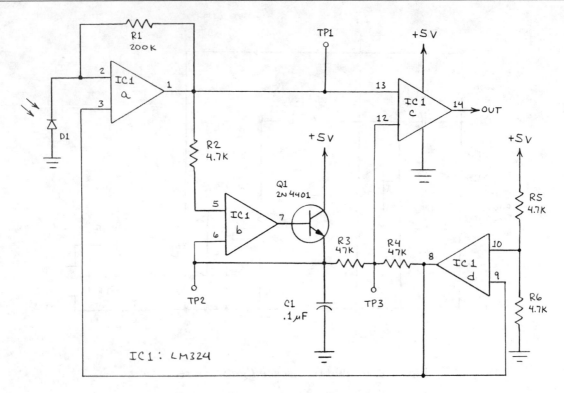

Fig. 5-4. EDU-LINK fiber optic receiver.

4. IC1 is an LM324N quad op amp, and D1 is a high speed PIN photodiode. In operation, D1 is a reverse biased and connected to one of the amplifiers (*a*) to form a voltage-to-current converter or, as it is commonly known, a *transimpedance amplifier*. The operation of the remainder of the circuit is more complex than it might at first appear.

Amplifier *d* is connected as a unity gain follower that provides, at pin 8, a buffered version of the voltage generated by the divider made from R5 and R6. Amplifier *b*, together with Q1 and C1, forms a peak detector that stores the amplitude of the incoming pulses. According to the EDU-LINK instructions, "This stored reference signal allows one to sample the incoming signal at its point of minimum distortion, thereby reducing pulse width distortion."

The output from the peak detector is halved, relative to the reference voltage, by the divider formed from R3 and R4. This signal, along with the amplified signal from the photodiode, is then applied to the output comparator (*c*).

The neatest feature of the EDU-LINK kit is the plastic connectors into which the LED and photodiode are installed. Both connectors are attached to their respective circuit boards by a pair of mounting screws.

Only a few minutes are required to terminate the plastic optical fiber supplied with the kit. After a quarter-inch of sheathing is removed from each end, the ends of the fiber are inserted into plastic ferrules. The exposed fiber emerging from the end of each ferrule is then cut with a hobby knife. The link between transmitter and receiver is completed when the plastic ferrules

at each end of the fiber are snapped into the LED and photodiode connectors.

All that's necessary to test the link is a 5-volt power supply and an audio amplifier or oscilloscope. I powered the unit I assembled with a pair of 6-volt batteries. (I dropped the voltage from the transmitter battery to about 5.4 volts by connecting a 1N914 diode between the positive battery terminal and the circuit.)

You can hear the test tone by connecting the output from the receiver to a small audio amplifier. Or you can monitor the signal with an oscilloscope as described in the instructions supplied with the kit.

Going Further

If you would like to learn more about lightwave communications, you may want to read my book *A Practical Introduction to Lightwave Communications* (Howard W. Sams & Co., 1982). For a variety of both amplitude and pulse-modulated circuits, see *The Forrest Mims Circuit Scrapbook* (McGraw-Hill, 1983).

Optical Data Communications

The colorful computer ads in magazines cleverly conceal the tangle of power cords and the maze of interconnection cables behind every computer system that includes more than just a keyboard and a display. True, computers with built-in peripherals, like the one I'm typing this column into, require fewer external cables and wires. But when peripherals are added, external cables are still required.

One way to reduce the power cord problem is for manufacturers to design a central power supply into their computers. Power for peripherals could then be taken from the central supply rather than wall sockets. Another is to power the computer and its peripherals with self-contained batteries. And one way to eliminate entirely the maze of interconnection and bus cables is to send data between a computer and its peripherals (or another computer) over beams of near-infrared radiation.

This subject was high on my mind a few days ago while I was reading a catalog that described an Atari wireless, remote control joystick system. My first reaction was that someone had finally replaced the clumsy joystick cable with an infrared emitting diode. Further reading, however, disclosed that the Atari system uses miniature radio transmitters. A photograph revealed a stubby antenna protruding from each joystick assembly. A longer, telescoping antenna emerged from a receiver unit that connects to the computer.

Since I've spent many hours using joysticks and other graphic input devices for genuinely useful applications (no, I'm not a video game freak), I'm delighted that Atari has seen the wisdom of replacing those bothersome joystick cables with a wireless link. But I'm puzzled why they and many other companies haven't yet introduced free-space infrared data links for computers and their peripherals.

Although many companies make fiber optic data transmission links, most electronic equipment manufacturers have been very slow to adopt near-infrared emitting diodes for transmission of data through free space. This is all the more puzzling when one considers that there are some similarities in both components and the operation of fiber optic and free-space near-infrared data.

Since both infrared emitting and detecting diodes and a host of modulation and demodulation methods were pioneered in the United States, we certainly cannot plead ignorance of the subject. Perhaps our love affair with radio is the culprit.

In any event, though audio was first transmitted over light in the United States (beginning with Alexander Graham Bell in 1880), infrared-coupled high-fidelity earphones and speakers were first manufactured on a commercial scale in West Germany. Similarly, while we and the Japanese were building radio-controlled toy cars, a German company introduced an infrared controlled toy vehicle with detection circuitry so sensitive it responds to signals bounced from ceilings, walls, people, and plants. Unlike its radio-controlled counterparts, it completely ignores transmission from CB operators and passing taxis. And it is subject to no rules regarding frequency, radiated power, and antenna size.

As for computers, several companies have finally begun to recognize the advantages of short-range, free-space infrared data links between computers and peripherals. In 1979, for instance, Fritz R. Geller and Urs Bapst of IBM's Zurich Research Laboratory in Ruschlikon, Switzerland, published a paper describing in detail the transmission of in-house data by means of reflected beams of near-infrared ("Wireless In-House Data Communica-

tion via Diffuse Infrared Radiation," *Proceedings of the IEEE*, Nov., 1979).

This excellent paper discussed virtually all aspects of practical in-house infrared data transmission. Block diagrams of typical systems, design equations, and detailed discussions concerning the near-infrared reflectance properties of various surfaces were included.

Never slow to adapt a useful technology, several Japanese computer firms have developed or introduced various kinds of near-infrared coupled in-house, wireless data transmission systems. One of the most ambitious is an infrared modem developed by Fujitsu, a major Japanese manufacturer of computers.

Designed specifically for office automation, Fujitsu's system optically links several terminals equipped with infrared transceivers to one of several *satellite* transceivers connected by conventional cable to a central processor. The system is RS-232 compatible at full or half duplex with data rates of up to 19.2 kbaud. The normal operating range of the system is about ten meters, but this is reduced to about three meters should any of the system's photodiodes be exposed to direct sunlight.

Incidentally, Fujitsu's system uses the new high-power AlGaAs "super" LEDs. Each terminal transceiver uses five diodes and each diode emits about fifteen milliwatts. The satellite stations use nine diodes.

Fujitsu's system shows what can be done on a large scale with infrared links. But I think you'll be more interested in a new handheld business calculator developed by Hewlett-Packard that can communicate with a printer by means of a midget infrared data link.

Design Tips for Optical Data Links

Possibly one reason for the slow development of free-space optical data links for computers and nearby terminals and peripherals is the special design requirements imposed by such a system. First, a communications format must be selected. For instance, will one LED channel be sufficient or will others be required? What form of multiplexing, if any, will be used?

Next, a suitable LED and LED driver must be selected. If high data rates are not a requirement, high power AlGaAs near-infrared emitters are the best choice. A power MOSFET makes an excellent driver for these and other infrared-emitting diodes. For more power, several LEDs can be connected in series.

Many kinds of receivers are possible. A typical design might employ a photodiode or phototransistor detector followed by a low-noise, high-gain amplifier. Low noise is important for high sensitivity. Immunity to noise from incandescent and, especially, fluorescent lights is essential. Infrared filters can help. So can a filter that rejects 60- and 120-Hz interference.

Over what range will the system operate? Two simple range equations can be used to give reasonable predictions. Both equations require a knowledge of such parameters as the optical output power from the transmitter (P_o), the receiver sensitivity (P_{th}), the area of the receiver's detector chip or lens (A_{rec}), the divergence or beam spread angle of the transmitter in radians (θ) and the transmissivity of the receiver's optics (τ).

The first equation, a simplified form of the optical communications range equations, is for line-of-sight operation where the infrared emitted by the transmitter's emitter has an unobstructed, direct path to the detector. In this case, the range is

$$\sqrt{\frac{P_o A_{rec}\tau}{P_{th}\pi\theta^2}}$$

Figure 5-5 summarizes this equation.

The second equation, a simplified form of the optical radar range equation, is for systems in which the transmitted beam is reflected from diffusely reflecting walls, ceilings and other surfaces before reaching the receiver's detector. In this case, the range is

$$\sqrt{\frac{P_o A_{rec}\rho\tau}{P_{th}\pi}}$$

This equation is summarized in Fig. 5-6.

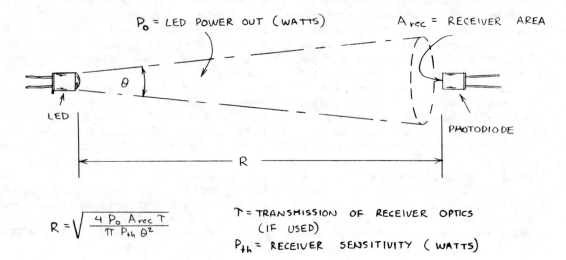

$$R = \sqrt{\frac{4 P_o A_{rec}\tau}{\pi P_{th}\theta^2}}$$

τ = TRANSMISSION OF RECEIVER OPTICS (IF USED)

P_{th} = RECEIVER SENSITIVITY (WATTS)

Fig. 5-5. Range equation for line-of-sight operation.

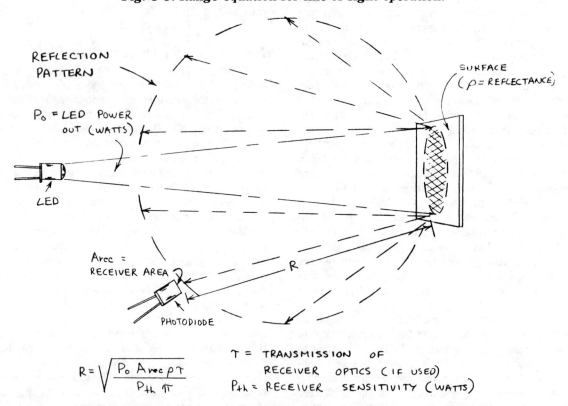

$$R = \sqrt{\frac{P_o A_{rec}\rho\tau}{P_{th}\pi}}$$

τ = TRANSMISSION OF RECEIVER OPTICS (IF USED)

P_{th} = RECEIVER SENSITIVITY (WATTS)

Fig. 5-6. Range equation for operation with a diffuse reflector.

The Greek letter ρ (rho) denotes the reflectance of the surface from which the transmitted beam is reflected. At 880 nm, plaster and unpainted pine have a typical reflectance of about 70 percent (0.7). Human skin and green vegetation have typical reflectances of from 50 to 60 percent (0.5 to 0.6). (In the near infrared, differences in skin pigmentation are barely discernible.)

Experimenting with Low-Cost Fiber Optic Links

At one time, high prices prevented most experimenters and hobbyists from experimenting with fiber optic data links. General Electric, Motorola, and other companies have now come to the rescue with low-cost fiber optic components including near-infrared emitters and phototransistor detectors. Many are installed in threaded plastic receptacles to which low-cost fiber optic connectors can be quickly mated.

Here, I will emphasize General Electric's inexpensive fiber optic components. Those made by other companies are similar in design and function. Figure 5-7 shows General Electric's side-looking package design as well as how the package is interfaced with a threaded Optimate plastic connector. Figure 5-8 is an internal view of a package that contains an infrared-emitting diode. Note the use of a reflector and lens to collect and direct toward the aperture in the package, hence a terminated fiber, the radiation emitted by the diode.

The Emitter

One of General Electric's inexpensive emitters is designated GFOE1A1. This device is a silicon-compensated, liquid-phase, epitaxial gallium arsenide diode that emits near-infrared radiation peaking at about 940 nanometers. At room temperature and 30- to 50-milliamperes forward current, the diode has a power conversion efficiency of about 4 percent. The conversion efficiency increases to from 5 to 6 percent at 200 milliamperes (pulse drive).

The diode is mounted behind a diffuse, molded epoxy lens that provides a 1.2-millimeter diameter source having nearly uniform intensity across its surface. The large source size assures good optical coupling between the LED and the wide variety of different fibers to which Optimate connectors can be terminated.

At 50-milliamperes forward current, a typical GFOE1A1 will couple more that 100 microwatts into a 1-millimeter diameter fiber. This is approximately 10 percent of the total power (i.e., a −20-dB loss) radiated by the chip and is comparable to the light injection efficiency of other low to moderately priced fiber optic links.

The silicon dopant added to the GFOE1A1 increases both power conversion efficiency and wavelength at the expense of the diode's *response time* (the sum of delay and rise times of storage and fall times). Conventional GaAs emitters, for example, have response times measured in tens or nanoseconds or even less. A GaAs:Si emitter like the GFOE1A1, however, has a response time of nearly a microsecond. This places an upper limit of about 400 kHz on the modulation bandwidth of the GFOE1A1. This, of course, is more than adequate for many kinds of telecommunications and data transmission applications.

The Detector

General Electric makes at least two low-cost detectors. One is an npn phototransistor (GFOD1A1), and the other is a photodarlington transistor (GFOD1B1). Both are installed in side-looking packages identical to those used for the GFOE1A1 emitter. Since the base of a phototransistor is not often used in optoelectronic circuits, neither device is provided with an external base lead.

The spectral response of both phototransistors peaks at about 850 nanometers. At the 940-nanometer wavelength emitted by the GFOE1A1 emitter, both phototransistors exhibit about 80 percent response efficiency.

When illuminated by radiation from a GFOE1A1 emitter transmitted to the phototransistor via a 1-meter length of Crofon 1040 fiber, the GFOD1A1 exhibits a minimum responsivity of

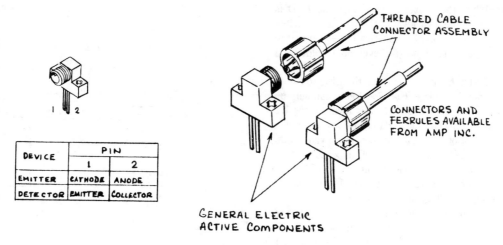

DEVICE	PIN	
	1	2
EMITTER	CATHODE	ANODE
DETECTOR	EMITTER	COLLECTOR

Fig. 5-7. GE active fiber optic components.

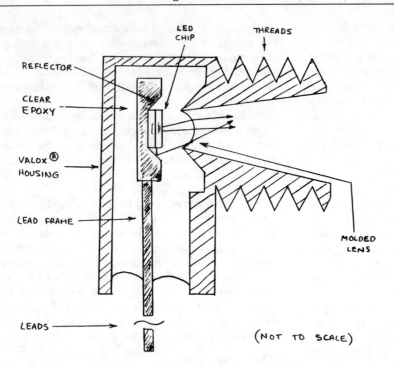

Fig. 5-8. Internal view of GE side-looking emitter.

(NOT TO SCALE)

70 microamperes per microwatt. The GFOD1B1 provides 1000 microamperes per microwatt under the same conditions. The turn-on and turn-off times of the GF0D1A1 are each 3 microseconds when the load resistance is 0 ohms. The turn-on and turn-off times for the GFOD1B1 are, respectively, 10 and 25 microseconds when the load resistance is 0 ohms.

Digital Logic Application Circuits

Short fiber optic links are ideally suited for transmitting digital data through noisy environments. The circuits that follow illustrate straightforward ways to send and receive signals through such a fiber optic link.

TTL Emitter Driver Circuits

Figure 5-9 shows a basic TTL LED driver made from a single NAND gate. When the enable input is *low*, the LED is turned off irrespective of the logic level at the TTL input. When the enable input is *high*, the LED is forward biased when the logic level input is high. When the logic level input is low, the LED is turned off.

Series resistor R1 limits current through the LED to a safe value. The output from a standard TTL 7400 gate can sink up to 16 milliamperes. To drive the LED at this level means R1 must be 312.5 ohms. (From Ohm's law, R1 equals 5 volts divided by 16 milliamperes.) The closest standard resistance value, 330 ohms, will provide an LED current of 15 milliamperes.

Incidentally, the maximum current output from an LS TTL gate is only 5 milliamperes. Therefore you should use standard TTL in the circuit shown in Fig. 5-9.

In applications where higher infrared emission levels are required, more drive current can be provided in adding a transistor driver stage as shown in Fig. 5-10. Note that the transistor inverts the signal from the gate. Also note that since the transistor and not the gate drives the LED, an LS TTL gate can be used.

R2 should be selected to limit the current through the LED (I) to the desired level. The combined voltage drop of Q1 and the LED is about 2 volts. Therefore the series resistance is $(V_{cc}2 - 2)/I$. If $V_{cc}2$ is 5 volts and the desired current level is 50 milliamperes, then R2 should have a resistance of 60 ohms. Higher current levels can be achieved by increasing $V_{cc}2$ or reducing R2's resistance. It is essential, of course, that Q1, R2, and the LED emitter be rated for the selected current level.

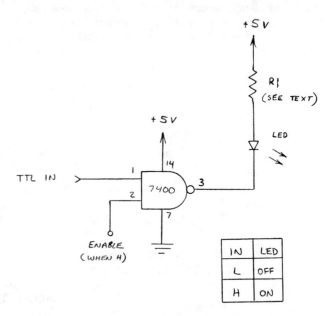

Fig. 5-9. Basic TTL compatible LED driver.

IN	LED
L	OFF
H	ON

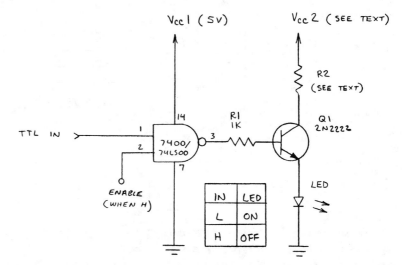

Fig. 5-10. TTL compatible fiber optic LED driver with gain stage.

IN	LED
L	ON
H	OFF

Detector Circuits

Figure 5-11 shows two basic phototransistor detector circuits. In Fig. 5-11A, the phototransistor is normally off and the voltage across R_L is high. When the phototransistor is turned on by an incoming light pulse, the output is brought low.

Although a basic phototransistor circuit is very sensitive, its response time is slowed by the RC time constant of the internal capacitance of the phototransistor and R_L. The delay induced by the load resistor is virtually eliminated by adding a common base stage as shown in Fig. 5-11B. The low resistance path provided by Q2 greatly speeds up the charge-discharge time of Q1's internal capacitance, thereby substantially improving its response time.

Figure 5-12 compares the response of the two basic phototransistor circuits in Fig. 5-11. Response *A* is the signal delivered to the TTL compatible LED driver shown in Fig. 5-10. The signal is inverted by the LED driver transistor. Response

B is the output of a basic phototransistor load resistor circuit (Fig. 5-11A), and response *C* is the output of a phototransistor circuit to which a common base stage has been added (Fig. 5-11B).

Note the slow response of the phototransistor circuit in trace *B*. The risetime of the common-base circuit in trace *C* is approximately ten times faster. Also note the phase reversal of the output from the two phototransistor circuits.

TTL Compatible Fiber Optic Receiver

The outputs from the phototransistor circuits in Fig. 5-11 are not TTL compatible. To provide a fully transparent TTL fiber optic link, that is one that accepts and outputs TTL level logic signals, amplification and pulse restoration is required.

Figure 5-13 shows a fiber optic receiver that provides a TTL compatible output signal. In operation, an optical signal received by Q1 is delivered directly to the inverting input of a 741 op-

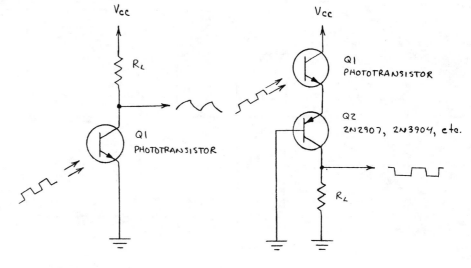

Fig. 5-11. Two phototransistor circuits.

(A) Phototransistor and load resistor.   *(B) Phototransistor and common base stage.*

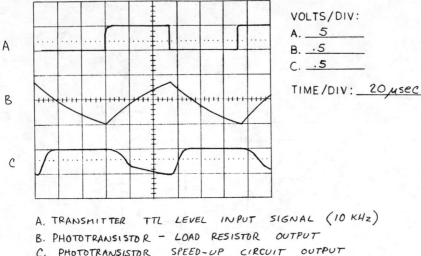

VOLTS/DIV:

A. __5__

B. __.5__

C. __.5__

TIME/DIV: __20 μsec__

Fig. 5-12. Received signal waveforms for basic phototransistor detector circuits.

A. TRANSMITTER TTL LEVEL INPUT SIGNAL (10 KHz)
B. PHOTOTRANSISTOR - LOAD RESISTOR OUTPUT
C. PHOTOTRANSISTOR SPEED-UP CIRCUIT OUTPUT

erational amplifier. R2 permits the gain of the 741 to be adjusted. The amplified signal from the 741 is coupled through C1 into a Schmitt trigger formed by a 555 timer.

Figure 5-14 is a set of oscilloscope waveforms that confirms the transparent nature of this receiver. Waveform A is the TTL signal delivered to a TTL compatible LED transmitter (Fig. 5-10), B is the signal at the output of the 741, and C is the TTL level signal at the output of the Schmitt trigger.

The 555 output is a phase reversed image of the TTL input at the transmitter. The phase reversal can be eliminated simply by following the 555 with a TTL inverter stage.

The circuit in Fig. 5-13 requires an initial gain adjustment via R2. It may also be necessary to alter the transmitted signal level. Too much infrared at the phototransistor may result in failure of the phototransistor to follow the transmitted signal.

Though the phototransistor-load resistor arrangement is very slow (see trace B in Fig. 5-12), this circuit has a surprisingly fast response of about 60 kilobits per second. This performance is made possible by the Schmitt trigger stage. The oscilloscope traces in Fig. 5-14 were produced when the circuit was receiving a pulse train of 50 kilobits per second.

General Purpose Fiber Optic Receiver

The circuit in Fig. 5-15 is a more versatile version of the preceding circuit. The Schmitt trigger stage has been replaced by a 555 connected as a monostable multivibrator. Optical pulses received by the phototransistor are amplified by the 741 and passed directly to the trigger input of the 555.

A negative-going pulse triggers the 555 into delivering a positive output pulse having a duration of 1.1 R3C2 seconds.

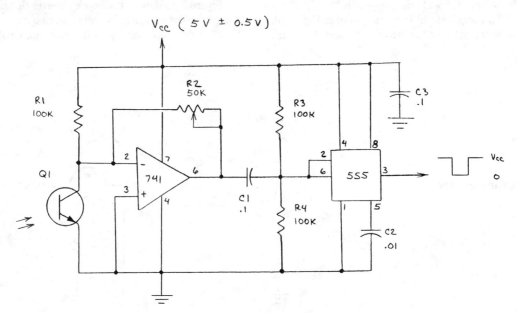

Fig. 5-13. Fiber optic receiver with TTL output.

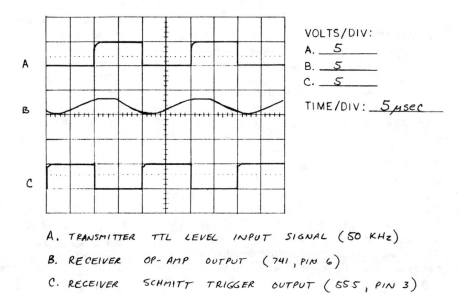

Fig. 5-14. Received signal waveforms for circuit of Fig. 5-13.

VOLTS/DIV:

A. _5_

B. _5_

C. _5_

TIME/DIV: _5 μsec_

A. TRANSMITTER TTL LEVEL INPUT SIGNAL (50 KHz)

B. RECEIVER OP-AMP OUTPUT (741, PIN 6)

C. RECEIVER SCHMITT TRIGGER OUTPUT (555, PIN 3)

With the values given in Fig. 5-15, the pulse duration is about 2.5 microseconds. This corresponds to an upper bandwidth of about 160 kHz. (Bandwidth is found by dividing 0.4 by the pulse duration.)

The circuit in Fig. 5-15 will provide a response greater than 100 kilobits per second. As with the circuit in Fig. 5-13, it is necessary to adjust the transmitted signal level and the op-amp gain potentiometer (R2) for optimum results.

Going Further

Though I used GFOE1A1 emitters and GFOD1A1 detectors in the test versions of the circuits described here, many other emitters and detectors can also be used. Many kinds of fiber optic cable can also be used. Indeed, over short distances, no fiber is required so long as infrared from the emitter can reach the detector.

The GFOE1A1 emitter and GFOD1A1 detector can also be used to transmit *analog* signals through a fiber optic cable. Several of these circuits appear in Chapter 2 of *The Forrest Mims Circuit Scrapbook* (McGraw-Hill, 1983). Additional circuits appear in *Engineer's Notebook* (McGraw-Hill, 1986).

Finally, the GFOE1A1 emitter can also function as a detector. This means half-duplex, bidirectional transmission over a single fiber optic cable can be achieved by placing a GFOE1A1 at each end of the link.

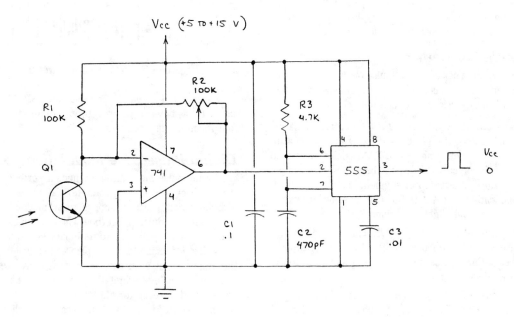

Fig. 5-15. Fiber optic receiver with pulse restorer output stage.

A Pulse-Frequency Modulated Infrared Communicator

If you have constructed and operated an amplitude or intensity modulated lightwave communicator, you have probably been impressed with both the simplicity and high-quality sound transmission of such a system. Several such projects have been described in *Popular Electronics* since 1970.

This magazine's first voice modulated laser communicator was described by C. Harry Knowles in a May 1970 cover story. Harry's breakthrough project used a low-power helium-neon laser he offered for a bargain price of only $50.50 postpaid! Harry's company, Metrologic Instruments, Inc. has since become a leading manufacturer of HeNe lasers and related products.

The first infrared LED communicator described in *Popular Electronics* also made the cover. The system was called the Opticom and it was designed primarily by H. Edward Roberts and described by Ed and me in the November 1970 issue. Incidentally, a kit version of the Opticom was one of the first products offered by MITS, Inc., the company that introduced the ALTAIR 8800 microcomputer in 1975.

Amplitude modulation is ideal for ultrasimple lightwave links through either fibers or free space. In free-space systems, however, it is susceptible to noise from both artificial and natural light sources. Furthermore, the transmission range controls the volume of the receiver's audio output unless some form of automatic gain control is provided. Finally, the average power consumption of an amplitude modulated system is high since the LED or injection laser source is continuously biased.

All these objections can be overcome by selecting one of the various forms of pulse or digital modulation. Though pulse modulation requires more complex transmitter and receiver circuitry than amplitude modulation, its advantages are significant. They include a very high degree of noise immunity, low average power and duty cycle of the optical source, and simplified multiplexing.

Pulse Modulation Methods

There are several major classes of pulse modulation suitable for lightwave communications, and they are compared in Fig. 5-16. Here is a brief description of each method:

1. *Pulse-Amplitude Modulation (PAM)*. In this modulation scheme the amplitude of the pulses is directly proportional to the amplitude of the modulating signal. PAM is closely related to amplitude modulation in that PAM can be achieved by simply sampling at a uniform interval brief segments of an AM signal. An obvious application of PAM is the transmission of two or more signals over a single lightwave channel.

2. *Pulse-Width Modulation (PWM)*. This method is also known as Pulse-Duration Modulation (PDM). The duration of individual pulses within a pulse train is made proportional to the amplitude of the modulating signal.

3. *Pulse-Position Modulation (PPM)*. Here the amplitude of the input signal controls the relative position of individual pulses in a pulse stream. Unlike PAM and PWM, *all* the pulses

in PPM have precisely the same amplitude and duration. This means the PPM receiver can be optimized for the processing of identically shaped pulses. This gives a higher degree of noise immunity than provided by PWM and especially PAM. Another advantage is that high peak power optical sources such as injection lasers can be used to full advantage.

The detection of PPM pulses by a receiver requires synchronization with the transmitter. This implies the necessity to transmit a clock signal along with the data or on a separate channel.

4. *Pulse-Frequency Modulation (PFM)*. This modulation method resembles that used in fm radio in that the transmitter emits a steady train of pulses called the *carrier*. Information is superimposed on the carrier by altering the carrier's frequency.

Detection of PFM is straightforward since, unlike PPM, no clock signal from the transmitter is required. Since the pulses have uniform duration and intensity, PFM offers many of the same advantages of PPM. PFM is particularly well suited for audio bandwidth lightwave links.

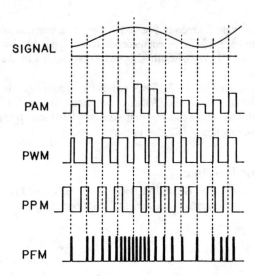

Fig. 5-16. Pulse modulation methods.

Later, we'll assemble and evaluate a PFM LED communicator. First, however, I want to explain briefly the most important form of pulse communications.

Digital Pulse-Code Modulation (PCM)

All the pulse modulation methods described thus far require that such critical pulse parameters as amplitude, duration, or position be altered in response to an analog signal. PCM is a true digital modulation method which involves the transformation of an analog signal into its binary equivalent.

Voice, for example, is sampled at a sufficiently fast rate, and the amplitude at each sample point is converted into a binary word by an analog-to-digital converter. The binary word is then transmitted in serial form a bit at a time.

In PPM, PFM, and PCM, the shape of each pulse is identical. PCM, however, offers two significant advantages over PPM and

PFM. First, the pulses remain fixed in time, and a pulse is either absent or present. This greatly enhances the noise immunity of the signal. Second, the predictable spacing between pulses greatly simplifies time division multiplexing. The major disadvantages of PCM are system complexity and bandwidth limitations.

Several pulse formats are used to implement PCM, the most important being return-to-zero (RZ) and nonreturn-to-zero (NRZ). A binary signal is either high (1) or low (0). A pulse, therefore, represents binary 1 and the absence of a pulse denotes binary 0.

Figure 5-17 illustrates some of the important pulse formats. In the RZ mode, all bits return to zero before the next bit is transmitted. In the NRZ mode, a 1 remains high and a 0 remains low for the duration of the bit transmission interval. This means two consecutive 1 bits merge into a pulse having twice the duration of an individual bit position. The RZ format is more efficient than the NRZ format since only half the time is required to transmit a single bit position. Both the NRZ and RZ modes require synchronized clocking at both the transmitter and receiver.

The Manchester format, also shown in Fig. 5-17, is a modification of the RZ format in which half of every bit position is denoted by a pulse. If the bit position contains a 0, the first half of the pulse is low and the second half is high. If it contains a 1, the first half is high and the second half is low.

Manchester coding eliminates the need for a separate clock channel. Like the NRZ format, however, it requires twice the time space of the RZ format.

In a typical, practical PCM system, one second of voice requires 8000 samples of eight bits each for a total of 64,000 bits/second (bit/s). Two voices require twice this figure or 128,000 bit/s. The two signals can be multiplexed onto the same channel by placing the individual bits in each voice signal in unique time slots within discrete pulse windows of the signal.

PCM greatly simplifies multiplexing. One established channel frequency, for example, is 44.736 megabits/second (the T3 rate). This permits the transmission of 672 voices over a single channel. The actual information content is $64,000 \times 672$ or 43.008 megabits/second (Mbits/s). The remaining bits provide demultiplexing information for the receiver.

A PFM Infrared Transmitter

Figure 5-18 is the circuit for a straightforward PFM transmitter designed around the popular 555 timer IC. The 7555 CMOS version of the 555 can also be used. In my experience the 7555 gives better performance.

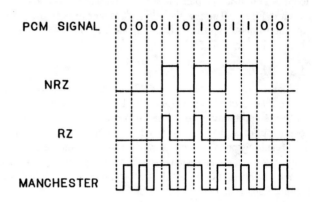

Fig. 5-17. NRZ, RZ, and Manchester formats.

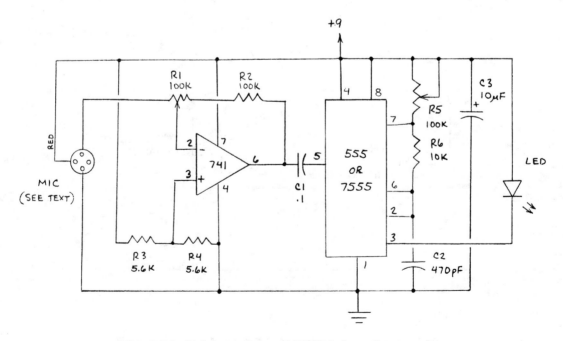

Fig. 5-18. Voice modulated PFM infrared transmitter.

In operation, the 555 oscillates at a center frequency determined by the time constant of R5 and C2. Typically the center frequency is 40 kHz. Low level audio signals appearing at the microphone are amplified by the 741 and passed into the modulation input of the 555 where they alter the chip's oscillation frequency. The pulse-frequency modulated signal appears at pin 3 where it is used to drive an infrared LED.

For best results use an electret microphone. A crystal microphone may also be used. Better quality op amps can be used in place of the 741 to give a lower transmitted noise level. The LED can be any GaAs, GaAs:Si, or AlGaAs infrared emitter. For high power operation, select one of the super LEDs made from AlGaAs such as Xciton's XC880 series or General Electric's F5D1/F5E1 series.

The peak pulse current supplied to the LED by the 555 is about 50–60 mA. The pulse width is about 5 microseconds. The circuit in Fig. 5-19 employs a VFET transistor to raise the pulse current to about 450 mA. Even higher current levels can be obtained by increasing the voltage at V_{ss}.

A PFM Infrared Receiver

Figure 5-20 is the circuit for a PFM receiver designed around a 565 phase-locked loop (PLL). The PLL is tuned via R4 and C4 to the approximate center frequency of the transmitter. Incoming optical signals are detected by the phototransistor (Q1). The resulting photocurrent is converted to a voltage by load resistor R1 and coupled via C1 into the 741 op amp. The signal is then amplified 1000 times (R3/R2) and passed into one of the phase comparator inputs of the PLL.

The phase comparator generates an error voltage proportional to the difference between the PLL's on-chip VCO center frequency and the instantaneous transmitter frequency. The error voltage is fed back to the VCO in a feedback loop which causes the VCO to track the transmitter frequency. The error voltage represents the demodulated analog of the transmitter signal, so it is tapped via pin 7 for power amplification by the 386. R5 controls the gain of the system by controlling the amount of signal which reaches the 386.

Several modifications can be made to the basic receiver circuit. To reduce the effect of sunlight upon the detector, Q1 can be replaced by a silicon photodiode. The diode should be connected in the reverse direction.

The gain of the 741 preamplifier can be altered by changing the ratio of R3/R2. And the center frequency of the PLL's VCO can be changed via R4 and C4. For best results, R4 should be at or near 4 kΩ, although Signetics observes R4 can range up to about 20 kΩ.

Testing the Communicator

Before applying power to the transmitter and receiver, carefully inspect both circuits for possible wiring errors or omissions. To avoid severe interference when using a phototransistor in the receiver, place an opaque hollow tube over Q1 to keep artificial light from striking Q1's active region. A photodiode may not require this kind of protection.

For initial tests, disconnect the transmitter's microphone and connect the audio output of a transistor radio to the circuit via a 1-microfarad capacitor. The negative lead of the capacitor should be connected to R1. The radio output should be connected to the positive lead of the capacitor and the transmitter's ground connection. For best results use a radio with an earphone jack and make your connections with the help of clip leads soldered to an appropriate plug.

Turn the radio on at low volume and select a station that gives clear reception. Then apply power to both the transmitter and receiver while pointing the transmitter's LED at the re-

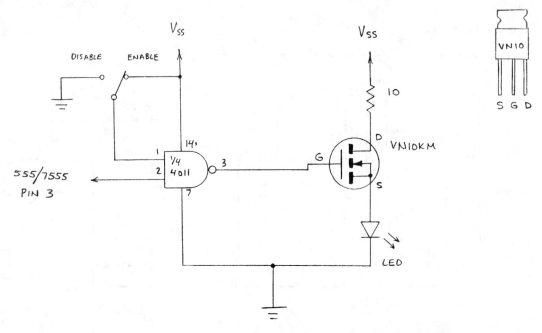

Fig. 5-19. LED power booster for PFM infrared transmitter.

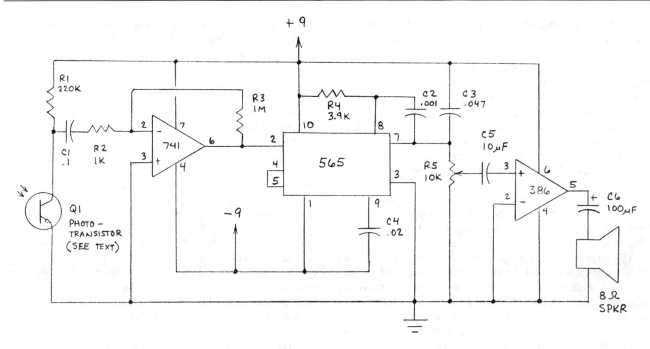

Fig. 5-20. PFM infrared receiver.

ceiver's detector. At this point you should hear noise, oscillation or, ideally, the sound of the radio from the receiver's speaker. In any event, carefully adjust the setting of R5 in the transmitter in an effort to match the center frequency of the VCO in the receiver's PLL, in this case about 40 kHz.

While adjusting R5 you will probably hear loud noises and whistles interspersed with very clear sounds of the radio. Select the best quality transmission point with R5 and then adjust the gain of the receiver (R5 in Fig. 5-20) for comfortable listening. You can then experiment with the setting of the transmitter's R1 to find the optimum gain of the transmitter's preamplifier.

When the transmitter and receiver are properly adjusted for good transmission of the radio signal, you can experiment with the units. The most interesting demonstration is to gradually point the LED away from the receiver's detector while listening to the received signal. The amplitude of the signal will remain constant until a threshold point is reached where a sudden noise level appears. As the LED is moved still more, the noise will stop and the receiver will be silent.

This demonstration illustrates the fade-free operation of a PFM communications link. In an actual field test, you will find that the transmitted signal from a PFM transmitter is much more pleasant to receive than an amplitude modulated transmitter. Artificial light sources which produce relatively low power densities at the receiver's detector are undetected by the PLL in the receiver. And the received signal has constant volume no matter where the receiver is located within the reception range.

It should be noted, however, that the signal from an amplitude modulated system can be detected at a greater distance than that from a PFM transmitter assuming both systems bias their LEDs at the same level. This occurs because of the brain's ability to pick out a low amplitude signal buried in noise. The

listening quality may not be good, but the signal can be understood.

Of course this advantage does not necessarily occur in real systems since it is a simple matter to pulse drive the LED in a PFM transmitter at much higher current levels than are possible with an amplitude modulated transmitter.

A CMOS PFM Transmission System

If you are interested in experimenting with PFM infrared transmission systems, be sure to refer to *The Forrest Mims Circuit Scrapbook* (McGraw-Hill, 1983) for information about a CMOS PFM transmission system I have designed. The system is described on pages 100–101.

By combining the information given there and here, you may be able to design a customized transmission system. For example, you may wish to employ a 7555 modulator-transmitter in conjunction with a CMOS 4046 PLL receiver-demodulator. You may also wish to try the 4046 PFM transmitter with a VMOS LED driver as shown in Fig. 5-19.

No matter which method you select, you can achieve transmission ranges of thousands of feet at night with the help of lenses or reflectors. You can also use PFM circuits to transmit through optical fibers.

An Infrared Temperature Transmitter

Obviously an ordinary thermometer provides the simplest way to measure temperature. For remote temperature measurements, a solid-state sensor can be connected to a monitoring station by means of a wire cable. Suitable sensors include thermocouples, thermistors, and silicon chips.

In some special applications, it's not practical or safe to connect a remote sensor to a monitoring station by means of a wire cable. For example, if the sensor is to be mounted upon a moving fixture, adjacent to moving parts, or near high-voltage circuits, a wire cable may get in the way or pose a shock hazard. Moreover, outdoor wires and cables are susceptible to lightning strikes.

Another special application is the measurement of temperature at various elevations in the earth's atmosphere. A temperature inversion occurs when a layer of cool air is trapped near the earth's surface by an overlying blanket of warm air. If the inversion persists over a few days, the trapped air can become heavily contaminated with automobile exhaust and other pollutants.

The measurement of temperatures at different altitudes is the application which led directly to the development of the system to be described below. For this application a radio or optical fiber link is preferred since it's not safe to connect a balloon- or kite-borne system to a ground station by means of electrically conductive wires. Lightning from thunderstorms is the obvious hazard. But it's important to remember that potentials as high as 50,000 volts have been measured on the lines of wire-tethered kites flown on perfectly clear days!

An Infrared Temperature Sensor System

Several circuit arrangements can be used to transmit temperature by means of an infrared beam. Figure 5-21 is a block diagram of the method I selected. Notice that the temperature sensing element is a thermistor, a temperature-dependent re-sistor. Temperature-sensing integrated circuits like the LM334 and LM335 can also be used. They provide an output which is highly linear with respect to temperature. Though thermistors are inherently nonlinear, they provide a faster response time.

Referring to Fig. 5-21, the thermistor is connected in series with a resistor to form a voltage divider. As the thermistor's resistance changes with temperature, the voltage appearing across the junction of the thermistor and resistor changes. The resistance of thermistors is typically inversely proportional to temperature. In other words, the resistance of a thermistor falls as its temperature rises. Therefore, the output from the thermistor-resistor voltage divider in Fig. 5-21 increases as the temperature of the thermistor increases.

The temperature-dependent voltage from the thermistor-resistor divider is applied to the input of a voltage-to frequency (V/F) converter. The V/F converter generates a pulse train whose frequency is dependent upon the temperature of the thermistor. The output from the V/F converter drives an LED. Note that both the thermistor and V/F converter are powered by a voltage regulator. This guarantees that the circuit will provide repeatable performance until the battery voltage falls below a usable point.

The infrared signal from the LED is detected by a receiver circuit, amplified, and coupled into a frequency meter. A calibration chart is then used to determine the temperature of the thermistor.

The receiver's detector can be a photodiode, phototransistor, solar cell, or LED identical to the one in the transmitter. The infrared signal can be coupled to the detector directly through the air or via a fiber optic cable. If free space coupling is employed, a lens at both the transmitter and receiver can increase

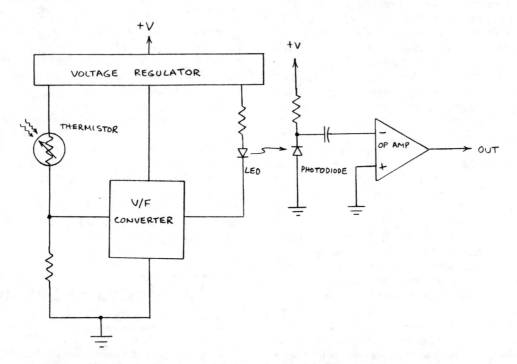

Fig. 5-21. Block diagram of infrared temperature sensing transmitter.

the transmission range from a few feet to hundreds of feet. Alternatively, the signal can be coupled via an optical fiber. A fiber will provide a high degree of reliability and will be unaffected by sunlight.

The Infrared Transmitter

Figure 5-22 shows a working infrared transmitter circuit based upon the block diagram in Fig. 5-21. I've built and tested several versions of this circuit. The final working version of the circuit is housed in a plastic enclosure measuring 0.6 × 1.2 × 2.3 inches. The entire circuit, including battery, weighs under an ounce.

Referring to Fig. 5-22, the thermistor and R3 form the temperature-dependent voltage divider. Many kinds of thermistors will work with the circuit. I used an unmarked glass bead thermistor that has a room temperature resistance of 2500 ohms. The current Newark Electronics catalog (Number 107) lists Fenwall Electronics glass bead and Yellow Springs Instruments Teflon coated thermistors with this and similar values (see pp. 202–203). They range in price from about $3.60 to $10.00.

Newark has some 160 local branch offices in the United States. Since they have a $25 minimum order requirement, you might be better off to call a local electronics parts distributor first.

The V/F converter is a National Semiconductor LM311, a chip that can also function as a frequency-to-voltage (F/V) converter. R8 provides the means to vary the output frequency of the LM311. It therefore serves as a calibration control.

Note that the LM311 directly drives the LED through R10, a 10-ohm series resistor. Almost any standard LED can be used. However, for best results, select a red or near-infrared AlGaAs or GaAs:Si device. I used a General Electric GaAs:Si GFOE1A1 fiber optic emitter. This device is housed in a plastic package equipped with a threaded coupler that mates with AMP Optimate fiber optic connectors.

Referring back to Fig. 5-22, note that an LM350T serves as the circuit's voltage regulator. The circuit I built is powered by a miniature 7-volt mercury battery (Duracell TR175 or equivalent). The LM311 will function at a power supply voltage as low as 4 volts, and that is the approximate output provided by the LM350T when R1 and R2 in Fig. 5-22 have the values shown. The regulator will provide a steady output until the battery voltage falls to about 5.5 volts. The output voltage from the LM350T can be changed by altering the value of R1. For details, see the data sheet for this chip.

The Infrared Receiver

A suitable receiver for the near-infrared signal from the transmitter can be fashioned from any op amp and a photodetector. Figure 5-23 shows the circuit I used. I tried a GFOE1A1 emitter identical to the one in the transmitter, a miniature solar cell and a photodiode. All worked well.

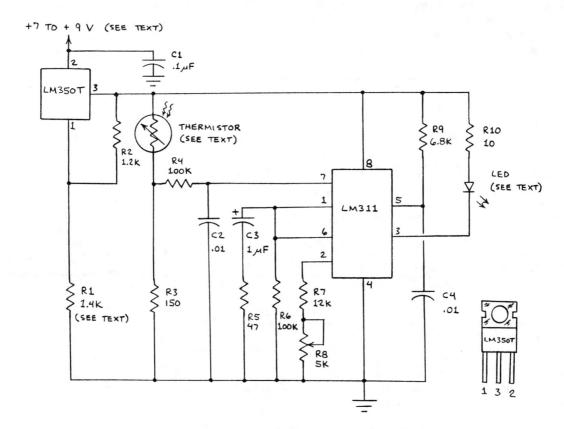

Fig. 5-22. Lightwave temperature transmitter.

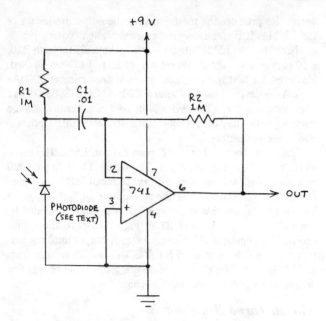

Fig. 5-23. Lightwave receiver for temperature transmitter circuit.

The output from the receiver can be coupled directly into a standard frequency counter or oscilloscope. If you don't have either of these instruments, you can make a simple but effective frequency meter from an LM311 or a 555. For details about an LM311 frequency meter, see "A Cassette Recorder Analog Data Logger" in Section 7 of this book. A frequency meter designed

around a 7555 (CMOS 555) is given in *Engineer's Notebook II* (Radio Shack, 1982, p. 104). A 555 version of this circuit can be found in *Engineer's Mini-Notebook* (Radio Shack, 1984, p. 17).

Calibrating the System

If your thermistor is the glass bead type, you'll need half a cup of boiling water, a full cup of ice water and a thermometer to calibrate the temperature transmission system. Use care to avoid spilling the hot water into your lap while performing the calibration!

Begin by recording both the temperature of the hot water and the frequency produced by the transmitter when the thermistor is immersed in the water. Be sure the leads from the thermistor are not immersed. Then add some ice water to the hot water and repeat the measurements. Continue adding cold water and making measurements until the cup is full. Then start making measurements of the cup of ice water.

For reliable measurements, the water should be stirred after water with a different temperature is added. Also, the thermometer should be allowed half a minute or so to stabilize before the readings are made. You'll probably find that the thermistor settles down much faster than the thermometer. Incidentally, avoid suddenly transferring the thermometer from the hot water to the ice water or vice versa. The sudden temperature change might fracture its glass housing.

Table 5-1 gives the results of the calibration procedure for the temperature transmitter I assembled:

You will have to modify the calibration procedure if your thermistor is uninsulated. Since you will be unable to immerse

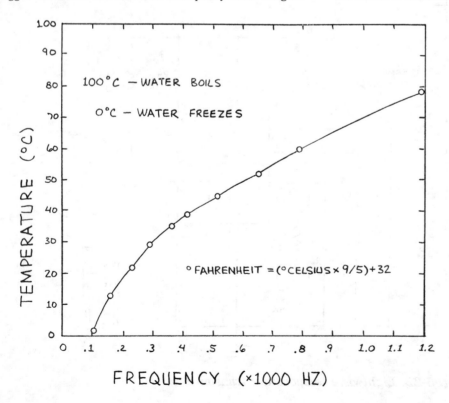

Fig. 5-24. Typical calibration curve for temperature sensor circuit.

Temperature Calibration Results

Temperature (Celsius)	Frequency (Hz)
2	103
4	114
13	166
21	217
28	285
35	369
39	421
45	510
53	649
60	795
78	1180

Figure 5-24 is a graph of these measurements. Since the transmitter uses a thermistor as a temperature sensor, the curve is nonlinear over its entire range. Keep in mind that other thermistors will provide different curves.

the thermistor into a fluid, one approach is to simply measure the temperature of the air at different times of the day. Then record the results and plot them on a graph. This method will not provide the range of the immersion method, but it will work.

Using the System

I designed this system to make temperature measurements from a helium-filled balloon flown to an altitude of up to 500 feet. For these tests, the temperature transmitter was dubbed the Fibersonde. It was connected to the receiver by means of a 200- meter length of 200-micron ITT silica optical fiber wound on a kite spool. The receiver was coupled to a tape recorder so the data could be saved and read back later. For details about saving analog data on cassette tape, see "A Cassette Recorder Analog Data Logger" in Section 7 of this book.

For previous experiments with airborne, remote-controlled cameras, I used commercial balloons. Unfortunately, such balloons are expensive and a good quality 4-foot diameter balloon costs as much as $18. Worse, such balloons are easily burst by inadvertent encounters with trees. Therefore, for flight tests with the Fibersonde, I used a cheap and rugged BPTB balloon.

(BPTB means *B*lack *P*lastic *T*rash *B*ag.) You can purchase a dozen BPTB balloons at any grocery store for under $2.00.

One cubic foot of helium lifts about 1.1 ounces. Allowing a 15 percent excess lift margin and assuming a total airborne package weight (balloon, Fibersonde, and fiber) of 4 ounces, then only about 3.6 cubic feet of helium are needed to fly the system. This is equivalent to 26.9 gallons which means a 30-gallon trash bag will suffice.

Helium can be purchased from some welding shops and party supply stores. You'll need to borrow a regulator to attach to the gas cylinder. Expect to pay about a quarter per cubic foot for the gas plus a deposit for the cylinder and regulator. Be sure to handle the heavy gas cylinder with care.

The optical fiber I used for flight tests has a breaking strength of 10 pounds and could easily have doubled as a tether line. Because of nearby trees, however, I attached a nylon tether line to the balloon. That proved to be a wise move since a slight breeze caused the balloon to drift into the top of a tree. The trash bag was so rugged that it could be pulled from the branches without a puncture.

Figure 5-25 is a photograph of the Fibersonde suspended from a flying trash bag. Unfortunately, the system can be flown only when the air is almost perfectly still. Otherwise the balloon drifts downwind and fails to reach a high altitude. An aerodynamically shaped balloon is required for flights when the wind is blowing. Unfortunately, such balloons cost hundreds of dollars.

Here are a few flying tips that will prove helpful when flying a balloon. Use only helium to inflate a balloon. Tie the free end of the tether to an object on the ground so, if necessary, you can release the line and run toward the balloon. Avoid jerking on the tether when the balloon is flying low over trees or a building. Both the tether and the balloon may be pulled into the obstacle below. Never fly a balloon near an airport, power lines, or high buildings. Finally, never use an electrically conductive wire between the balloon and the ground.

Going Further

If you're interested in flying instrumented packages from helium-filled balloons or kites, you will want to refer to the discussion of balloon- and kite-borne radio-controlled disc cameras elsewhere in this book.

If you want to transmit the infrared beam from the tem-

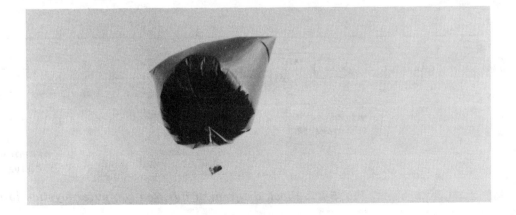

Fig. 5-25. Fibersonde suspended from helium-filled trash bag.

perature transmitter directly through the atmosphere, you'll need to use a lens at both the transmitter and receiver. I briefly covered this subject in *Getting Started in Electronics* (Radio Shack, 1983, p. 65).

An Experimental Infrared Joystick Interface

Radio Shack's TRS-80 Color Computer can be equipped with two joysticks. Each joystick assembly includes two mechanically linked potentiometers and a "fire" switch. Figure 5-26 shows the internal circuitry of each joystick.

The *Color Computer Technical Reference Manual* (Radio Shack, 1981) gives complete details about the operation of the Color Computer's joysticks. Briefly, the two potentiometers in

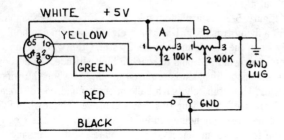

Fig. 5-26. TRS-80 Color Computer joystick circuitry.

each joystick function as voltage dividers. As the wiper of a potentiometer is rotated, the voltage appearing across the wiper and ground varies from ground to +5 volts. This voltage is applied via joystick input ports to a 6-bit digital-to-analog converter circuit in the Color Computer. A built-in software routine uses a successive approximation method to find to the 6-bit accuracy of the D/A converter the voltage equivalent to the position of the joystick. Since there are either two or four joystick potentiometers, a multiplexer is required to direct the selected potentiometer to the D/A converter.

One way to replace the cable between the joysticks and the Color Computer with an infrared link is to employ a pair of LEDs, each driven by a pulse generator with a repetition rate determined by the resistance of its respective joystick potentiometer. A pair of receivers would detect and amplify the signals from the LEDs and pass them to respective frequency-to-voltage converters. The resultant output voltages would then be applied to the joystick input ports of the Color Computer. Figure 5-27 is a block diagram of this method.

A Working Circuit

To test this idea, I assembled a working version of half the block diagram in Fig. 5-27. The transmitter is shown in Fig. 5-28. The circuit is a straightforward pulse generator designed around a 555 timer whose pulse repetition rate is determined by joystick potentiometer R1 and timing capacitor C1. Pulses from the 555 (pin 3) switch Q1 on and off, thus applying current to the infrared-emitting diode (LED1). R4 limits current through the diode to less than 100 milliamperes.

Figure 5-29 shows one of several simple receivers I tested

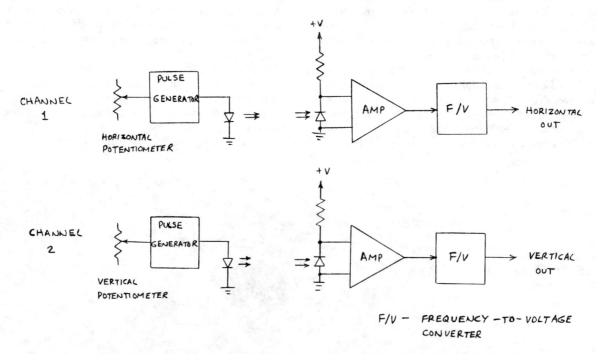

Fig. 5-27. Block diagram of infrared computer-joystick interface.

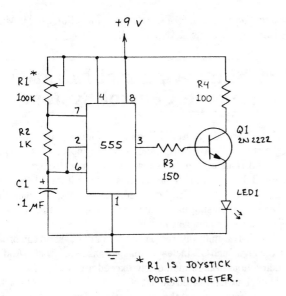

Fig. 5-28. Single-channel near-infrared joystick transmitter.

that can detect the signal from the transmitter, amplify it, and convert the variable pulse rate into its respective voltage.

For best results, especially in the presence of ambient light, a pin photodiode such as the Texas Instruments TIL413 should be used. This particular photodiode includes a built-in infrared filter that substantially improves performance in the presence of incandescent and fluorescent indoor lighting. A phototransistor can also be used, but noise immunity may be a problem.

If you use a phototransistor, connect its collector to the junction of C1 and R1 and its emitter to ground. You will also need to reduce R1 to about 100 kΩ. Keep in mind that although the phototransistor has built-in gain, the photodiode is less susceptible to the effect of ambient light.

I used a pair of 741 operational amplifiers to amplify the received signal, but many other amplifier arrangements can also be used. A 60- or 120-Hz notch filter can also be included.

The frequency-to-voltage converter is designed around the familiar 555. R9 and C5 at the output of the 555 form an integrator that transforms the pulses from the 555 into a variable dc voltage.

Although I've had good results with the 555 in this role, you can also use the 9400 or LM311 in their frequency-to-voltage converter roles for superior results. Both these chips were described in *Popular Electronics* (November 1979) and reprinted in *The Forrest Mims Circuit Scrapbook*, (McGraw-Hill, 1983).

Testing the Interface

I tested the experimental infrared joystick-computer link by disassembling a Color Computer joystick and unsoldering the yellow wire from the wiper terminal of one of the two potentiometers. I then connected this wire and the joystick's black wire (ground) to, respectively, the output of the receiver and its ground.

I then entered the following joystick test program into the Color Computer:

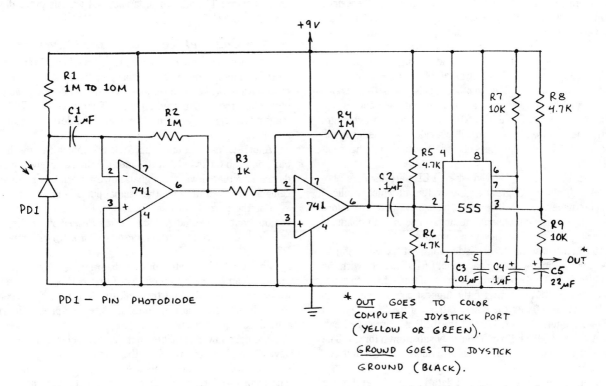

Fig. 5-29. Single-channel infrared-coupled joystick receiver.

```
10 CLS
20 PRINT @0, JOYSTK(0);
30 PRINT @5, JOYSTK(1);
40 GOTO 20
50 END
```

This simple program places the coordinates (0 to 63) of the right joystick on the upper left corner of the monitor's screen. Joystk(0) is the horizontal potentiometer and JOYSTK(1) is the vertical potentiometer.

For initial tests, the transmitter's LED and the receiver's photodiode should be closely spaced and pointed directly at one another. When the system is operating properly, move the transmitter away from the receiver to determine the maximum range. The prototype circuit I built gave a maximum range of about four feet when a photodiode detector was used at the receiver. A phototransistor at the receiver gave a maximum range of about three feet. If the coordinates of JOYSTK(0) do not reach the full range, you may need to alter the value of C1 in the transmitter.

You can monitor the output of the receiver with a high-impedance voltmeter. And you can gain a better understanding of the entire transmitter-receiver system by observing (on an oscilloscope screen) pin 3 of the transmitter's 555 and the outputs of the various stages of the receiver. If you are able to perform these tests, be sure to experiment with the position of the transmitter's LED with respect to the receiver's photodiode.

Going Further

The simple infrared joystick-computer interface described here links only one of the potentiometers in the joystick with a Color Computer. One way to link both potentiometers is to assemble two identical transmitters and receivers as suggested in Fig. 5-27. Use an 880-nm LED for one transmitter and a 950-nm LED for the second transmitter. Replace the receiver photodiodes with identical LEDs. In other words, the 880-nm transmitter should be coupled with a receiver that uses an 880-nm LED as a photodiode. Since the LEDs can function as wavelength selective detectors, the system will provide a wavelength-multiplexed, two-channel link between the joystick and the computer.

The range of the system in which LEDs are used as photodiodes will not be as great as the system in which a silicon PIN photodiode is used. Therefore, you may wish to explore various electronic multiplexing methods to send both channels of information from a joystick to a computer.

You can increase the transmission range of the infrared joystick-computer link by adding additional LEDs in series with the transmitter's LED. Also, select the transmitter's LED current-limiting resistor (R4) to provide the highest possible current to the LED consistent with the power ratings of the LED, Q1, and R4. For maximum power, Q1 and the LED may require heat-sinks. A power MOSFET can be directly substituted for Q1. The equivalent connections are base-gate, emitter-source, and collector-drain.

For more details about infrared data links in general and a derivation of the line-of-sight range equation, see *A Practical*

Introduction to Lightwave Communications (Forrest M. Mims, III, Howard W. Sams & Co., 1982).

A Single-Channel Infrared Remote Control System

Near infrared radiation is well suited for use as a carrier of trigger signals in miniature remote control applications. No governmental rules or regulations apply to remote control system triggered by radiation from nonlaser near-infrared emitting diodes. Such systems are less susceptible to false signals than similar systems using ultrasonic sound. And infrared transmitters can be very compact in size.

True, a 100-milliwatt radio remote control system can broadcast through such obstacles as foliage, haze, and walls. And its omnidirectional range may easily exceed a city block.

Though near-infrared cannot penetrate many such obstacles, the use of small lenses at the transmitter and receiver can extend the range of such a system to many hundreds of feet. In some applications the pointing problems associated with the narrow beam of such a system are a distinct disadvantage. In applications requiring a high degree of security, however, a narrow beam can be a major asset.

In most instances, radio edges out near-infrared long range remote control systems. Infrared, however, can be the clear winner in applications where the distance is under a few tens of feet. Typical applications include remotely controlled garage door openers, TV sets, toys, lamps, and various other devices and appliances.

Single-Channel Remote Control Systems

There are several approaches to single-channel remote control. The most common is the analog of the momentary contact push-button switch. The remotely controlled device is actuated only when the transmitted signal is being received. When the transmitted signal is absent, the controlled device is no longer actuated.

Another, less common approach resembles the mechanical push-on/push-off switch. In this method the remotely controlled device is actuated when a signal, however brief, is received. It remains actuated until another signal pulse is received.

For the purpose of this discussion, let's designate the first remote control method *Actuated when Pushed* or AP for short. We'll call the second method *Push-On/Push-Off* or simply PO/PO.

Figure 5-30 illustrates in block diagram form how both these methods can be implemented. Note that both systems can use the same tone-modulated transmitter. Also note the receivers for both methods share a common detection-preamplification-tone decoding front end.

A simple power amplifier that drives a relay, lamp, or other device completes the receiver that uses the AP approach. The PO/PO receiver requirements are more complex since false signals from inadvertent multiple input signals must be ignored. This problem is analogous to the well known contact bounce

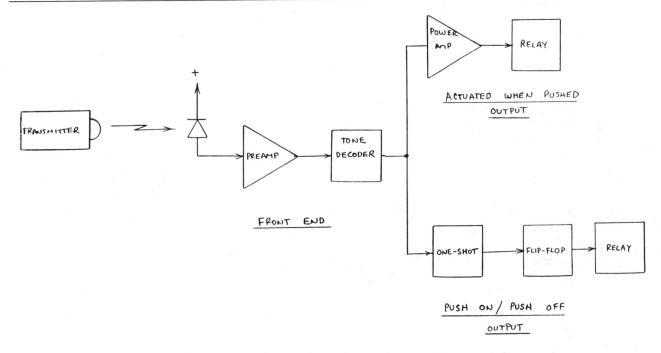

Fig. 5-30. A single-channel remote control system.

phenomenon that accompanies the opening and closing of most mechanical switches.

The output from the tone decoder in Fig. 5-30 is fed into a one-shot multivibrator that effectively stretches the incoming pulses burst from the transmitter into a logic pulse having a duration of several seconds. This pulse sets the flip-flop.

During its timing cycle, the one-shot ignores subsequent input signals from the transmitter. When the timing cycle is complete, the one-shot will trigger on the next arriving pulse and, in turn, reset the flip-flop. The set-reset cycle of the flip-flop provides the desired PO/PO action.

Tone Modulated Transmitters

Figure 5-31 shows a simple tone-modulated near-infrared transmitter suitable for short range remote control applications. When S1 is pressed, the LED emits a train of pulses at a frequency determined by R1's setting.

R3 controls the current through the LED. Its resistance can be reduced for more current per pulse, hence higher infrared output, as long as the LED's peak current rating is not exceeded. If the LED's forward voltage is 1.5 volts, a value typical of some gallium arsenide devices, then the peak current is given by

$$I = \frac{V_{in} - 1.5}{R3 + R_{DS}}$$

where,

V_{in} is the power supply voltage,
R_{DS} is the drain-source ON resistance of Q1.

The R_{DS} of the VN10KM specified in Fig. 5-31 is about 5 ohms. Therefore, the peak current through the LED in Fig. 5-31 is (9 − 1.5) / (100 + 5) or 71 milliamperes. This is *well*

within the safe operating range for pulses operation of most near-infrared LEDs, many of which can be driven to a couple of amperes by microsecond pulses so long as a high duty cycle is avoided to prevent excessive heating.

Incidentally, a CMOS 7555 timer can be substituted for the 555. Similarly, a 2N2222 or similar npn driver transistor can be substituted directly for the VN10KM power FET. Connect the base to pin 3 of the 555, the collector to R3, and the emitter to the LED's anode.

Figure 5-32 shows an alternative transmitter may wish to try. It's exceptionally efficient and will easily deliver 1.1 ampere pulses to the LED. It may not be as reliable, however. And it may consume excessive current. For additional details, refer to *The Forrest Mims Circuit Scrapbook* (p. 57).

A Single-Channel Receiver

The receiver whose circuit is shown in Fig. 5-33 will provide both AP and PO/PO operation when triggered by either of the near-infrared transmitters described here. The circuit is reasonably sensitive and will not trigger in the presence of line-powered, hence 60-Hz modulated, incandescent and fluorescent lamps.

A tungsten desk lamp placed within a few millimeters of the circuit's photodiode failed to trigger the circuit. The full flash from a Vivitar 283 photographic strobe unit placed 15 centimeters from the photodiode also failed to initiate a false trigger. Moving the strobe closer, however, did give a false trigger when it was flashed.

In operation, near-infrared from the transmitter is detected by reverse-biased photodiode PD1. The resultant photocurrent is then amplified by an LM308 op amp connected as high gain

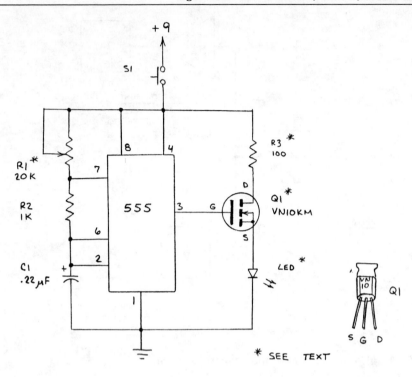

Fig. 5-31. Infrared remote control transmitter.

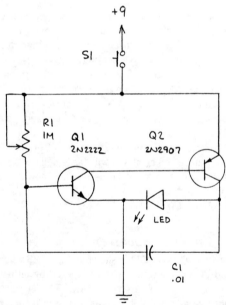

Fig. 5-32. Alternative infrared transmitter.

current-to-voltage converter. The amplified signal is coupled via C3 into a 567 tone decoder tuned to a center frequency of about 3 kilohertz.

When the 567 receives an in-band signal, pin 8 goes low. This turns on LED 1 and provides an AP output signal that can be used to drive a small relay or turn on a transistor. In Fig. 5-33, however, the signal from pin 8 is used to trigger a CMOS 4528 one-shot multivibrator. The 4528 then issues an output

pulse that turns on LED 2 and sets the 4027 flip-flop. The flip-flop's Q output then turns on LED 3 and Q1, which, in turn, pulls in relay RY1.

The time delay provided by the 4528 is determined by R6 and C7, either or both of which may be increased to further stretch the incoming tone burst into a longer logic pulse. The values given in Fig. 5-33 give a timing interval of a few seconds.

To preclude triggering on switching transitions and other noise generated within the circuit, it is essential to include capacitors C8 and C9. Both should be placed as close as possible to the 567 power supply pins. If false triggering occurs or if the circuit appears to operate erratically, it may be necessary to install additional 0.1-μF decoupling capacitors directly across the power supply pins of the LM308 and the two CMOS logic chips. Incidentally, be sure to ground all unused inputs of both CMOS chips. Both chips are dual versions, and floating inputs to the unused side may cause excessive current consumption, overheating, and erratic circuit operation.

The LM308 is ideally suited for this circuit. In a pinch you can substitute a 741 or other op amp, but for best results use an LM308. If you have to order yours from a mail order company, you might want to buy a few extras. Use them to make low-noise, high-gain amplifiers having a high input impedance.

For best results, avoid substituting a phototransistor for PD1. Although phototransistors will work very well in subdued light, they quickly saturate in the presence of even moderate light levels.

I used a Texas Instruments TIL413 photodiode for PD1. This low cost photodiode is equivalent in quality to photodiodes that cost considerably more. It incorporates a built-in epoxy lens designed to filter out visible radiation while transmitting near-infrared.

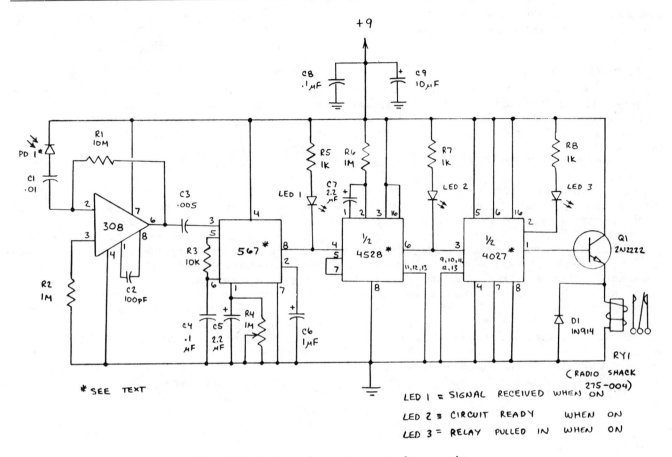

Fig. 5-33. Infrared remote control transmitter.

Testing the System

Proper operation of the receiver requires that the tone frequency of the transmitter match closely the center frequency of the receiver's 567 tone decoder. The values shown in Fig. 5-33 give a center frequency of about 3 kHz. If this frequency is acceptable in your application, *slowly* tune the transmitter while pointing its LED toward the receiver's photodiode.

LED 1 will flicker as you tune the transmitter through the 567's center frequency. When this occurs, carefully tweak the transmitter's tone frequency to give a bright, steady glow from LED 1. Then move the transmitter LED away until LED 1 *just* stops glowing. Again tweak the transmitter's tone frequency until LED 1 glows. This optimizes the tuning for the receiver. The circuit should now operate as follows:

1. Initially, LED 2 glows to indicate the receiver is ready to receive a signal. LEDs 1 and 3 are off.

2. LED 1 glows when a signal is present *and* being received.

3. LED 2 turns off immediately after LED 1 turns on. Note that even the slightest flicker from LED 1 is sufficient to insure that LED 2 will be extinguished.

4. LED 3 switches on or off to indicate the status of the relay. When LED 3 is glowing, the relay is pulled in.

5. LED 2 glows again after the time delay is complete. The receiver is now ready to again receive a signal.

Note that if the transmitter is pointed at PD1 for an interval longer than the time delay of the one-shot, LED 2 will turn back on to indicate the receiver is ready to receive another command. If the transmitter is still sending infrared signals to PD1 and is then moved away or turned off, the receiver will be triggered a second time. The flip-flop will then be reset and the device just actuated will be deactivated. In other words, for true push-on/push-off operation, the transmitter should be operated for only a moment *or* used to sweep a flash of infrared pulses across PD1.

If the circuit appears unreliable, the problem may be associated with properly pointing the transmitter at the receiver. For example, since the 567 requires a minimum number of pulses to acquire lock, a very brief sweep of the transmitter beam across the receiver's photodiode may *not* trigger the circuit, particularly at longer ranges. For this reason it's a good idea to stay with the 3-kHz operating frequency of this system by using the parts values given in Fig. 5-33.

Low battery voltage and temperature changes can also cause problems. Weak batteries, for instance, may alter the frequency of the transmitter and the center frequency of the receiver. This program can be alleviated by making sure fresh batteries are

131

used. Alkaline batteries will provide good service. Or you can design a line-powered supply or use a voltage regulator chip.

Construction Tips

Before building a permanent version, be sure to assemble a test version of the circuit on a solderless breadboard to test and evaluate its operation and to correct any bugs. This step is important since properly tuning and operating the system can be a little tricky.

Be sure to consult the 567 data sheet/application note before attempting to make major changes in the receiver's detection frequency. For instance, the 567 may require a second or more to respond to very low frequency signals. The data sheet/application note clearly explains this and other operating idiosyncrasies of the 567.

My system gave a range of more than eight feet without external lenses. Doubling the diameter of the photodiode's collection surface will double the range. Narrowing the beam of the transmitter with a suitable lens will give an even greater improvement in performance.

A range of hundreds of feet should be possible with patience and careful attention to detail. But make sure the system works well in the breadboard stage *before* trying such an ambitious range test. A light shield such as a hollow tube lined with black paper or coated with flat black paint might be helpful when the

receiver is used in the presence of bright sunlight. Place the tube over PD1 and avoid pointing PD1 at brightly illuminated objects and clouds. An infrared filter can also be used.

Applications

I have used this system to turn the sound of a television off when loud commercials interrupt news programs. The receiver and an external speaker box are connected to the TV phone jack by a short cable. This automatically disconnects the internal speaker of the TV. The relay in the receiver then switches the external speaker on or off.

I plan to use this or a similar system to remotely control a toy car and a camera. You can probably think of many other applications. In any case, be sure to follow appropriate safety procedures should you use the receiver to actuate line-powered devices. You should also avoid using this system in any application which might endanger people or property. For example, using it to control a garage door would require the inclusion of appropriate limit switches and other safety precautions to prevent accidents resultant from erratic operation.

Also, bear in mind the limitations of any remote control system. Say you use this system to control a toy boat. Should the boat exceed the reception range of the system, you will have one of two problems: If AP operation is used, the boat will simply ignore the transmitter. If PO/PO operation is used, the boat will continue to follow the last command it received.

SIX

Radio Control and Remotely Triggered Cameras

- Experimenting with a Servomechanism
- A Multifunction Radio-Control System
- Experimenting with a Touch Tone DTMF Receiver
- Experimenting with Kodak's Disc Camera
    - Part 1. Modifying the Camera for Electronic Triggering
    - Part 2. Controlling the Camera Remotely
    - Part 3. Radio Control and Aerial Photography

Radio Control and Remotely Triggered Cameras

Experimenting with a Servomechanism

A servomechanism is an automatic control system. The word servomechanism comes from the French *servo-moteur* which literally means *slave-motor*. The terminology is quite accurate, for under the proper conditions a servomechanism is totally obedient to any control or input signals applied to it.

In operation, a servomechanism develops an output signal which is continually compared to the input signal. Any difference or error between the two signals is applied back to the input, thus causing the output signal to move toward and eventually match the input signal.

The feedback loop responsible for the operation of a servomechanism is shown in Fig. 6-1. This simplified diagram illustrates the operation of servomechanisms which are entirely electronic, such as phase-locked loops, and those which incorporate such electromechanical devices as motors or solenoids.

Various kinds of phase-locked loops are described in detail in *The Forrest Mims Circuit Scrapbook.* You may also want to refer to Howard Berlin's excellent book *Design of Phase-Locked Loop Circuits, With Experiments* (Howard W. Sams & Co., 1978).

The servomechanisms covered here are those which are electromechanical in operation. Considering the increasing interest in robotics, radio-control, and computer control, electronics experimenters should be fully aware of the operation and capabilities of such servomechanisms.

The Electromechanical Servomechanism

Many kinds of electromechanical servomechanisms exist. Most incorporate a motor, a comparator, and a position sensing transducer connected as shown in Fig. 6-2. In this illustration, the input voltage is provided by one potentiometer of a two-axis joystick. The position sensing transducer is a potentiometer whose wiper is rotated by the motor's armature. Both potentiometers are connected as voltage dividers.

To understand the operation of this servomechanism, assume the voltages applied to both inputs of the comparator are equal. The output of the comparator will then be low and the motor will be off. When the joystick is moved so that the voltage applied to the noninverting input of the comparator exceeds the voltage from the position sensing potentiometer the comparator output will be high and the motor will be activated. This will cause the position sensing potentiometer to deliver a progressively higher voltage. When the two voltages are again equal or when the voltage from the position sensing potentiometer is slightly higher than that from the joystick, the comparator will switch off and the motor will cease rotation. The end result is that the motor has *tracked* the position of the joystick. That is to say the physical movement of the motor's armature is directly proportional to the movement of the joystick.

The simple servomechanism shown in Fig. 6-2 is not suitable for most practical applications since no provision is made for reversing the motor. Instead, the motor must complete up to a full revolution to provide an armature position *behind* an initial position.

Reversing a motor can be achieved under mechanical control by using the dpdt switch arrangement shown in Fig. 6-3. Forward-reverse operation can also be achieved by substituting transistors for the switches, as in Fig. 6-4. The circuit in Fig. 6-4 is actually the motor driven portion of Signetic's NE544 Servo Amplifier, a chip designed specifically for controlling a dc servo motor.

Integrated Servo Amplifiers

The NE544 and similar servo amplifier chips respond to a train of variable duration pulses by causing a small dc motor to assume a rotational position proportional to the average duration of incoming pulses. Pulse duration (or width) modulation is employed to simplify the use of a servomechanism in remote control applications. It's relatively straightforward, for example, to transmit pulse duration modulated signals via radio or infrared. Even though this method is not a true digital control system, it's commonly referred to as digital proportional control.

Figure 6-5 is a block diagram and connection network for the Signetics NE543 Servo Amplifier. In operation, pulse duration modulated pulses from a receiver are applied to the input. If the pulse duration *exceeds* that of an internally generated pulse, the difference is stretched and applied to the appropriate output stage. This causes the motor to rotate feedback potentiometer R3's wiper. R3 governs the duration of the internally generated pulses. When R3's value has been altered so that the duration of the internally generated pulses matches that of the incoming pulses, the motor is deactivated.

On the other hand, if the duration of the incoming pulses is *less* than that of the internally generated pulses, then the difference is stretched and applied to the output stage that rotates the motor in the opposite direction. When the value of R3 causes

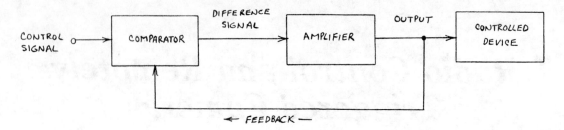

Fig. 6-1. A basic closed loop servomechanism.

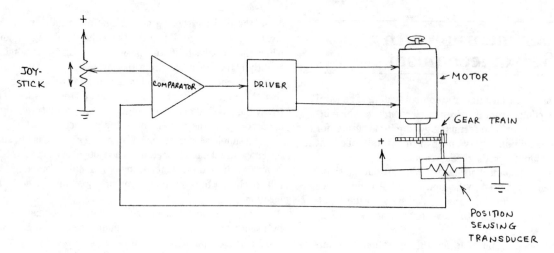

Fig. 6-2. The basic components of an electromechanical servomechanism.

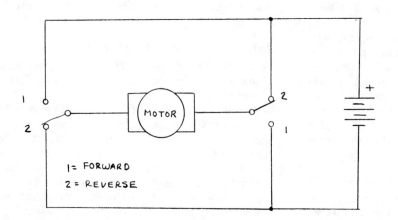

1 = FORWARD
2 = REVERSE

Fig. 6-3. A dpdt motor reverser.

the duration of the internally generated pulses to match the incoming pulse duration, the motor is deactivated.

The net result of this feedback operation is that the motor armature position follows the duration of the input pulse. If the input pulses are received from the transmitter in which the duration of pulses is controlled by the position of a joystick, then the armature position accurately tracks the joystick position, even though the two are separated by a considerable distance.

The successful operation of a servo amplifier such as the one in Fig. 6-5 involves some important subtleties about which you should be aware. Assume, for example, that the motor has

rotated the feedback potentiometer's wiper to the point where the duration of the internal reference pulses matches that of the incoming pulses. Though the motor drive transistor is switched off, the inertia of the motor's armature uses it to rotate *beyond* the null point. Now the duration of the reference and incoming pulses again differs, so the servo amplifier switches the motor into reverse in an effort to find the null point. The motor may again *overshoot* the null, thus requiring a series of command pulses before it assumes the desired position.

The forward-reverse oscillation of the motor around the null point is called *hunting* or *seeking*. In most cases, the damping

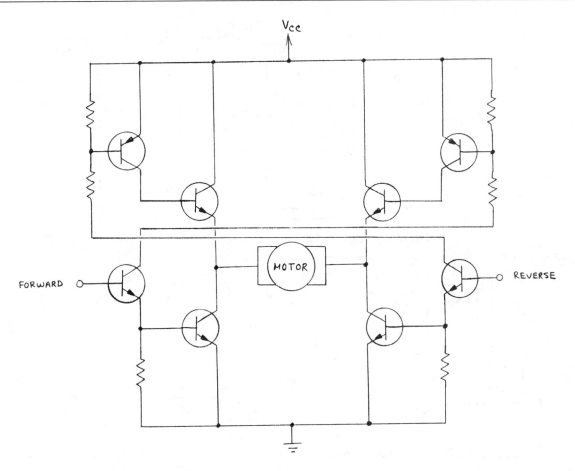

Fig. 6-4. Motor direction control circuit in Signetics NE544 servoamplifier.

effect of the motor's gears will cause progressively less hunting per cycle, thus allowing the motor to eventually stop at the desired position. In very sensitive systems, however, the motor may never settle down and will hunt continuously.

The hunting problem can be controlled by purposely designing a margin of error or, as it is sometimes called, *slop* into the system. In Fig. 6-5, this is accomplished by the inclusion of resistors Rd-1 and Rd-2. They establish what is called a *deadband* of about 5 microseconds. The motor will not be switched on until the input pulse duration differs from that of the internal reference pulses by *more* than the deadband. If hunting still occurs, the deadband can be widened by increasing the values of Rd-1 and Rd-2.

The NE544, whose motor drive section is shown in Fig. 6-4, is a second generation servo amplifier which features more precise operation that the NE543. You can find out more about both these chips be referring to their data sheets in the Signetics *Analog Data Manual*. The data sheets for the NE544 include two PC board layouts suitable for use in miniature servos.

Servo amplifier chips are also made by Texas Instruments and Exar. TI's SN76602 and SN76604 both feature an adjustable deadband and bidirectional motor drive.

Exar's XR-2264 is a pulse proportional servoamplifier designed to directly drive a motor. The XR-2265 is a similar chip

designed to drive relays, optical couplers, and triacs. The XR-2266 is designed specifically to provide two control channels for radio-controlled cars. One channel controls the direction and speed of travel. The second channel controls the steering. The direction/speed control channel can also directly drive a pair of backup lights to provide another degree of realism.

Improved Servomechanisms

The feedback potentiometer used in most closed-loop servomechanisms is a major source of problems. For instance, in radio-controlled aircraft, system failures attributed to dirty servo wiper arms are exceeded only by those caused by loss of transmitter or receiver battery capacity.

This problem can be checked by replacing the traditional feedback potentiometer with a mechanism less susceptible to contamination. In past years, some servos designed for radio-controlled aircraft used variable capacitance or inductance to provide a contamination-proof feedback signal. A better solution is to employ purely digital feedback. This can be achieved by a rotating coding disk. A miniature lamp and photocell array provide a means for detecting the position of the clear and opaque patterns encoded on the disk.

Unfortunately, the digital feedback method is more expensive that the traditional potentiometer. And it, too, is subject to

137

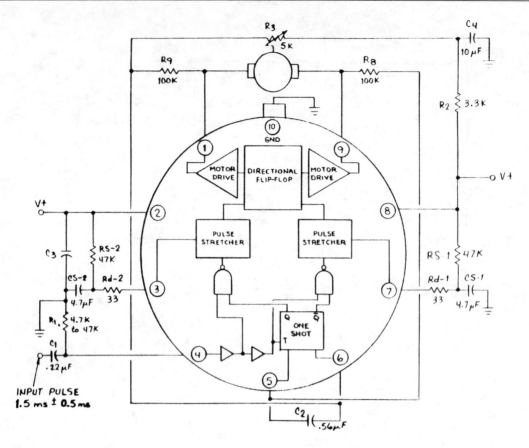

Fig. 6-5. Block diagram and operating circuit for NE543 servoamplifier.

failure. For these reasons, most servomechanisms still use a feedback potentiometer.

Experimenting with a Servo

Though it's possible to assemble a servomechanism from scratch, you can buy many different preassembled servos ranging in price from about $12 to $50 from suppliers of radio-controlled cars, boats, and planes. Nearly all these servos are installed in tough plastic cases no larger than two side-by-side 9-volt batteries.

A typical inexpensive servo is shown in Fig. 6-6. For a device of such compact size and relatively low price (around $20), a servo such as this is quite sophisticated. Typically, it contains a high quality miniature motor coupled to a train of three or four reduction gears. The feedback potentiometer is mechanically coupled to the gear train. The electronics are installed on a small circuit board adjacent to the motor and below the gear train. The axle of the output gear passes through a bushing or ball-bearing assembly in the top of the servo case. Various wheels and arms can be attached to the axle by means of a small screw.

A servo like the one in Fig. 6-6 weighs from about 1.25 to 2 ounces and can deliver several *pounds* of thrust. It can travel through its rotational limit of 90 degrees in about half a second. Typical no-load current is around 80 milliamperes and the stall current is around 450 milliamperes.

Figure 6-7 is a simple variable pulse duration pulse generator

I've used to directly drive an Aero Sport GS-ICR Servo. This servo, which is typical of those used in radio-controlled cars, boats, and planes, responds to a pulse duration of from 1 to 2 milliseconds. The servo is at its center (neutral) position when the incoming pulse has a duration of 1.5 milliseconds. At 6 volts, it consumes around 25 milliamperes at idle and around 80 milliamperes when traveling. Its stall current is around 250 milliamperes.

Resistance R3 in Fig. 6-7 can be a potentiometer in a joystick assembly or a standard potentiometer. It can even be a variable resistance transducer to give a mechanical movement in response to some environmental change. For example, a fixed resistor in series with a pair of electrodes inserted into soil will provide a transducer which gives a voltage output dependent upon soil moisture content. The servo to which this transducer is connected could open and close a small water valve to keep the soil moisture at any desired level. Numerous other possibilities exist.

Applications for Servomechanisms

By now you have probably thought of several applications for servomechanisms in addition to those mentioned thus far. Servos are used in remotely controlled television cameras, self-focusing film cameras, toys, industrial robots, and military hardware.

Robotics is a fertile field for servo applications. Industrial

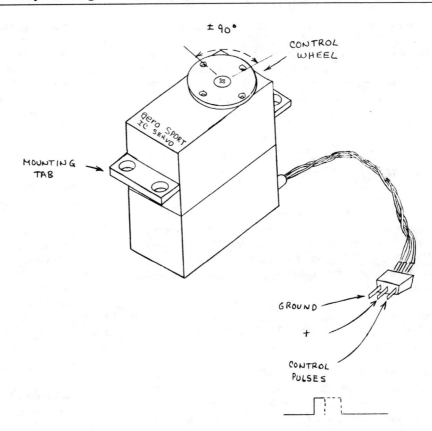

Fig. 6-6. A typical miniature servomechanism.

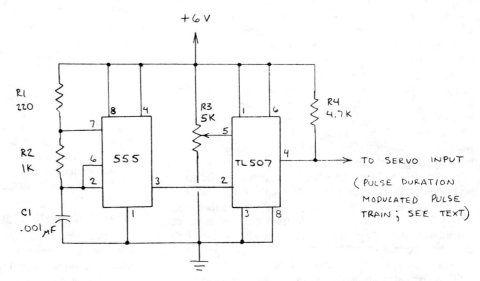

Fig. 6-7 Pulse duration modulator for miniature servomechanisms.

robots are already at an advanced stage of development. With suitable circuitry to provide the brains and servos to provide the muscles, one can envision various kinds of compact toys that provide a crude electronic analog to household pets.

Going Further

Radio-control car, boat, and plane enthusiasts use more servos than any other group. Their publications carry articles about servomechanisms and advertisements about commercial servos

and other radio-control equipment. One such periodical is *Radio Control Modeler.*

Several firms publish excellent catalogs describing various kinds of commercial servos. One such company is Ace R/C, Inc. Their catalog is $2.00. Another is Hobby Shack. Their catalog is also $2.00.

Many books describe the operation and construction of servos. One that includes considerably more detail than most is Fred M. Marks' *Getting the Most from Radio Control Systems*

(Kalmbach Books, 1980). It's available at hobby shops for $8.95.

If you're interested in robotics, you'll want to see Martin Bradley Weinstein's interesting book entitled *Android Design*. This book is filled with useful tips about robot design. It also includes many addresses for suppliers. *Android Design* is published by Hayden Book Company, Inc. (1981, $11.95).

A Multifunction Radio-Control System

Remote control is a field which offers many interesting opportunities for the electronics experimenter. For example, I've used various remote-control methods to control the volume of a TV set, the joystick inputs of a computer, a miniature guided rocket, and motor-driven cameras suspended from kites and balloons.

In the following I will describe an easily assembled, single-channel radio-control system with many on-off control applications. First, here's a case study that illustrates the versatility of such a system.

A Radio-Controlled Telephone Bell

My brother, Keith, called to tell me he was scheduled to give an important speech to officials at the head office of the company for which he works. The subject of his speech was telephone etiquette, and Keith had devised a clever way to catch the attention of his audience. During a key part of the speech, he planned to trigger the bell of a telephone placed on the podium. He would then answer the phone and simulate a conversation with the nonexistent caller.

After explaining his plan, Keith asked if I could design for him an infrared or radio remote-control unit to trigger the bell in the phone. An infrared system would be very compact and easy to conceal. But it might be difficult to guarantee a line-of-sight to the receiver. A radio system would be bigger, but it would work without a direct line-of-sight. Both methods would be susceptible to possible noise that might cause the bell to be triggered without a signal from Keith. The principal noise source that might affect the infrared receiver would be artificial lights, particularly the fluorescent variety. Noise sources that might trigger a radio receiver include electromagnetic disturbances transmitted by nearby electric motors, light switches, and lightning. Interference from CB and other radio-frequency transmitters might also cause false triggering.

Our discussion about the relative merits and demerits of infrared and radio remote-control systems quickly became moot when Keith asked if I could assemble a working system during a brief visit he planned to make to my office. Visions of a sophisticated, noise-immune, remote-control system vanished from my mind as I asked Keith to stop by a Radio Shack store on the way to my office and buy the cheapest radio-controlled toy car in stock.

Keith arrived with a radio-controlled, 1/16th scale U.S. Army jeep purchased for the sale price of $9.88. I immediately switched on my oscilloscope and began studying the operation of the jeep's noncrystal-controlled receiver and control system. The telephone Keith planned to use during his speech was an antique model with a nonfunctioning bell. Therefore, we decided to use an ordinary 2½-inch doorbell from a hardware store (Eagle Electric Manufacturing Company number 292).

Most of the receivers in the RC toy cars I've disassembled use driver transistors to switch the cars' motors off and on. The doorbell we planned to use, however, consumed much more current than the drive motor of a toy car. Fortunately, the receiver in the Radio Shack jeep incorporated an output relay. This greatly simplified the assembly of Keith's system since the relay could directly control the bell.

Within two hours or so, I managed to install the receiver from the toy jeep inside a plastic cabinet. The bell was mounted on the outside of the cabinet, and the batteries for both the receiver (9 volts) and the bell (6 volts) were installed inside the cabinet. The pocket-size transmitter reliably triggered the bell each time its button was pressed. Unfortunately, so did the CB rig in my pickup. Even though the toy receiver was far from being noise immune, Keith managed to use the remote-controlled bell quite successfully. He reduced the chance for an unwanted ring by switching on the concealed receiver just before his speech.

Practical Radio-Control Systems

The Federal Communications Commission regulates radio-control transmitters. Prior to 1983, users of RC transmitters whose output power exceeded 100 milliwatts were required to secure a license from the FCC. The license requirement was eliminated in 1983 as part of the Reagan administration's program to reduce federal regulations.

Today anyone can operate an RC transmitter that has been tuned and certified according to FCC regulations. The transmitter must operate at one of the frequencies authorized by the FCC. And its output power must not exceed 4 watts.

Some day I plan to design an ultraminiature, crystal-controlled RC transmitter having an output power of less than 100 milliwatts. Until then, I'll continue to use commercial RC equipment. Commercial RC gear is relatively inexpensive, readily available, and easy to use.

Types of Radio-Control Systems

Most commercial RC systems are intended for use by model car, boat, and plane enthusiasts. These systems range from inexpensive single-channel units to those having seven or even more channels.

Most commercial systems use "digital" proportional modulation. This method is not a true digital modulation method wherein control signals are transmitted as sequences of binary words. Instead, control signals are embedded as one or more variable duration pulses, one for each channel, within a constant interval period called the frame. The duration of the individual pulses within the frame determines the position of the servos connected to the receiver's output. The signal from the receiver of this kind of RC system is not suited for directly driving a relay. However, a relatively simple decoder circuit can be designed to accomplish this function.

A second kind of commercial system uses pulse proportional modulation. Here the transmitter sends a series of pulses to the receiver at a rate of about 6 Hz. The receiver is connected to an electromagnetic actuator with a movable arm that flips back and forth each time a pulse is received. Varying the pulse parameters varies the average position of the actuator's arm. Pulse proportional modulation is used in very simple RC installations for model planes in which only the average position of the rudder is controlled. This kind of RC system will directly drive a relay, but the relay will chatter instead of being either on or off.

A third kind of commercial RC system, which is rarely used to control models, uses a simple on-off method of control which is well suited for driving a relay. When a switch on the transmitter is pressed, a tone is transmitted. The receiver decodes the tone and actuates a relay, motor, or other electromechanical device. This kind of system is ideal for single-channel applications such as actuating a camera, pager, intrusion alarm, or bell.

Commercial Radio-Control Systems

For experimenters with a small budget or those who need a "quick-and-dirty" RC system, a single-channel unit salvaged from a toy RC car will suffice in some applications. Often the major drawbacks of such systems are limited range (a few hundred feet) and susceptibility to noise and interference from signals on nearby frequencies. The interference problem usually occurs when the receiver is not crystal controlled.

A good time to look for salvageable single-channel RC toy cars is during the first few months after Christmas. Unsold systems are sometimes available on sale for as little as $10 and returned systems, particularly defective ones, are available for even less.

Some relatively low-cost toy cars use the same digital proportional modulation method of more sophisticated systems. The receivers of such systems are usually crystal controlled. These systems enable multichannel operation. If you want to control

more than one device with a single transmitter, you might want to consider such a system.

A disadvantage of RC systems designed to control toy cars is their susceptibility to interference and electrical noise. If your budget permits, you can achieve better results by purchasing an RC system designed specifically for controlling more sophisticated model planes and boats. One system with which I have experimented extensively is the Aero Sport Two. This two-channel RC system operates in the 72-MHz RC band and is therefore unaffected by signals in the 27-MHz CB and RC bands. The receiver is enclosed in a plastic housing, weighs about two ounces, and measures $1\%_{16} \times 1^{23}/_{32} \times \frac{3}{4}$ inches. A complete system is available from Hobby Shack. The system includes a transmitter, receiver, two servos, and a battery box for four AA cells.

A Multifunction Single-Channel RC System

The preceding commercial RC systems each have important advantages and disadvantages. Without modification, however, none is well suited for the simple task of single channel, go/no go (on/off) control. Ace R/C, Inc., a manufacturer and supplier of RC equipment since 1953, sells a single-channel transmitter/receiver pair ideally suited for general purpose, go/no-go remote control applications. The transmitter, which is called the "Wee 1," has an output power of nearly 500 milliwatts. The Wee 1 measures only $5\frac{5}{8} \times 2\frac{3}{4} \times 2\frac{1}{16}$ inches and is available as a kit (11K16) or factory assembled (11K17).

Two versions of Ace's Commander superheterodyne receiver are designed specifically for use with the Wee 1. One is designed for a 2.4-volt nickel cadmium power supply (12K12) and the other for a 3-volt supply (12K13). Each is available assembled but without an enclosure. The Commander weighs less than an ounce and measures $1\frac{5}{16} \times 1\frac{3}{4} \times \frac{9}{16}$ inches.

Figure 6-8 is a block diagram that illustrates the operation of the Wee 1 transmitter and Commander receiver. In operation, the transmitter's crystal-controlled oscillator generates a radio-frequency carrier signal that is amplified and coupled to a 52-

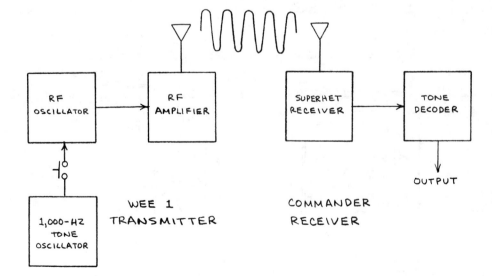

Fig. 6-8. Block diagram of Wee 1/Commander RC system.

inch telescopic antenna. The transmitter includes a normally off 1000-Hz multivibrator coupled to the rf oscillator through a buffer transistor. When the switch that supplies power to this tone oscillator is closed, the carrier is modulated by the 1-kHz tone. The receiver incorporates a crystal-controlled superhet circuit and a tone decoder. When the receiver detects a carrier signal having the proper rf and tone frequencies, it supplies an output current capable of directly driving a small relay, opto-isolator, or external circuit.

Any of five FCC-authorized frequencies can be selected for this transmitter/receiver pair. The frequencies are 26.995, 27.045, 27.095, 27.145 and 27.195 MHz. Selecting a frequency requires some planning, particularly if others may be using RC systems within half a mile or so of your location. Interference from CB radios might also cause problems. The 40-channel CB band extends from 26.965 to 27.405 MHz and encompasses all five frequencies at which the Wee 1 transmitter is available. The 27.145-MHz RC frequency is close to CB channel 14 (27.125 MHz), a frequency used in many older 100-mW toy transceivers. The 27.195-MHz RC frequency is even closer to CB channel 19 (27.185 MHz), the popular trucker's channel. Therefore, you

may reduce potential interference problems by selecting one of the three other frequencies.

Incidentally, the frequency of the crystal in a superhet receiver does *not* match the frequency of the crystal in the transmitter whose signal the receiver is intended to receive. The receiver crystal is always ground for a frequency 455-kHz *lower* than that of the transmitter's crystal. Thus the crystal of a superhet receiver designed to detect the signal from a transmitter operating at a frequency of 27.145 MHz will have a frequency of 26.690 MHz (27.145 − 0.455 = 26.690).

Figure 6-9 shows how Ace's Citizenship receiver can be connected to an optoisolator which, in turn, can drive an external circuit or a relay. R1 limits current to the infrared-emitting diode in the optoisolator. R2 limits current to a red indicator LED which is included to show when the receiver has received a signal. A General Electric H11A10 GaAs diode/npn phototransistor optoisolator was used in the prototype circuit I assembled. However, any similar optoisolator should also work.

Figure 6-10 shows how the receiver and the external optoisolator circuit can be installed in a compact plastic enclosure that measures only 4 × 2 × 13/16 inches (Radio Shack 270-

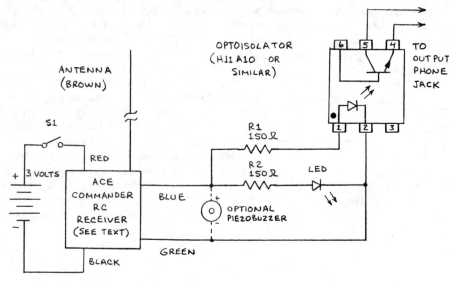

Fig. 6-9. Multifunction RC receiver unit.

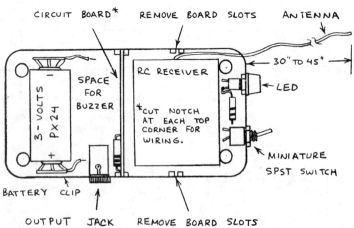

Fig. 6-10. Internal layout of general-purpose RC receiver system.

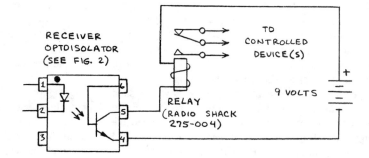

Fig. 6-11. Relay interface for general-purpose RC amplifier.

220). By installing the optoisolator and R1 on a perforated board, space is made available for the later installation of additional components.

Using the Single-Channel RC System

Before connecting the battery to the system, be sure to carefully check the wiring to make sure there are no errors. Also make sure none of the exposed printed circuit connections on the receiver board touch the phone jack or other exposed terminals. Each time the power switch is closed, the indicator LED will flash to indicate the receiver is functioning.

There are many practical applications for this single- channel RC system. The simplest is to install a piezobuzzer (Radio Shack 273-065 or similar) inside the receiver's case and use the system as a pager. There's ample space for the buzzer between the optoisolator and the 3-volt battery (see Fig. 6-10). Connect the positive terminal of the buzzer to the receiver's blue lead and the negative terminal to the receiver's green lead.

I installed a piezobuzzer in the prototype unit for this purpose. However, I connected its leads to the circuit by means of a miniature connector so the buzzer can be easily removed when not needed.

The output transistor inside the optoisolator will drive many circuits directly. It will also function as a low-current switch for activating a camera such as Kodak's disc camera. Figure 6-11 shows how the receiver can drive an external relay. This permits the system to control devices that exceed the switching capability of the optoisolator alone.

To test the immunity of the system in Figs. 6-8 through 6-10 to interference from CB units, I tried keying the 40 channel rig in my pickup at the frequencies near that at which the RC system operates while whistling into the microphone. Even though the receiver was only a few feet away, in no case did false triggering occur. Incidentally, the FCC prohibits using a CB station to whistle for entertainment purposes. You can, however, transmit audio tones that last no longer than 15 seconds for the purpose of making contact.

Though the receiver is apparently immune to rf interference, it is susceptible to electromagnetic pulse phenomena. For example, a nearby lightning bolt will trigger from one to several quick flashes of the indicator LED. The receiver LED will also flash when an electric space heater switches on and off. Fortunately this occurs only when the receiver antenna is within a few feet of the heater. If these sources of false triggering pose a problem, you can connect a timer circuit to the optoisolator

that will trigger an output relay only when the received signal exceeds a preset interval. See *Engineer's Mini-Notebook: 555 Timer IC Applications* (Radio Shack, 1984) for timer IC design tips.

CAUTION: Because of the possibility of interference, *never* use an RC system to control any circuit or system under circumstances that might endanger life or property. For example, an RC system should never be used to control a life support system. Also, when using an RC system, you must abide by the radio service rules and regulations of the FCC. For detailed information about these regulations, write the FCC (Gettysburg, PA 17326).

Going Further

The catalogs available from ACR R/C, Hobby Shack, and other RC dealers include abundant information about radio control. Fred Marks has written two of the best books available on the subject. They are *Getting the Most from Radio Control Systems* and *Basics of Radio Control Modeling*. Both these books are published by Kalmbach Publishing Company. They are usually available from hobby shops that specialize in radio controlled models.

Experimenting with a Touch Tone DTMF Receiver

Several semiconductor companies have introduced various integrated circuits capable of generating and decoding the Touch Tone signals used in push-button telephones. Since Touch Tone signals can be easily transmitted over wires, radio waves, beams of light, or through the air as sound waves, these new chips permit Touch Tone signals to be used in many kinds of remote control applications.

For example, integrated circuits that decode Touch Tone signals make possible receivers capable of responding to signals sent over the telephone line. Lights or appliances at any location equipped with a telephone and such a receiver can be switched on or off by pressing buttons on a second telephone equipped with a Touch Tone keypad. If the second telephone is not equipped with a Touch Tone pad, then the signals can be transmitted by means of a commercial or homemade Touch Tone circuit placed next to the telephone's microphone.

The dual-tone principle of a Touch Tone system requires that two specified tones be simultaneously present before an output signal is generated at the receiver. This greatly reduces the impact of interfering signals and means Touch Tone encoders and decoders can be used in nontelephone applications which might be subject to interference from external signals.

Several years ago experimenters and hobbyists who wished to experiment with Touch Tone signals were forced to assemble the required circuits from scratch. Integrated circuits capable of generating Touch Tone signals, such as the crystal-controlled MC14410 tone encoder, greatly simplified the assembly of such circuits.

Today, it's possible to buy preassembled, pocket-size Touch Tone encoders for less than the cost of assembling a do-it-yourself unit. For example, Radio Shack sells a compact Touch Tone generator complete with batteries for less than $20. More expensive units include such features as memory, digital display, and crystal-controlled clock.

Commercial Touch Tone decoder systems are not as widely available or as economical as encoder units. At least not yet. Therefore, the emphasis here will be upon the operation of a particularly versatile decoder chip suitable for do-it-yourself circuits, Teltone Corporation's M-957. Before looking at this chip in some detail, let's quickly review the basics of the Touch Tone system.

Touch Tone Basics

The Bell System invented the Touch Tone system specifically for push-button telephones. The keypad on a typical push-button phone includes twelve buttons, ten labeled 0 through 9 and two special function keys, known as spares, marked * and #. The system can be expanded to include four additional keys.

Pressing a Touch Tone button generates two simultaneous audio-frequency tones. Figure 6-12 shows the frequency pairs assigned to each button. The four additional keys would control an eighth tone having a frequency of 1633 Hz.

The technical term for the various Touch Tone frequencies is *Dual-Tone Multifrequency* (DTMF) signals. If you refer to Fig. 6-12, you'll observe that the seven frequencies seem arbitrary and rather oddly distributed. Actually, the frequencies were very carefully selected to reduce to a minimum interference from voice, the dial tone, and the harmonics from alternating current power lines.

The frequencies are divided into a low group (697 to 941 Hz) and a high group (1209 to 1477 Hz). Pressing any button simultaneously selects one frequency from each group. Since both tones must occur simultaneously, the possibility of a false signal is virtually negligible. This is why a DTMF system should be considered for remote control applications that might be subject to false triggering from noise or interfering signals.

For instance, a few years ago I designed a radio-controlled camera system for making aerial photographs from kites and balloons (see elsewhere in this book). Using this system, which was triggered by single-frequency audio tones superimposed on the carrier of a low-cost radio-control transmitter, I have obtained many good-quality aerial photographs. Unfortunately, the system is very vulnerable to false triggering, as have been most remote control systems I've built that use a single frequency. In other words, it can sometimes be triggered by ship-to-shore radios and CB units in passing cars and trucks. Now that I've experimented with DTMF circuits, I plan to modify my aerial photography system for Touch Tone operation. The system should be virtually immune to false triggering and it will provide the added bonus of up to sixteen channels.

Integrated Circuit DTMF Receivers

Several companies make chips that receive and decode DTMF signals. Among them are Mitel Semiconductor, Silicon Systems, Inc., and Teltone Corporation.

The DTMF receiver chips made by these companies all incorporate switched-capacitor filters to detect the transmitted tones. The filter stages are followed by various kinds of amplitude detection circuitry and logic that determine when two detected tones are present. Each chip includes an output decoder that transforms the detected tone into a binary bit pattern. CMOS circuitry is generally used to provide low-power operation.

Teltone's M-957 DTMF Receiver

Teltone Corporation has for several years made available to hobbyists and experimenters a line of reasonably priced DTMF receiver kits. The latest is the TRK-957 DTMF Receiver Kit. This kit includes an M-957 CMOS DTMF receiver, a 3.58-MHz crystal, a 1-megohm resistor and a 22-pin DIP socket. You can order the kit by writing the company at the address given in the Appendix.

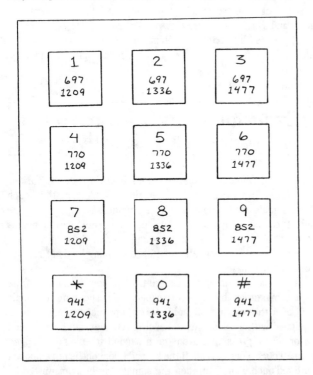

Fig. 6-12. Simultaneous tone frequencies for a Touch Tone keypad.

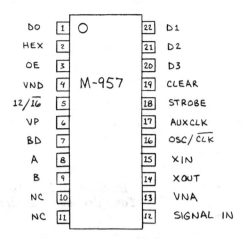

Fig. 6-13. M-957 pin outline.

Figure 6-13 is the pin diagram of the M-957. The pin placements are similar, though not identical to, those of the M-947, an earlier DTMF receiver which was once sold in a Teltone kit. The M-957 is very easy to use so long as operating requirements and precautions given in the manufacturer's data sheet are followed. Since the M-957 is a CMOS chip, it's important to observe proper handling procedures to avoid damaging the chip with static electricity.

Referring to the pin diagram in Fig. 6-13, the positive supply, which should be from 5 volts to an absolute maximum of 16 volts, is applied to the M-957 at pin 6 (VP). Pins 4 (VND) and 13 (VNA) should be at ground potential.

The DTMF signal is applied to the M-957 at pin 12 (SIGNAL IN). The input signal may be ac coupled through a capacitor. If the signal is dc coupled, the peak signal voltage must not exceed the positive supply voltage. Therefore, the input signal should always be removed *before* power to the M-957 is switched off.

A 3.58-MHz crystal and a 1-megohm resistor are both connected across pins 15 (XIN) and 14 (XOUT) to provide a precise time base for the chip's internal oscillator. If the internal oscillator is not used, pin 15 should be tied to logic 1.

Pin 16 (OSC/CLK) is the time base control. When pin 16 is at logic 1, the internal oscillator is selected. When pin 16 is at logic 0 (and pin 15 is at logic 1), a signal applied to AUXCLK is selected as the time base. Pin 17 (AUXCLK) should be left open when the internal time base is selected (pin 16 at logic 1). If an external signal is used for a time base, its frequency must be 3.58 MHz divided by 8 or the M-957 will not decode signals properly.

The M-957 has four output pins (1, 22, 21, and 20) conveniently grouped at one end of the chip. As is explained in the following, these pins provide two kinds of binary bit patterns that correspond to the detected DTMF signal.

Several control and output pins add to the M-957's versatility. The 12/16 input (pin 5) determines which range of DTMF signals will be detected. When pin 5 is at logic 1, the standard 12 DTMF signals of the push-button telephone will be detected. When pin 5 is at logic 0, all 16 DTMF signals are detected.

The A and B inputs (pins 8 and 9) control the sensitivity of the M-957 to the input signal. Applying various combinations of logic states to these two pins adjusts the sensitivity in steps to a maximum of −31 dBm.

The OE input (pin 3) controls whether the output pins are enabled or placed in the high-impedance or so-called third state. When pin 3 is at logic 1, the output pins are enabled and represent the contents of the M-957's output register. When pin 3 is at logic 0, the output pins are placed in the high-impedance third state.

The HEX input (pin 2) controls the format of the four output pins. When pin 2 is at logic 1, the output pins provide a standard 4-bit binary bit pattern. When pin 2 is at logic 0, the output pins provide a 2-of-8 binary code. Table 6-1 summarizes the two output modes of the M-957.

Table 6-1. Summary of M-957 Output Modes

Signal	Low Tone	High Tone	Binary Output	2-of-8 Output
1	697	1209	0001	0000
2	697	1336	0010	0001
3	697	1477	0011	0010
4	770	1209	0100	0100
5	770	1336	0101	0101
6	770	1477	0110	0110
7	852	1209	0111	1000
8	852	1336	1000	1001
9	852	1477	1001	1010
0	941	1336	1010	1101
*	941	1209	1011	1100
#	941	1477	1100	1110
A	697	1633	1101	0011
B	770	1633	1110	0111
C	852	1633	1111	1011
D	941	1633	0000	1111

The STROBE input (pin 18) indicates when a valid frequency pair is present at the input. Normally pin 18 is at logic 0. When a valid frequency pair has been detected, pin 18 goes to logic 1 until the signal ends or the CLEAR input (pin 19) is placed at logic 1. If the CLEAR input is not used, it should be tied to ground (VNA or VND) to prevent stray signals from causing inadvertent clear operations.

Output BD (pin 7) provides an early indication of a possibly valid DTMF signal at the input pin. Normally BD is at logic 0, but it goes to logic 1 when a signal has been received and is being validated. The BD output responds to an input signal within about 18 milliseconds but the STROBE output requires about 40 milliseconds to verify a correct signal.

Using the M-957

Figure 6-14 is a straight test circuit for the M-957. The input signal may be supplied directly from a Touch Tone pad, tape recorder, amplifier, or radio receiver so long as the signal

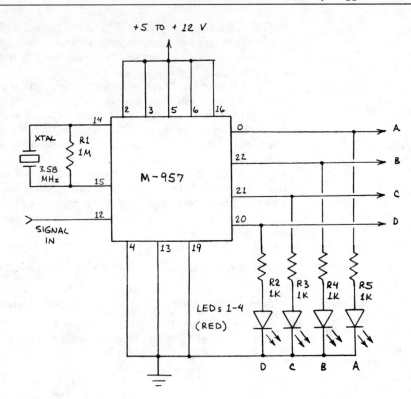

Fig. 6-14. M-957 Touch Tone decoder test circuit.

amplitude does not exceed the positive supply applied to the M-957. The signal may be coupled directly (dc) or through a 0.01-μF capacitor (ac).

The circuit in Fig. 6-14 includes four optional LEDs connected to the outputs to provide a visual indication of the received signal. Indicator LEDs can also be connected to other outputs of the M-957.

Generally it's desirable to decode the binary output from the M-957 into a 1-of-16 format. Figure 6-15 shows how to connect a 74C154 4-line to 16-line decoder to the outputs of the M-957 to achieve this purpose. Keep in mind that the 74C154 is a CMOS chip and should be handled accordingly. Other CMOS decoder chips can also be used.

Figure 6-16 shows how to drive a relay from one of the outputs from the 74C154 or a similar decoder. Diode D1 absorbs reverse voltage generated by the collapsing field in the relay's coil when the relay is switched off. The relay is normally de-energized. When the base of Q1 is placed at logic 0, the relay is energized. If the circuit doesn't switch consistently, try making slight changes in the value of R2.

I have had excellent results experimenting with the M-957 in a variety of nontelephone remote control applications. Those readers wishing to connect the chip directly to a telephone line will want to first carefully review the M-957 data sheet. Figure 6-17 is adapted from a suggested telephone line interface circuit given in the M-957 data sheet. If this single-supply circuit doesn't work properly, try using a dual-polarity supply. In other words, connect pin 4 to the negative counterpart of the positive supply instead of to ground.

It's very important that you understand the interfacing requirements imposed by your local telephone company before

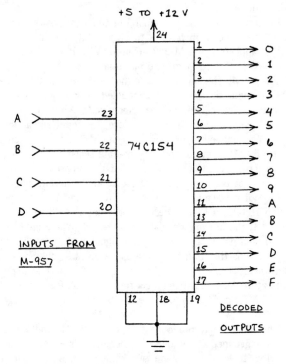

Fig. 6-15. A 1-of-16 decoder for the M-957 tone decoder.

connecting a do-it-yourself circuit to their lines. The Federal Communications Commission allows customer-provided equipment to be connected to telephone lines if the equipment meets

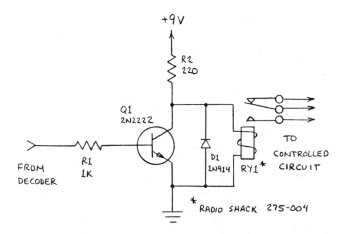

Fig. 6-16. A relay driver circuit

FCC guidelines. In every case the equipment must be connected to the lines with standard four-prong or modular telephone plugs and jacks. Check your telephone directory for guidelines and call the company if necessary.

Going Further

As you can see by now, the familiar Touch Tone DTMF signals have far more applications than merely dialing telephone numbers. For more information, write the manufacturers listed previously at the addresses given in the appendix and request data sheets and application notes. Also, look for articles on DTMF applications and new encoder and receiver chips in the various electronics and communications magazines available at most good university libraries.

If your primary interest is connecting circuits to the telephone line, be sure to thoroughly research the topic and proceed with caution. An excellent book for hobbyists is *Electronic Telephone Projects, Second Ed.* by Anthony J. Caristi (1986, No. 22485, Howard Sams & Co.).

Experimenting with Kodak's Disc Camera

Part 1. Modifying the Camera for Electronic Triggering

Kodak's widely imitated system of disc photography was once acclaimed as the most important development in snapshot photography since Polaroid introduced instant picture taking. Several years after its introduction, however, a new generation of compact 35-millimeter cameras attracted far more customers. The chief reason for this switch was the small size of the disc camera's photographic negative. Even though Kodak developed a new film for the disc camera, the much larger 35-mm negatives provide much better photographs.

Its small negative size notwithstanding, the automatic operation of the disc camera opened up a wide range of applications for experimenters who add electronic accessories to a basic camera. Kodak has discontinued production of its disc cameras, but they are still widely available. Several other companies manufacture similar disc cameras.

Each of the cameras in the disc family combines a motorized film advance, built-in strobe, batteries, and totally electronic triggering in a rugged package that weighs only six ounces and costs as little as $45 or even less. These features make the disc cameras ideally suited for many fascinating assignments that otherwise require hard to find equipment and accessories costing hundreds, or even thousands, of dollars.

Many electronic accessories for the disc camera family can be designed. Some of the more obvious include a variable speed sequence controller and circuits that remotely trigger the camera in response to light, sound, or radio signals. Applications for a disc camera and one or more accessories such as these are wide ranging. A sound-activated disc camera can photograph wildlife or an unwanted intruder. A light-triggered disc can record lightning or serve as a combination slave flash *and* camera.

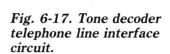

Fig. 6-17. Tone decoder telephone line interface circuit.

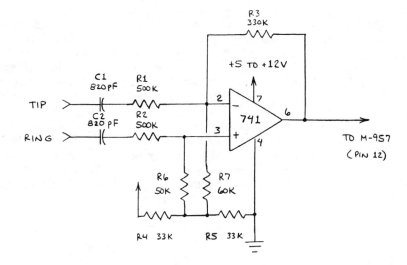

A sequentially triggered disc can take elapsed time photos of flowers opening, cloud movements, and traffic patterns. It can also be used to provide a timed photographic record of an instrument panel or a sequence of zones through which a vehicle or aircraft equipped with the camera has passed. A radio- or infrared-trigger disc camera has numerous applications in remotely controlled photography.

Until a disc camera with an external socket or jack suitable for connecting various triggering devices is introduced, it's necessary to modify an existing disc camera. I'll describe how that is accomplished later. First, let's find out more about the design and operation of a disc camera.

The Disc Camera Family

This discussion is limited to the four original Kodak disc cameras, the models 2000, 4000, 6000, and 8000. All four cameras share many common features with disc cameras made by other companies. Each is about the size of a pocket calculator ($1 \times 3 \times 4.5$ inches) and includes motorized film advance and a built-in strobe. All four cameras accept a 15-exposure flat film cartridge that contains a unique rotating disc of Kodacolor film. All the cameras also include a fixed-focus, four-element, all-glass lens system with a focal length of 12.5 mm and a fully open aperture of f/2.8. According to William H. Price, manager of optical engineering at the Kodak Apparatus Division, the performance of the lens approaches "the diffraction limits posed by the wave nature of light."

The Disc 2000 is a low priced version of the camera. It is powered by a replaceable 9-volt alkaline battery and lacks the fully automatic features of the three other disc cameras.

The Disc 4000 includes a built-in light sensing integrated circuit with a threshold of 125 footlamberts. Above that value, the camera provides an exposure speed of $1/200$ second at a lens aperture of f/6. Below that value, the exposure speed is automatically dropped to $1/100$ second and the lens aperture is opened fully to f/2.8. Furthermore, the electronic strobe always flashes when the light level is below 125 footlamberts.

A second 16,000-square mil integrated-injection logic chip housed in a miniature 18-pin DIP makes the timing and control decisions necessary to charge the flash capacitor, fire the strobe, select the lens aperture and exposure speed, and advance the film disc. The chip drives the camera's 6-volt slot car-type motor at an average power of 2 watts and a peak current of 2 amperes.

The flash capacitor is fully charged in less than a second. When combined with the automatic film advance feature, which rotates the film disc to the next frame in 0.4 second, this means the camera can take flash photographs at intervals of only $1\frac{1}{3}$ seconds!

The Disc 4000 is powered by a pair of 3-volt lithium polycarbon monofluoride batteries made by Panasonic, thus making it one of the first consumer products to be powered by this exceptional energy source. These batteries have a shelf life in excess of five years and a capacity of 1200 milliampere hours. Consequently, Kodak guarantees the *entire* camera, including the batteries, for a full and unprecedented *five* years. Assuming no electronic or mechanical problems, the camera should provide well over 2000 exposures before the batteries are exhausted.

The Disc 4000 is housed in a robust plastic and silver anodized aluminum case. A sliding lens and viewfinder cover automatically actuates the strobe capacitor charging circuit when the camera is readied for use.

The Disc 6000 is identical to the Disc 4000 in every respect with two exceptions. The first is a folding cover that protects the entire front of the camera when it is not in use. When opened, the cover serves as a handle. It also automatically actuates the strobe capacitor charging circuit.

The second addition to the Disc 6000 is a close-up lens that can be quickly slid into action by moving a small protrusion under the lens opening. The close-up lens reduces the minimum picture taking distance from 4 feet to 18 inches.

The Disc 8000 is the most sophisticated of the disc family. It incorporates the close-up lens and cover of the Disc 6000 plus a self-timer, a rapid sequence film advance, and a digital alarm clock. The self-timer provides a 10-second delay before the camera automatically takes a picture, thus allowing the user to be included in a photograph. The timer activates a blinking red LED on the front of the camera and an audible, pulsating tone. The tone sequence speeds up during the final two seconds before the exposure is made to notify the user the camera is about to be triggered.

The rapid sequence feature of the Disc 8000 permits the camera's user to take photos at a rate of three per second in daylight simply by holding down the shutter button. If the flash is needed, the camera will take a picture once every $1\frac{1}{3}$ second when the shutter button is held down.

The Film Disc

Kodacolor disc film has an ISO speed of 200 and is coated on a thick 7-mil Estar base to provide the flatness and stiffness necessary to assure consistently sharp exposures. The film has twice the speed and a finer grain than Kodacolor II film.

The disc includes frame numbers and both alphanumeric and bar coded identification codes. This data as well as the individual frame numbers is preflashed on the film when it is manufactured and made visible during development. The individual disc identification code permits the use of highly automated processing equipment.

Individual film discs are installed in a flat plastic cartridge. The light-tight cartridge, whose label contains the disc's identification number, is 0.2 inch thick, 3 inches long, and 2.8 inches wide. The cartridge includes a dark slide which rotates away from the aperture opening after the disc is inserted in a camera. When the camera is opened, the dark slide is moved back over the aperture.

Disc Film Image Quality

Serious photographers may wonder about the image quality of Kodak's disc film. Since an individual disc negative measures 8.2×10.6 millimeters, about a tenth the area of a 35-mm negative, there's good reason for concern.

Because disc imagery must be enlarged much more than equivalent 35-mm imagery to obtain the same size print, the grain is definitely more apparent. On the other hand, disc film produces very sharp, highly resolved imagery. For example, photographs taken by a radio controlled Disc 4000 camera suspended from a helium filled balloon 145 feet above my backyard garden reveal individual cucumbers, rocks, and a 1-inch diameter hoe handle.

The disc camera's high quality glass lens system is one key factor in the sharpness of disc imagery. The other, of course, is the film itself. Because of its exceptionally small negative size, development of an acceptable disc film required simultaneous improvements in speed, grain, and sharpness. According to Dr. Dave Nelander, who coordinated development of the new film at Kodak, "We had never before been asked to make such a huge improvement in one film."

Kodak's researchers solved the sharpness problem by reducing the thickness of the film's emulsion. This reduced the optical scattering that causes fuzzy, poorly defined images. Changing the film's chemistry reduced grain size and doubled its speed. Furthermore, exposure latitude was broadened to permit scenes having a wide range of lighting to be properly photographed. The new film can produce acceptable pictures when overexposed by up to three f-stops or underexposed by up to two f-stops.

Although prints made from 35-mm Kodacolor II negatives are superior to equal size prints made from Disc Kodacolor film, I've found disc prints to be perfectly acceptable in many applications. Amateur photographers will find them at least as good as those made with 110 cameras. And many of us who prefer 35-mm photography may find the advantages offered by a *modified* disc camera to outweigh any loss in image quality.

Modifying a Disc Camera

Until a disc camera with an external shutter jack is introduced, it's necessary to modify existing cameras if you want to control them electronically. You have at least two options. One is to employ a servo or solenoid to electromechanically trip the existing shutter button. The other is to gain access to the camera's circuitry and attach a set of external connection leads.

The advantage of the electromechanical approach is there is no need to open the camera, thus protecting its warranty. On the other hand, the electromechanical approach requires more space, is heavier, consumes more power, and is less reliable than purely electronic triggering.

I've found two ways to gain access to the shutter connections inside the Disc 4000. The simplest is to gently pry off the plastic shutter button to reveal two metal contacts and a flexible metal plate. A third contact is hidden beneath the metal plate.

Unfortunately, I've not found a practical way to connect leads to these contacts. The plate can be removed with tweezers, but attempts to solder leads to the contact below the plate melt the surrounding plastic.

For these reasons I've modified two Disc 4000 cameras by removing the front panel and soldering connection leads directly to the cameras' circuit boards. I'll describe how this is done next, but first here are a few precautions you must heed and read *before* opening a disc camera.

1. Most disc cameras come with a warranty. In Kodak's case, the warranty is voided " . . . if the camera is damaged by misuse or other circumstances beyond Kodak's control . . . ." Since the manual provided with the disc cameras specifically states that the camera should not be disassembled, opening and modifying the camera might be grounds for voiding the warranty. On the other hand, if a malfunction is not associated with a modification, the warranty might stand. In any event, you risk voiding your camera's warranty if you open and modify it.

2. Unless you are careful and follow the instructions given in the following, you might damage the camera you are attempting to modify. You must avoid touching or manipulating the complex and fragile mechanical parts of the camera. You must also avoid bridging solder across adjacent terminals on its circuitboard.

3. The camera's built-in strobe circuitry constitutes a potential shock hazard. In this connection, the Kodak Disc Camera manual includes the following warning notice: "WARNING: Do not disassemble or attempt to repair this camera. The voltages associated with the power source in this camera will, under certain conditions, present a potential shock hazard."

The primary shock hazard is a 160-μF photoflash capacitor which is almost always charged to about 180 volts. Even *weeks* after the camera is last used, this capacitor retains a hefty charge! The discharge from this capacitor across a finger or hand can cause an involuntary jerk that may dump a soldering iron in your lap or jam your elbow into a wall. A discharge through your body (as from one hand to the other) may cause a more severe reaction. Therefore, you should open the camera *only* if you know what you are doing *and* if you plan to use the proper precautions.

For example, when the camera is open, *never* touch any part of the circuit board or any electronic parts or components with your fingers or an uninsulated tool. There's no need to touch anything inside the camera to make the modifications to be described. Furthermore, you should always keep one hand *away* from the camera to avoid a possible shock through your body. Of course you should not attempt to open the camera if you have had no prior electronics experience.

4. Finally, keep in mind that the discussion here pertains primarily to early models of Kodak's disc camera line. Subsequent cameras may incorporate design changes that may preclude the modifications given. Likewise, disc cameras made by other manufacturers differ in *many* ways from those made by Kodak. Therefore modifying these cameras requires considerable care. I have found that some such cameras cannot be modified for the applications described.

Opening a Kodak Disc Camera

Opening a disc camera requires a *clean* work area and a steel implement about half a millimeter thick and a centimeter or so wide. It should be at least 10 centimeters long. A 15-centimeter stainless steel pocket rule like those available at hardware stores works reasonably well. Avoid the temptation to use a screwdriver! It will damage the case and may slide up inside the camera.

You want to remove the aluminum front cover with its attached black plastic lens and viewfinder door assemblies. Along the bottom of the camera there is a narrow gap between the aluminum front cover and the camera's plastic body. Look closely, and you'll see two slots in the gap on either side of the camera. With the lens facing away from you, the widest of the two slots is to your left.

Make sure the camera's lens door is fully *closed*. Then insert the steel tool onto the widest of the two slots and twist the tool from side to side until the aluminum cover begins to give. Repeat this procedure with the slot on the right side of the camera's bottom. Be patient. Several cycles of twisting and prying may be necessary to remove the cover. Above all, don't force the tool or push it up inside the camera's body where it might damage the circuit board or the delicate moving parts or even cause a shock.

Eventually you will be able to lift the cover from the camera. Avoid getting dust on the lens and do not touch any of the camera's internal parts.

Connecting External Shutter Leads

When the camera is opened, make a 1-millimeter hole in the bottom of the camera at the location shown in Fig. 6-18. Use a small drill or simply twirl a sharp hobby knife into the plastic. Remove any protrusions or cuttings from inside the case. Incidentally, if you select a different location for the shutter leads' access hole, make sure it does not interfere with the protruding lips of the camera front panel.

Next, notice the square opening in the yellow plastic circuit board protective cover. The three rectangular pads visible through the opening are the shutter contacts. You many solder external connection leads directly to them, but to avoid complications you may then have to remove the flexible contacts from the shutter switch on the back side of the camera's cover panel. Of course this will permanently disable the camera's manual shutter switch.

Alternatively, you can do as I have done and temporarily remove the yellow circuit board cover in order to solder the leads to the terminal points of the lands leading away from each of the three shutter switch pads.

CAUTION: To avoid being shocked by the strobe capacitor, do *not* touch the exposed circuit board! See the safety remarks given previously.

Since early versions of the disc camera have used at least two entirely different circuit board layouts, I've not included a photograph of my modified cameras, each of which employed a different circuit board. All you have to do is follow the land leading away from each shutter switch pad to its end point and carefully solder an 8-inch length of wrapping wire to each terminal. See Fig. 6-19 for the color arrangement you should use.

Only a few millimeters of insulation need be removed from one end of each wire. Do *not* remove any insulation from the opposite end of each wire. Use a low wattage soldering pencil to make the connections and be careful to avoid bridging solder between the closely spaced terminals on the board. A rubber bulb solder slurper will remove solder bridges.

After the leads are in place, inspect the board to find and remove any solder balls or bits of wire. Make sure the soldered

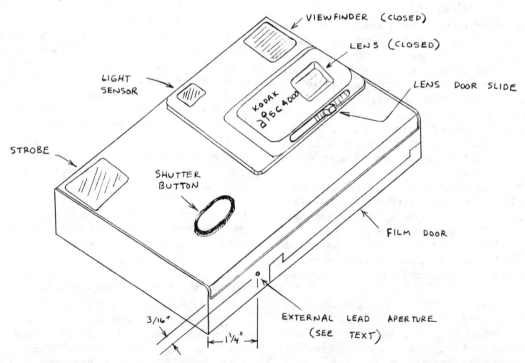

Fig. 6-18. Location of external leads outlet aperture in the Disc 4000.

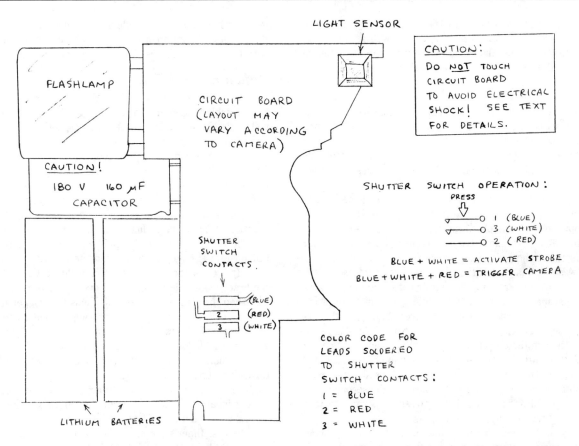

Fig. 6-19. Connecting external leads to the shutter switch terminals of the Disc 4000.

ends of each wire do not extend away from the terminals and contact any nearby terminals or lands.

Next, clip off any exposed wire from the end of each lead. Then thread the three leads through the hole in the bottom of the case. Pointed electronic tweezers will be very helpful. Again, do not touch the circuit board! If necessary, insulate the tweezers with vinyl tape. Pull the wires so they extend in a gentle curve over the two batteries and then carefully replace the adhesive yellow cover over the circuit board. There's no need to touch the board. Just allow the cover to fall into position on the board and press it in position with the eraser end of a wood pencil.

Finally, replace the camera's cover. First, make sure the lens door is fully *closed*. Then insert the upper edge of the cover into the top side of the camera's body. When it is aligned, press the bottom edge of the cover down into position. Press along all four corners to snap the cover in place. The modification of the camera is now complete.

Triggering the Camera Externally

Figure 6-19 shows how the shutter contacts in Kodak's disc camera are arranged. Caution: The connections in Fig. 6-19 are subject to possible change!

When the shutter is lightly pressed or merely touched, the upper two contacts close to activate the strobe capacitor charging circuit. Opening the lens door has the same effect.

You can hear a brief high pitched hum from inside the camera when this occurs. For more volume, place the camera near an am radio and touch the shutter button. The speaker will emit a brief but noisy hum (or a few clicks if the capacitor is already charged).

To trip the shutter, the upper two contacts must make contact with the lower contact. The exposure is then made and the film disc is advanced to the next frame. You can accomplish this with your modified camera by removing some insulation from the end of each lead, twisting the blue and white strobe charging leads together and touching them to the red lead. Disconnect the blue and white leads to save a few mils of current drain. You can use a pair of miniature switches to manually trigger the modified camera. Add longer connection leads to take pictures from across a room.

Part 2. Controlling the Camera Remotely

Now that we've covered the operation and modification of Kodak's Disc 4000, you should be prepared to connect a variety of external control circuits to the camera. If you haven't yet modified your camera in accordance with the instructions given in Part 1, it is essential that you refer to that discussion before attempting to open and modify a Disc 4000 or any other disc camera. Follow the detailed instructions as well as important precautions concerning the camera's warranty and the potential shock hazard posed by the camera's flash capacitor.

About the Circuits

I used a 9-volt battery to power test versions of the various circuits described here. For special purpose applications or when space and weight requirements are critical, you may wish to consider powering the circuit you are using with the camera's internal 6-volt lithium battery. I'll describe how this can be accomplished in Part 3.

Many of the trigger circuits with which we'll be experimenting are coupled to the disc camera by an LED-phototransistor optoisolator. The phototransistor's collector is connected to the blue and white leads from the modified camera. Its emitter is connected to the red lead.

As was noted in Part 1, the camera's shutter release switch has three contacts. Pressing the shutter button lightly activates the strobe capacitor charging circuit (blue plus white). Pressing the shutter button with a little more pressure begins the exposure/flash/film advance sequence (blue plus white plus red).

Even though the current drain is low, you should connect a switch between the white lead and the junction of the blue lead and the optoisolator (or whatever component is used in its place). I prefer to use a dpst switch for this purpose, the second set of terminals serving as a power switch between the positive battery terminal and the circuit.

Finally, if you spend a lot of time bench testing any or all of the following circuits, you can save money by recycling the first film disc you expose during initial testing. Simply pop off the back side of the plastic holder and rotate the disc so frame number one shows in the window. Since the camera will not

work without the disc installed, this procedure has saved me the expense of a dozen or so fresh film discs which might otherwise have taken pictures of the top of my work bench.

Adjusting Self Timer

The circuit in Fig. 6-20 will trigger a modified Disc 4000 an adjustable time interval after S1 is pressed and released. This will allow you to make special purpose photographs or permit you to be a part of your own pictures.

The circuit consists of two monostable multivibrators connected so that the output from the first triggers the second following a time delay controlled by R1 and C1. After the delay interval, which can be varied from a few seconds to a minute or so by means of R1, the second monostable generates a fixed duration pulse having sufficient duration to trigger the camera.

The monostables are designed around the two timers in a 556 dual timer chip. S1 can be connected directly to the trigger input (pin 6) of the first timer. I've included C2, however, so the first monostable will time out even if S1 remains closed.

When the circuit is at rest, R2 pulls the trigger input high to prevent inadvertent false triggering from extraneous electrical noise. R3 acts as a bleeder across C2. Without R3 to discharge C2 following the closing and opening of S1, the charge on C2 would have to be shorted to ground or drained away through natural leakage paths.

The output from the first timer (pin 5) is coupled to the trigger input of the second timer (pin 8) via C5. R4 pulls the second input high to prevent false triggering.

The first timer's output is also coupled, via R6, to LED 1. This LED glows when S1 is closed and remains glowing for the duration of the first timer's cycle. It therefore serves as a ready light to indicate that the self timer has been actuated.

The fixed duration pulse from the second timer must be long

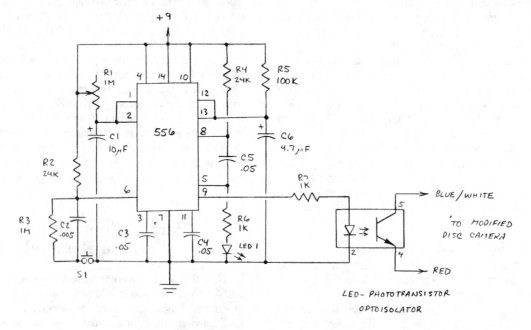

Fig. 6-20. Self timer for the disc camera.

enough to trigger the camera, 100 milliseconds or more. This time interval is controlled by R5 and C6.

The trigger pulse is coupled to the camera by means of an LED-phototransistor optoisolator. R7 limits current to the LED in the optoisolators.

You can adjust R1 to provide a time delay of up to a minute or so. For longer delays, increase C1 to 100 microfarads, and you should be able to obtain delays of at least several minutes.

Interval Timer

The circuit in Fig. 6-21 will convert a modified Disc 4000 into an elapsed time camera. Simply by adjusting the resistance of R1, the camera can be caused to make exposures at intervals ranging from seconds to minutes. The camera can then be used to record on a single film disc fifteen sequential images of such time dependent subjects as an opening flower, a busy intersection, passing clouds, playing children, sports events, and many others.

Like the self timer, the interval timer requires both timers in a 556 dual timer. The first is connected as an astable multivibrator whose period of oscillation is controlled by R1 and C1. The output from the oscillator triggers a monostable whose fixed duration output pulse is controlled by R3 and C4. For each cycle of the astable, the monostable provides a pulse having sufficient duration to trigger the disc camera via the optoisolator.

You'll find many interesting applications for this interval timer. For example, when R1 is set to provide a trigger pulse about every 1.3 seconds (you may need to reduce C1 to a few microfarads), the minimum recycle time for the Disc 4000, you can record about 20 seconds or so of an athletic event, experiment, or other fast moving event on a single film disc. Longer delays are well suited for recording slower events such as those described previously.

Triggering the Disc Camera with Light

Many simple circuits can be devised which will permit a modified disc camera to be triggered by light. A light-triggered camera can be used to photograph lightning, people or objects breaking a light beam, or even intruders. Of course, such a camera can be triggered from a distance of tens or even hundreds of feet by pointing a visible or infrared source at its sensor.

A Light-Activated SCR Trigger Circuit

The light-activated SCR (LASCR) has long been used as a light sensor in slave flash units. In this role, one or more LASCR controlled slave flashes are placed around an area to be photographed. The flash from the camera's strobe then triggers the LASCR equipped strobes to provide additional illumination.

Figure 6-22 shows a simple LASCR circuit for triggering a modified Disc 4000. Most strobes provide too brief a flash to allow the output from this circuit to trigger the camera. The circuit does, however, work well when triggered by flash bulbs or the beam from a flashlight or a visible or infrared laser.

The LASCR I used is a Motorola MRD920. It's available from Motorola suppliers or Radio Shack (Cat. No. 276-1095A). It's necessary to open normally closed push-button switch S1 after each operation since the current through the camera's switch leads is sufficient to keep the LASCR turned on after it has been illuminated. This means the camera will respond to an initial flash of light and ignore subsequent flashes, at least until S1 is pressed.

A Phototransistor Trigger

The simplest light activated trigger for a modified Disc 4000 is a single phototransistor connected across the shutter switch contacts as shown in Fig. 6-23. Many different silicon photo-

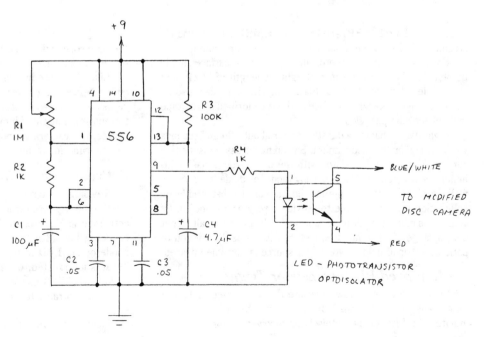

Fig. 6-21. Interval timer for disc camera.

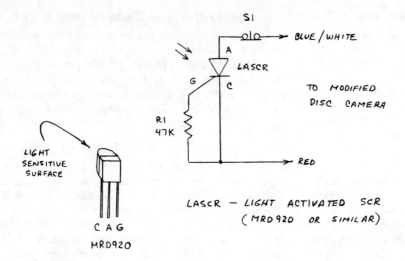

Fig. 6-22. Light activated SCR trigger for disc camera.

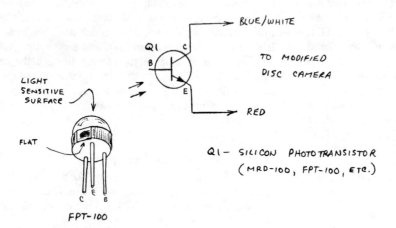

Fig. 6-23. Light sensitive trigger for modified disc camera.

transistors (Fairchild FPT-100, Motorola MRD-310, Texas Instruments TIL-414, etc.) can be used in this application.

For best results, the phototransistor's active surface should be shielded from direct external light. A length of black heat shrinkable tubing works well. Black electrical tape can also be used, but the sticky inner surface of the tube formed by the tape will collect dust particles.

Light flashes having a duration under about 100 milliseconds will not trigger the disc camera. Sweeping across the sensitive surface of the phototransistor with the beam from a flashlight, helium-neon laser, infrared emitting diode, or diode laser works well. In all cases, the range can be increased substantially by collimating the light source to provide a very narrow beam. Of course it's more difficult to point a narrow beam at a small target over a range of a few hundred feet. But it can be done if you're patient. A tripod helps if your light source is invisible infrared.

An Improved Phototransistor Trigger

Figure 6-24 shows how to isolate the disc camera from the phototransistor in Fig. 6-23. In operation, when Q1 is not illuminated the LED in the optoisolator receives no forward bias.

When Q1 is turned on by an external light source, the LED in the optoisolator is biased through R1, the phototransistor in the optoisolator switches on and the camera is triggered.

Like the preceding phototransistor circuit, this trigger circuit will fire the camera every time you sweep a light beam across the phototransistor. The only restriction is that the phototransistor must be illuminated for 100 milliseconds or so. Of course, the camera cannot be triggered during the 1.3 second recycle time following the making of an exposure.

A Break-Beam Phototransistor Trigger

Revising the circuit in Fig. 6-24 permits the modified Disc 4000 to be triggered when a continuous beam illuminating Q1's active surface is broken. Figure 6-25 gives this circuit. In operation, when Q1 is turned on by a continuous light source, the anode of the LED in the optoisolator is pulled low. When the beam is interrupted, Q1 switches off, thereby allowing the optoisolator's LED to be forward biased through R1. This switches on the phototransistor in the optoisolator and triggers the camera.

Incidentally, very brief interruptions in the beam (less than

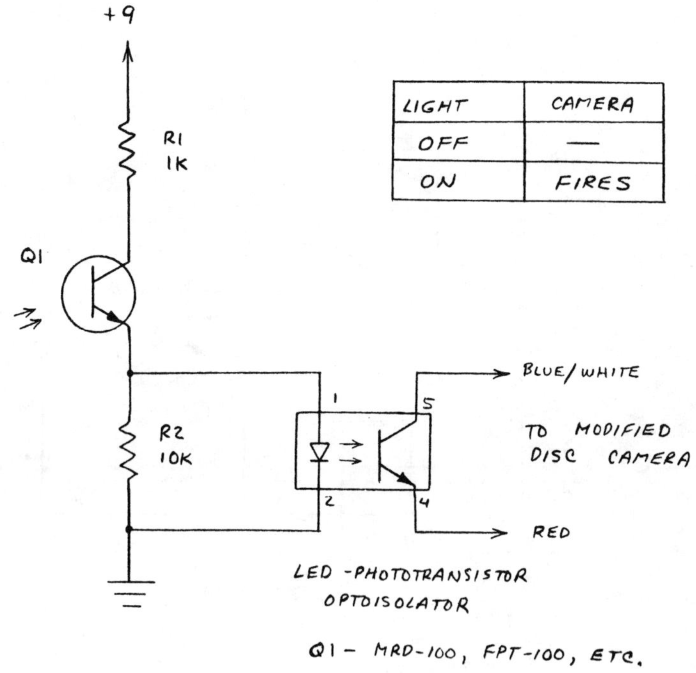

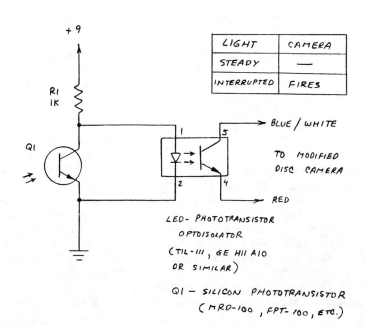

Fig. 6-24. Improved light sensitive trigger for disc camera.

Fig. 6-25. Break-beam trigger circuit for disc camera.

about 100 milliseconds) will not trigger the camera. This provides good protection from false triggering from falling leaves when the system is used out of doors.

Xenon Strobe Activated Trigger

For applications in which it's necessary to trigger the modified Disc 4000 with a very brief flash of light, it's necessary to stretch the incoming pulse. Figure 6-26 shows one way this can be accomplished.

In operation, a 555 timer is configured as a monostable multivibrator which outputs a pulse of 100 milliseconds or more when it has been triggered. A phototransistor connected to the trigger input of the 555 initiates the timing sequence when a brief light flash occurs. The output from the 555 is coupled to the modified Disc 4000 via an optoisolator.

This pulse stretching method can be used in other circuits designed to trigger the camera with a very brief event. For example, very intense pulses are produced by SH diode lasers

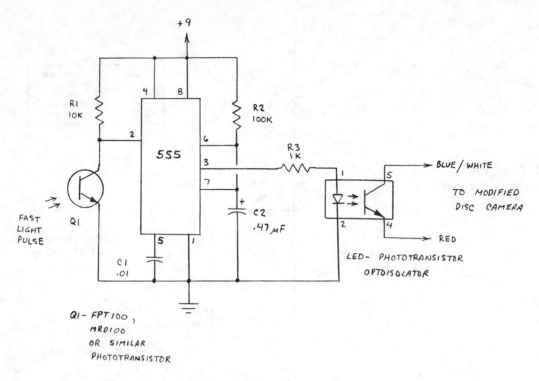

Fig. 6-26. Xenon strobe activated trigger circuit for disc camera.

driven with high current pulses. Substituting a fast risetime PIN photodiode (TIL413 or similar) for the phototransistor would allow such a laser to trigger the camera from a considerable distance. The PIN diode should be connected in the *reverse* biased mode (anode to ground). The beam from the laser should be collimated with a small f/1 lens for best results.

Going Further

The circuits presented in this column are merely representative of methods for triggering a modified disc camera. You may wish to utilize one or more of the circuits in conjunction with other circuits to provide special purpose triggering circuits. For example, a sound triggered camera can be achieved by coupling the output from an audio amplifier into the 555 pulse stretcher shown in Fig. 6-26. Another possibility is a disc camera triggered by a tone modulated light beam.

Perhaps the most interesting method of triggering the camera remotely is by means of radio control. This is the subject of Part 3.

Part 3. Radio Control and Aerial Photography

Aerial Photography was born in 1858 when 262 feet over the Valley of Bievre near Paris Gaspard Felix Tournachon ("Nadar") made the first photograph ever taken from the basket of a gas-filled balloon. Prior to the invention of the airplane, many photographers made photos from both manned and unmanned balloons as well as kites.

The largest aerial photo ever taken was made by George R. Lawrence in 1906 with a piano-size camera flown 2000 feet over San Francisco Bay by a team of seventeen kites. The negative, which gave a spectacular view of San Francisco in ruins after the great earthquake, measured 48 by 18¾ inches.

Over the past fifteen years, I've enjoyed making both still and moving pictures with small cameras flown in model rockets. I've even been able to take photos from helicopters, airplanes, and the baskets of several hot air balloons. Now I've found an updated way to take aerial photos from small balloons and kites using equipment costing under $150.

Obtaining aerial photos using a disc camera modified in accordance with the procedures and precautions outlined in Part 1 of this series is the principal subject that follows. Before getting to the fun part, however, let's find out how to add radio control (RC) to the modified camera.

Radio Control

A modified disc camera can be easily triggered from afar by a suitable RC system. I've tried three different systems, each having relative advantages and disadvantages.

By far the most economical RC system for a modified disc camera is one salvaged from a toy RC car. The output from such receivers usually directly drives one or more small dc motors and can therefore be connected directly to a relay (Radio Shack 275-004 or similar) whose contacts then control the camera. To save space and weight, the relay can be replaced with an LED-phototransistor optoisolator so long as a current limiting series resistor of a few hundred ohms is inserted between the receiver's output and the optoisolator.

After Christmas, you can often purchase for as little as ten dollars or even less returned, defective, or discontinued RC toy cars. Having actually tried a salvaged RC system from a toy car to control an airborne disc camera, I've found the major drawbacks to be limited range (a few hundred feet) and susceptibility to false triggering, particularly in or near metropolitan areas. On the positive side, the economic advantages of this approach cannot be disputed.

For better and more reliable results, you will want to consider more sophisticated RC equipment such as the two systems described next. Both are less susceptible to interference and have considerably more range. They also weigh less.

Ace R/C Special Applications RC System

Ace R/C, Inc., a long-time manufacturer and supplier of RC equipment, makes a single channel transmitter-receiver pair ideally suited for remotely actuating a modified disc camera. The Wee 1 transmitter, which has an output power of nearly half a watt, transmits a 1-kHz tone. Available as a kit (11K16) or factory assembled (11K17) the Wee 1 measures only $5\frac{5}{8} \times 2\frac{3}{4} \times 2\frac{1}{16}$ inches. Available frequencies are 26.995, 27.045, 27.095, 27.145, and 27.195 MHz.

Two versions of Ace's Commander superhet receiver are available for use with the Wee 1. One is designed for a 2.4-volt nicad power supply (12K12) and the other for a 3-volt supply (12K13). Each is available assembled but without an enclosure. The Commander weighs less than an ounce and measures $1\frac{5}{16} \times 1\frac{3}{4} \times \frac{9}{16}$ inches.

Figure 6-27 shows how the Commander receiver can be connected through an optoisolator to a modified disc camera. Any standard LED-phototransistor optoisolator can be used.

The 40-channel CB band extends from 26.965 to 27.405 MHz and encompasses all five FCC allocated RC frequencies at which the Wee 1 is available. The 27.145-MHz RC frequency is close to CB channel 14 (27.125 MHz), a frequency used in many older 100-mW toy transceivers. The 27.195-MHz RC frequency is even closer to CB channel 19 (27.185 MHz), the

popular trucker's channel. Therefore, you may reduce potential interference problems by selecting one of the three other frequencies.

Should you order an Ace Special Applications System, be sure to specify an identical frequency for the transmitter and receiver. The frequency stamped on the crystal of the receiver you receive will be 455-kHz *below* the transmitter's frequency in order to be compatible with the receiver's 455-kHz local oscillator.

You can reduce the interference problems which plague the 27-MHz RC band by using equipment designed for the 72-MHz RC band. The three frequencies in this band not allocated specifically for controlling model aircraft are 72.160, 72.320, and 72.960 MHz.

CAUTION: Certain radio-control frequencies are designated "aircraft only." These frequencies must *never* be used for other purposes since transmissions may interfere with the flight of an RC aircraft. The potential danger posed by such interference is very real and should not be minimized. In 1988 the FCC will restrict the use of the existing 72-MHz RC frequencies in favor of a new band of frequencies at 75 MHz. For the latest information about these changes, write the Academy of Model Aeronautics. You might also want to discuss the new changes with local hobby shop dealers who specialize in radio controlled models.

Ace R/C and many other firms make 72-MHz digital-proportional and pulse-proportional RC systems. Digital-proportional systems are designed specifically to control the position of a servo. In such systems the transmitter broadcasts a train of duration-modulated pulses. The signal is not suited for triggering directly a modified disc camera.

Figure 6-28 shows a straightforward circuit that, in effect, decodes a pulse duration modulated signal and provides an output suitable for triggering a modified disc camera through an LED-phototransistor optoisolator. In operation, Q2 and the 555 function as a missing pulse detector. Q1 buffers and inverts the signal

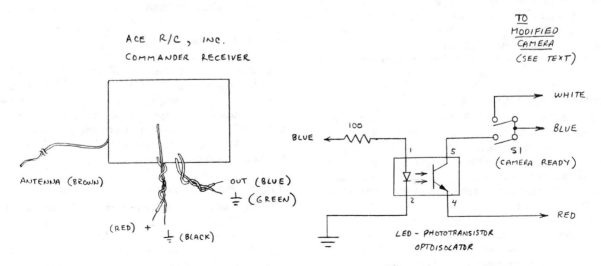

Fig. 6-27. Interfacing an RC tone receiver to a modified disc camera.

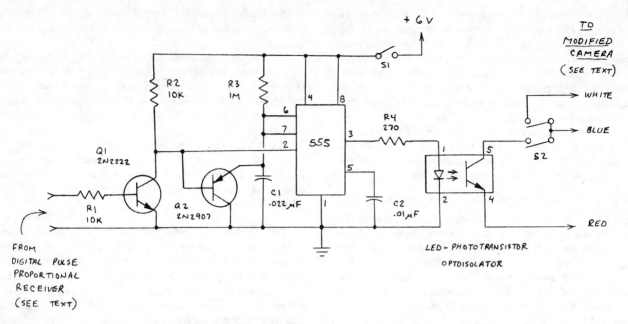

Fig. 6-28. Pulse proportional RC receiver decoder circuit.

from the RC receiver. When pulses do arrive from the receiver, the 555 turns on the LED in the optoisolator and the camera is triggered.

I've used the circuit in Fig. 6-28 to enable an Aero Sport Two Two-channel RC system to remotely trigger a disc camera. The Aero Sport transmitter includes a two-axis joystick with trimmers, but all that's necessary to trigger the camera is to switch the transmitter on. For this reason, I've modified an Aero Sport transmitter by disconnecting the red battery connector lead from the unit's on-off slide switch and connecting a normally open push button between the lead and the switch. The transmitter is readied for use by closing the slide switch. The push button is pressed to actuate the camera.

The Aero Sport Two system operates in 72-MHz RC band (see the caution statement given previously). The receiver, which is enclosed in a plastic housing, weighs about two ounces and measures $1^{9}/_{16} \times 1^{23}/_{32} \times {}^{3}/_{4}$ inches. A complete system is available from Hobby Shack. The system includes the transmitter, receiver, two servos, and a plastic battery box for four AA cells.

Selecting an RC System

If your budget is severely limited, you will want to give strong consideration to salvaging your RC equipment from a toy RC car. If you can afford better equipment, the Ace R/C Special Applications Systems is inexpensive, simple to use and compact in size. If you want to use your RC system in other projects (e.g., robotics), a digital proportional system like the Aero Sport Two is the best choice.

You can find considerable information about various RC systems by referring to books on the subject and model airplane magazines. Some hobby dealers, incidentally, sell for bargain prices used but perfectly functioning RC equipment.

Applications for a Radio Controlled Camera

Photographers have long used radio controlled cameras to photograph wildlife, hazardous events, and otherwise inaccessible locations. All these and many other applications can be accomplished with a radio controlled disc camera.

One of the more frustrating aspects of RC photography is not knowing if the camera is responding to your signals when it's too far away for you to hear the sound of an exposure being made. One solution is to add a light or tone generator that responds each time an exposure is made.

When the lighting is low, the disc camera winks back with a brilliant flash each time an exposure is made, a particularly reassuring sight when the camera is tied to a kite flying high above the ground. This brings us to the most interesting application for a radio controlled disc camera, aerial photography.

Aerial Photography

I have spent much of my spare time flying radio controlled disc cameras from various kinds of kites and helium-filled balloons and trash bags. My original goal was to develop a low-cost method of obtaining aerial photographs of my house and garden. However, the experience of making aerial photos in this fashion is so interesting and entertaining that I've flown my camera from many different sites and assembled an album of hundreds of photos.

The Aerial Package

To fly your radio controlled disc camera from a kite or balloon you must first install the apparatus within a suitable package. Plastic refrigerator boxes such as the Superseal line made by Eagle Affiliates make ideal enclosures for airborne camera packages. The plastic is resilient and does not shatter or break.

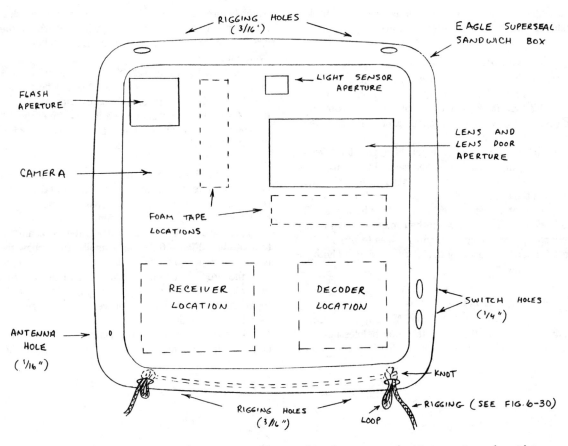

Fig. 6-29. Bottom view of camera package showing approximate aperture locations.

Openings for lenses and switches can be easily formed with a sharp knife and a drill.

Figure 6-29 is a bottom view of the second of two airborne disc camera packages I've assembled and flown from kites and balloons. The system employs an Aero Sport receiver connected to the decoder circuit shown in Fig. 6-28. It's installed in a Superseal No. 3427 sandwich box.

Note the use of foam insulation tape to provide vibration protection for the camera. A single 2–56 screw and nut secures the decoder board to the box. The camera and receiver are held in place by strips of foam plastic inserted between them. It is essential that the top of the plastic enclosure be held securely in place to prevent the camera and receiver from being dislodged should the package strike the ground. I use a heavy rubber band for this purpose.

I powered the receiver and decoder in the original version of this package with a miniature 6-volt lithium battery (Duracell PX28L). Later, to save weight, I removed the battery and its holder and connected the receiver and decoder to the 6-volt lithium battery in the camera. Since the combined current consumption of the receiver and decoder is only 13 mA and the camera's batteries are rated at 1200 mAh, the additional load is not significant so long as the power is switched off between flights.

Together with four nylon suspension lines and some snap

swivels, the package in Fig. 6-29 weighs 12 ounces. The first system I assembled, which used an RC system salvaged from a toy car, required a 9-volt battery and weighed a full pound.

Ace R/C's Special Applications receiver weighs less than the Aero Sport receiver and can trigger a disc camera *without* a decoder circuit. It requires a 3-volt (or 2.4-volt) supply, however, and cannot be directly powered by the camera's 6-volt supply.

Borrowing Power from the Camera

There are two ways to connect leads to the 6-volt lithium battery in the disc camera. One is to carefully solder wrapping wire to the exposed portions of the wires to which the two series connected lithium cells are terminated. Solder the leads to the bare wire in the small gap between the insulated portion and the terminal.

A better method, the one I've used, is to solder wrapping wire to the termination points of the battery leads on the back side of the circuit board. The points can be located visually and confirmed with a voltmeter.

CAUTION: Should you choose to power your system by borrowing current from the camera, it is *absolutely essential* that you follow *all* the procedures and safety precautions given in Part 1. You must also abide by the precautions printed on the lithium batteries: "CAUTION! May explode, leak, and/or

159

flame if crushed, cut, soldered, short-circuited, connected backwards, recharged, heated, or disposed of in fire."

Rigging the Airborne Package

After experimenting with various rigging arrangements, I've settled upon the straightforward approach shown in Fig. 6-30. Variations of this four line rigging can be attached to a kite line to provide vertical and two kinds of oblique photos. It's also well suited for use with balloons.

Referring to Fig. 6-30, form each pair of rigging lines from a single high-strength, braided-nylon line several feet long. Tie two knots, each with an extension loop, about 5 inches apart equidistant from the ends of each line. Pass the ends of the lines through the inside of adjacent holes in the box. The knots will secure the lines in place. Later, when the camera is flown from a kite, the loops can be pulled through the holes to provide tie down points for additional rigging that permits oblique photography toward the kite flier.

Flying the Camera from a Kite

Words alone cannot express the thrill of flying a camera from a kite hundreds of feet in the air or the total helplessness of watching it suddenly dive to within inches of hard rocks or deep

water before zooming back to its former altitude. Best of all is the ability, when the wind is right, to maneuver the camera directly over a sailboat mast, palm trees, tall signs, and even flying birds!

You'll need a sturdy, reliable kite to accomplish these aerial feats. Many different homemade kites can be fashioned from readily available materials, and you can find excellent books on their design and construction at a library. An early (1929) book reissued by Dover Publications is Leslie L. Hunt's *25 Kites That Fly*. Mr. Hunt was formerly a kite maker for the U.S. Weather Bureau.

I have used with excellent results a 3 × 6 feet nylon Spinnaker delta kite manufactured by Spectra Star Kites. The kite is available in six color patterns. When properly rigged, these kites will lift a full pound in a 15-mph wind.

There are several ways to attach the disc camera package to a kite line. Figure 6-31, for example, shows how the camera package is rigged to provide oblique photos in the direction *away* from the kite flier. This simple rigging arrangement resists camera swing and is easy to fly. The rigging holes must be near the top of the camera package to prevent the package from flipping over high in the air.

The attachment in Fig. 6-31 is ideal for aerial views of the horizon. Many of the photos I've made resemble those taken

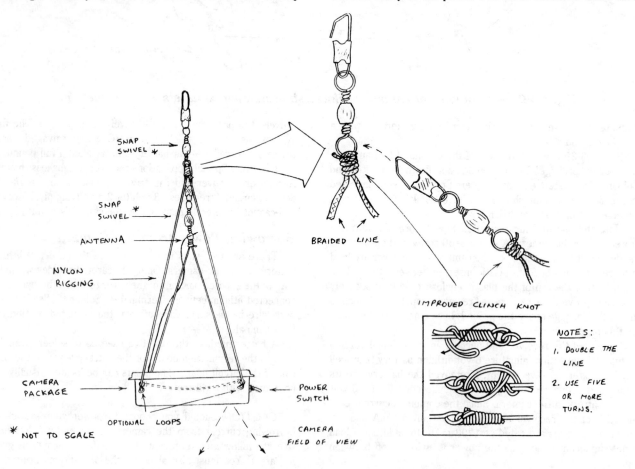

Fig. 6-30. How to rig the airborne camera package.

from the window of an aircraft. The major drawback is the guess-work required to "aim" the camera.

Figure 6-32 shows how the camera is rigged to provide oblique photos in the direction *toward* the kite flier. The camera package's four nylon lines are connected together and to the kite's leader. A halter made from four clear fishing leaders having a test strength of at least twenty pounds each is attached via snap swivels to the loops on the camera package rigging (see Fig. 6-30). All four leaders are terminated at a single size 6 or larger snap swivel which is then connected to the main kite line.

To prevent the camera package from rotating or providing tilted photos, it may be necessary to use a counterweight. I slip a ¼ × 36-inch wood dowel under the heavy duty rubber band that secures the camera package's cover. A length of brightly colored plastic streamer slipped through a small eyescrew in the end of the dowel provides a convenient visual reference of the camera's orientation when it is high in the air.

I've used the rigging in Fig. 6-32 to photograph my family feeding seagulls from the vantage point of the gulls. I've also used it to photograph the stern of a tour boat from the stern of the same boat and my car from an altitude of 545 feet. Surprisingly the barely visible halter through which the photos are taken does not detract from their appearance.

Unfortunately, it was not possible to include in this book examples of the spectacular color photographs made by my airborne disc camera system. If your library carries back issues of *Computers & Electronics*, you can see four such photos by referring to the January 1983 issue (p. 33).

Figure 6-33 shows how the camera is rigged for vertical photographs. The stabilizer boom prevents camera rotation. It does not, however, eliminate camera swing. To avoid blur, trigger the camera only when it is not swinging.

Kite Flying Tips

I've flown my RC cameras from kites as high as 550 feet over both land and water and have obtained hundreds of aerial photos. Here are some important kite flying tips I've learned:

1. Flying a camera kite can be both tricky and busy at times. It helps to have a helper to hold the RC transmitter and to stay with the equipment while you rescue a downed kite or walk one down.

2. Unless you design a kite to which the camera package is directly attached, insert a twenty feet leader between the camera and the kite. This will allow the kite to gain altitude and stability before it's required to lift any weight.

3. Always use quality nylon line. Braided line is best. I always use at least 50 pound test line with a Spectra Star delta.

4. Flying a payload from a kite requires a stiff breeze. I've found that the camera package will swing wildly and likely strike the ground if the wind is under 12-15 mph.

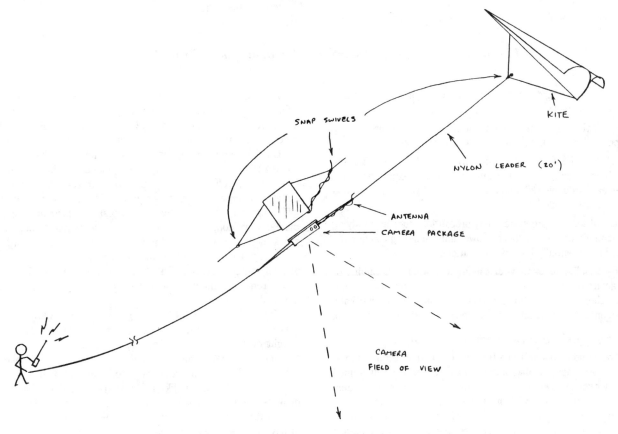

Fig. 6-31. Rigging the camera for oblique-away photos from the kite.

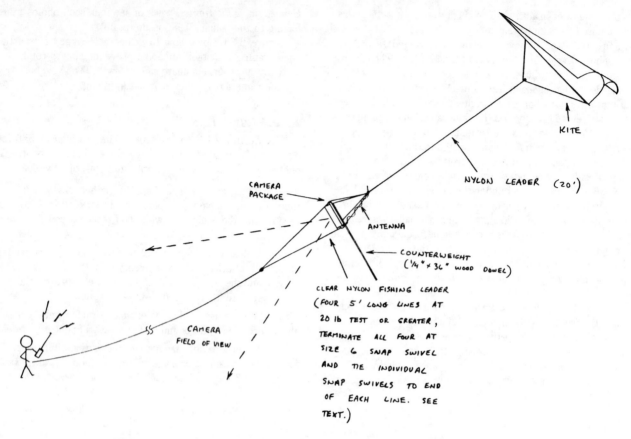

Fig. 6-32. Rigging the camera for oblique-in photos from the kite.

5.   Avoid flying downwind from hills, buildings, bridges, signs, and other obstacles that cause turbulence.

6.   If the wind is strong, be sure to protect your hands from the possibility of painful burns by wearing gloves.

7.   Practice flying with a dummy payload before flying the real thing. Flying a working kite is a unique experience, and you'll learn valuable skills that may later save your camera package from a possible crash.

8.   Observe appropriate safety precautions such as not flying your camera kite near airports, power lines, buildings, highways, and other potentially hazardous locations.

9.   Finally, never fly your camera kite when it is raining or with a wire tether or a two conductor remote switch attachment. Researchers have measured potentials as high as 50,000 volts on the lines of wire-tethered kites flown on perfectly clear days!

Flying the Camera from a Balloon

I've found a balloon to be the best lifting device for pinpoint aerial photography. Unlike a kite, a balloon can be easily guided directly over an area of interest without guesswork or the help of an assistant to tell you where the camera is pointing.

Unfortunately, balloon flying is riskier and requires almost perfectly calm air. Furthermore, preparing for a flight entails considerably more expense and time than does a kite launch.

First, a source of helium must be found. Welding and party shops sell the gas and, depending upon the volume of the cylinder, you can expect to pay $.23 or more per cubic foot plus a deposit for the cylinder and regulator.

Good balloons are more difficult to find than helium. I've spent $18 each for heavy duty 4-foot diameter rubber balloons. One worked well for a week until I tried to refill it after it was deflated. It promptly burst in my face. The second blew its top half 20 feet in the air before it was half filled. Small 2–3-foot diameter balloons cost a few dollars or so at party shops that stock them, but more than one will be required to lift a camera.

To fly a camera balloon in a breeze requires an aerodynamically shaped balloon. The minimum cost for such balloons, which require a hundred or more cubic feet of gas, is several hundreds of dollars.

Helium lifts about 1.1 ounces per cubic foot. Allowing a 15 percent excess lift margin and assuming a *total* airborne package weight of 16 ounces (camera system, balloon, parachute, and tether), you'll need about 17 cubic feet of gas. This implies a spherical balloon having a diameter of about 3.2 feet.

Although this seems like an ideal arrangement for a 3-foot diameter balloon, I've learned the hard way *never* to fill a balloon to its rated capacity. You can avoid the unpleasant experience of having an expensive balloon full of equally expensive helium burst in your face by filling your balloons to only half to two-

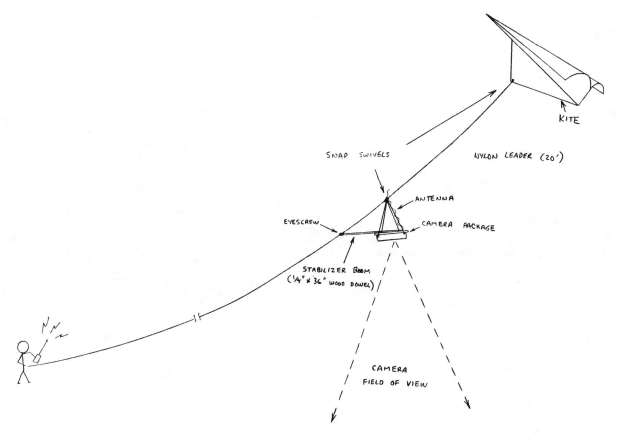

Fig. 6-33. Rigging the camera for vertical photos from the kite.

thirds their rated capacity. This implies using a 4- or 5-foot diameter balloon or two or more smaller balloons.

The advantage of using two or more balloons is that, should one burst, the camera package will descend to earth relatively slowly rather than crashing outright. I discovered this firsthand when one of two balloons popped a hundred feet over my house.

The advantage of a single, large balloon is simplicity. In case of catastrophe, however, you will need to include a lightweight parachute between the balloon and camera to protect the camera and any people below it. I use a 3-foot diameter flare chute similar to those sold by Edmund Scientific. Of course you can make your own parachutes. Figure 6-34 shows the arrangement.

You can even make your own balloons! An ordinary iron set to low heat will nicely weld polyethylene film. Enclosed cylinders made from 1- or 2-mil polyethylene can be easily formed.

Balloon Flying Tips

My balloon flying experiences have brought home some important lessons about flying a camera from a tethered balloon. Here are some key pointers:

1. Fly only when the air is *perfectly* calm. Otherwise, the balloon will drift with the breeze and begin to descend and vibrate when it reaches the end of its tether.

2. Check the atmospheric conditions with a small pilot balloon before inflating and launching the camera rig. Keep the small balloon tied nearby to a 20 foot tether. If the breeze blows it down, do not fly your camera rig.

3. Use only helium to inflate your balloons. Hydrogen and other lighter-than-air gases are highly flammable.

4. Tie the tether between the parachute and the camera package. Be sure to tie the free end of the tether to a heavy object. If you need to run toward the balloon, you can drop the tether without fear of losing the balloon.

5. Avoid jerking the tether when the balloon is flying low over trees. You may pull the tether and the balloon down into the branches.

6. If you use multiple balloons, rig them so the suspension lines do not rub against any balloon. Otherwise, as I learned the hard way, the balloon may burst.

7. Never use a metallic tether or a two conductor remote switch arrangement.

8. Avoid flying a balloon near airports, power lines, and high buildings.

Interpreting Aerial Photos

Dividing the altitude of a disc camera by its focal length (12.5 mm or 0.041 feet) provides a scale factor for calculating the size of objects in an aerial photo. Assume the camera makes from 200 feet, a photo showing a rock wall measuring 0.017

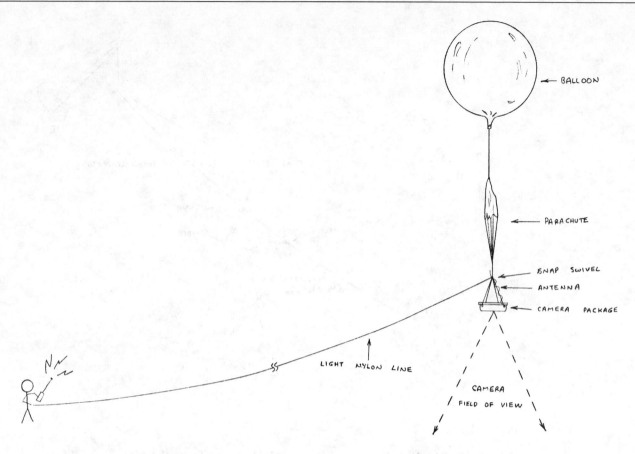

BALLOON

PARACHUTE

SNAP SWIVEL

ANTENNA

CAMERA PACKAGE

LIGHT NYLON LINE

CAMERA FIELD OF VIEW

Fig. 6-34. Rigging the camera for vertical photos from a tethered balloon.

feet (0.2 inch) on the disc negative (*not* the print). The scale is 200/0.041 or 4878. The length of the wall is then 0.017 × 4878, or 82.9 feet.

It's difficult to measure objects on tiny disc negatives, and often the altitude from which the photo was taken is unknown. But you can still make measurements by using as a scale reference any object of known size in a photo. You can even plan to include such objects in areas you plan to photograph from above.

Conclusion

Space precludes an account of my many adventures with airborne disc cameras. For example, there isn't room to relate how I managed to retrieve a kite and camera package undamaged after an unplanned landing in a pasture full of cattle!

While I have very much enjoyed my experiences with modified disc cameras, in recent years I have also enjoyed adding external control devices to low-cost 35-mm cameras. For example, the evening before this was typed I flew two radio-controlled 35-mm cameras from a 12-feet long helium inflated blimp 500 feet above my office. The aerial photographs provided by this system are truly spectacular. By the time this revised edition of this book is published, this work with modified 35-mm cameras will have been published in several of the magazines for which I write.

To find out more about early kite and balloon aerial photography, see *Airborne Camera*, by Beaumont Newhall (Hastings House, 1969). In the meantime, may the wind be in your favor and may your camera make soft landings.

SEVEN

Sensors and Sensing Systems

- Fiber Optic Sensors
- Homemade Pressure-Sensitive Resistors
- More About Pressure-Sensitive Resistors
- Adjustable Threshold Temperature and Light Alarms
- Detecting Sound
- Audio Amplifier Experiments
- Measuring the Flow of Air
- Nuclear Radiation
- Experimenting with an Ultrasonic Rangefinder
- A Cassette Recorder Analog Data Logger
- Electronic Aids for the Handicapped

Sensors and Sensing Systems

Fiber Optic Sensors

Most references to fiber optics relate to lightwave communications. Optical fibers, however, have *many* applications as sensors that are completely unrelated to communications. Moreover, unlike conventional electronic sensors, fiber optic sensors can be made immune to electromagnetic interference and electrical shock hazards.

In this discussion I'll describe the most important kinds of optical fiber sensors. Then I'll explain how to make several kinds of sensing systems based upon fiber optics.

Optical Fibers

Figure 7-1 shows how a light ray travels through the simplest kind of optical fiber. This fiber consists of a central *core* surrounded by a *cladding*. So long as the index of refraction of the core is *higher* than that of the cladding, the light ray will always be reflected away from the core-cladding interface and back into the fiber. Therefore the ray can travel through a fiber which is bent or coiled. Since there is a sharp difference in index of refraction between the core and cladding, the fiber in Fig. 7-1 is called a *step-index* fiber.

Over long distances, step-index fibers can introduce signal distortions that limit their usefulness in lightwave communications. This occurs when light rays take different paths through the fiber, thereby emerging from the opposite end of the fiber at different times.

Fibers intended specifically for communications often lack an abrupt core-cladding boundary. Instead, the core material gradually merges into the cladding material, thereby providing a gradual change in the index of refraction. This causes light to travel through the fiber as shown in Fig. 7-2.

A fiber like the one in Fig. 7-2 is called a *graded index* fiber. Since light near the cladding travels faster than light in the core, several rays injected simultaneously into one end of the fiber will arrive at the opposite end at the same time even if they follow different paths. Consequently, light travels through a graded-index fiber with considerably less distortion than light that travels through a step-index fiber.

Both stepped-index and graded-index fibers can be used to make fiber sensors. Interestingly, some of the very same factors that limit the usefulness of these fibers in communications make them ideally suited as sensors.

Fiber Optic Sensors

There are surprisingly many ways to make sensors that employ optical fibers. Most such sensors can be placed into one of two broad categories. The first category, which is by far the largest, includes all applications wherein optical fibers simply carry light to and from detectors and emitters. Since in these applications the fiber plays only a passive role, sensors in this category can be called *indirect-mode fiber optic sensors*.

The second category includes those applications in which the light passing through a fiber is in some way altered by mechanical stress or some other force or influence upon the fiber. Since the fiber plays an active role in the sensing process, sensors in this category can be called *direct-mode fiber optic sensors*.

If you've never before worked with fiber optics, these explanations probably seem more complicated than they really are. Actually, many kinds of fiber optic sensors, particularly those in the indirect-mode category, are very easy to make.

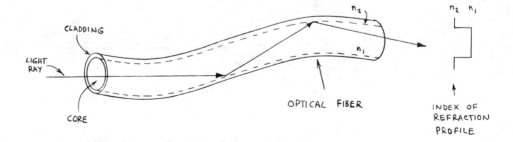

Fig. 7-1. How a light ray travels through a step-index fiber.

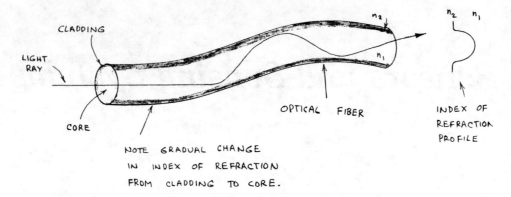

Fig. 7-2. How light travels through a graded-index fiber.

Indirect-Mode Fiber Optic Sensors

Figure 7-3 shows three basic kinds of indirect-mode fiber optic sensors. All of these sensors are easily assembled and have many applications. They can each use low-cost plastic or glass fiber. Suitable light sources include light-emitting diodes and small incandescent lamps. If the light source is operated at a constant dc bias (i.e., unmodulated), the detector can be a low-cost cadmium sulfide photoresistor, solar cell, phototransistor, or photodiode. If the light source is modulated, the response time of the cadmium sulfide detector will be too slow to permit its effective use.

The in-line sensor in Fig. 7-3A is by far the simplest. In operation, a light source couples light into a fiber separated from a second fiber by a narrow gap. Normally, light crosses the gap and arrives at the detector. If, however, the end of the first fiber is moved, the light level at the detector will be reduced proportionally. If the end of the first fiber is caused to vibrate, the light arriving at the detector will be modulated at the frequency of the vibration.

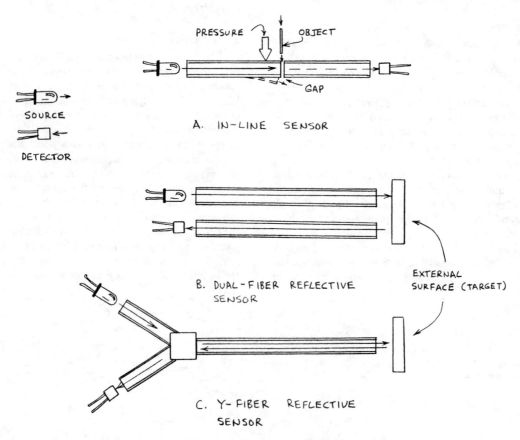

Fig. 7-3. Indirect mode fiber optic sensors.

In-line fiber optic sensors can be used as vibration sensors, accelerometers, microphones, and pressure sensors. They can also detect the presence of very small objects passing through the gap between the two fibers.

In some cases, only one fiber is required to make an in-line sensor. For example, the second fiber may be omitted if the free end of the first fiber is placed close to a light source or detector having a small emitting or detecting region.

The dual-fiber reflective sensor in Fig. 7-3B can be used to sense the presence or absence of objects. It can also detect the presence or absence of markings on paper or other material.

The dual-fiber sensor is particularly useful when there's a possibility of electrical shock. It's also very useful when the object to be detected is submerged in a liquid, perhaps one that's very hot and corrosive. Conventional electronic detectors require careful shielding, insulation, and encapsulation before they can be used in harsh environments like these.

The Y-fiber reflective sensor in Fig. 7-3C is a modification of the sensor in Fig. 7-3B. Here the two fibers have been spliced into a single fiber that carries the light to and from the target or object being detected.

All the sensor systems in Fig. 7-3 can be made using bundles of fibers instead of individual fibers. Bundles can carry more light, but their detection resolution is not nearly as good as that of a single fiber. When a bundle is divided into the Y configuration shown in Fig. 7-3C, the result is called a *bifurcated* optical fiber cable.

Direct-Mode Fiber Optic Sensors

Strange things can happen when an optical fiber is bent. If the bend has a large radius of curvature, the light will travel through the fiber relatively unaffected. But if very tiny bends, so-called *microbends*, are present, the light passing through a fiber may be attenuated. This may occur when some of the light passing through the core is coupled out of the core and into the cladding. Other phenomena may also take place.

Figure 7-4A shows a very simple arrangement for detecting pressure using an optical fiber. This is the principle behind the fiber optic guitar designed several years ago by Dynamic Sys-

tems, Inc. of McLean, VA. Instead of the usual strings, six optical fibers are stretched over the bridge of this guitar. Light from an LED is passed through the fibers and detected by a photodetector. When the fibers are plucked or strummed, the intensity of the light passing through them is modulated. The resulting signal variations are amplified and transformed into sound by a conventional amplifier and speaker.

I have no idea how well the sound from a fiber optic guitar compares with that from a conventional instrument. At least the fiber optic instrument has the potential of being shock-proof, a fairly important advantage over standard electric guitars.

A more advanced pressure sensing application has been studied by the United States Navy. The fiber sensor consists of a coil of fiber immersed in water. Sound waves passing through the water distort the fiber and the effect upon light passing through the fiber can be detected by a photodetector.

Figure 7-4B is an interesting sensor that uses an uncladded fiber to detect the presence of a liquid. Normally, the fiber transmits light directly to a detector. When the uncladded fiber is surrounded by liquid, however, some of the light is coupled out of the fiber and into the liquid. This occurs because the index of refraction of a liquid is higher than that of air. A detector senses the reduction in light level that indicates the presence of the liquid.

Other direct-mode fiber optic sensors are also in use. Among the most sophisticated is the fiber optic gyroscope, a sophisticated sensor that employs a fiber coil as part of an interometer that senses rotation. The fiber gyroscope may some day provide a compact, all-solid-state substitute for traditional mechanical gyros.

Do-It-Yourself Fiber Optic Sensors

You can make many kinds of inexpensive fiber optic sensors from readily available components like LEDs, phototransistors, photodiodes, and cadmium sulfide photoresistors. Black heat shrinkable tubing is handy for securing fibers in place and blocking ambient light. You can find these components and supplies at most electronic parts suppliers.

Plastic optical fibers are available from some hobby and craft shops. Though these fibers aren't as transparent as glass or silica fibers, they are very inexpensive and can be cut with a hobby knife.

Glass and silica fibers are available in large quantities from more than a dozen manufacturers, but small quantities may be harder to find. Some mail order and hobby electronics dealers stock plastic, glass, and silica fibers.

Dolan-Jenner Industries, Inc. makes various kinds of fiber optic sensors and a wide variety of glass and quartz (silica) multifiber (bundle) light guides suitable for use with do-it-yourself sensors. The company has a minimum order of $50, a requirement which shouldn't be hard to meet in light of the prices charged for its multifiber light guides. I've used a well-made Dolan-Jenner bifurcated light guide in various sensing roles for many years.

For some applications it's desirable to use one or two single fibers rather than a bundle. Keep in mind, however, that single fibers can be more difficult to work with than larger diameter

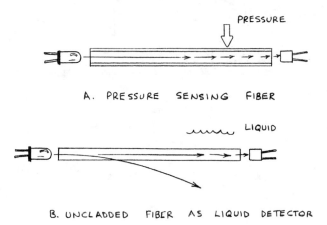

A. PRESSURE SENSING FIBER

B. UNCLADDED FIBER AS LIQUID DETECTOR

Fig. 7-4. Direct fiber optic sensors.

bundles. For example, silica fibers may have a diameter not much greater than that of a human hair.

Glass and silica fibers transmit the near infrared radiation emitted by GaAs:Si and AlGaAs LEDs far better than do plastic fibers. However, they are more difficult to prepare for use since their exposed ends must be perfectly flat for maximum light transmission.

One way to obtain flat ends is to cleave the fiber by lightly scoring it with a carbide tool while stretching it around a cylinder such as a plastic film container. If the tension is just right, the fiber will separate at the cleavage point and leave both exposed ends perfectly flat.

Learning to cleave glass and silica fibers requires experience. For more information, see *The Forrest Mims Circuit Scrapbook* (McGraw-Hill, 1983, pp. 38–40).

CAUTION: Small bits of glass and silica fibers are like invisible, very sharp splinters. *Always* protect your eyes with clear goggles when cleaving such fibers. Pick up small bits of discarded fiber with a piece of masking tape to be sure no unpleasant surprises befall vulnerable elbows or bare feet.

A Fiber Optic Vibration Sensor

Figure 7-5 shows an ultrasimple fiber optic vibration sensor. The key part of the sensor is a short length of plastic fiber cemented into a small hole bored into the top of an epoxy encapsulated red LED.

Figure 7-6 shows one of many possible transmitter-receiver circuits that can be used with the sensor in Fig. 7-5. In operation,

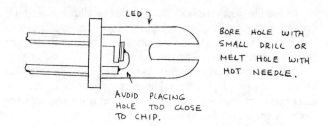

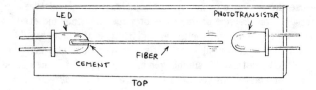

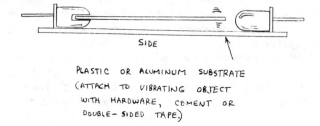

Fig. 7-5. Simple fiber optic vibration sensor.

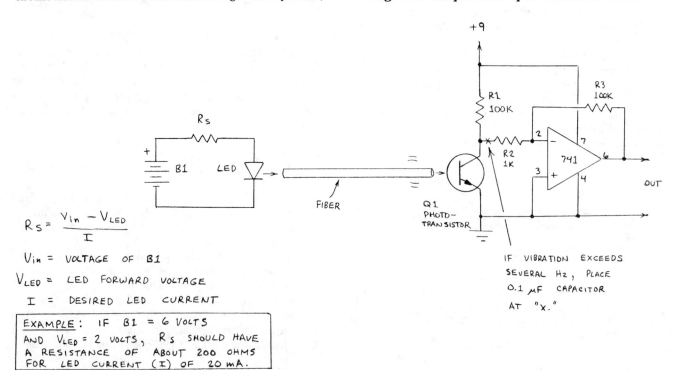

$$R_s = \frac{V_{in} - V_{LED}}{I}$$

V_{in} = VOLTAGE OF B1

V_{LED} = LED FORWARD VOLTAGE

I = DESIRED LED CURRENT

EXAMPLE: IF B1 = 6 VOLTS AND V_{LED} = 2 VOLTS, R_s SHOULD HAVE A RESISTANCE OF ABOUT 200 OHMS FOR LED CURRENT (I) OF 20 mA.

Fig. 7-6. Transmitter and receiver circuitry for the fiber optic vibration sensor.

light emerging from the fiber strikes the sensitive surface of phototransistor Q1. When the fiber is displaced, less light reaches Q1. Therefore, vibration of the fiber causes the signal from Q1 to be modulated at the frequency of modulation. The signal can be viewed on an oscilloscope or made audible by connecting an external speaker amplifier to the output.

Many modifications to the sensor in Fig. 7-5 are possible. For example, adding a fixed fiber to the phototransistor might provide much more vibration sensitivity since the ends of the two fibers could be adjusted for any desired tolerance. This method might also increase the light level at the phototransistor, thereby providing a stronger signal.

The side of the fixed fiber can be cemented to a small nut. Very fine adjustments of the relative positions of the ends of the two fibers could then be made by rotating a screw inserted in the nut.

A Multipurpose Fiber Optic Sensor

Dual-fiber reflective sensors like the one shown in Fig. 7-3B have many applications. They can detect objects as well as measure their reflectance and indicate if the object is moving, vibrating, or stationary. They can also detect markings on paper or other surfaces.

Figure 7-7 shows one of many ways to make a reflective fiber optic sensor using a bifurcated light guide. I have used this set-up along with the circuits in Fig. 7-6 to measure the reflectance of various surfaces in order to predict the detection range of miniature infrared travel aids for use by the blind.

Though the LED circuit in Fig. 7-6 is operated at a dc bias, it can also be operated in a pulsed mode to reduce interference from external light sources. This requires that the receiver's detector be capacitively coupled to its amplifier to eliminate the passage of dc signals.

Going Further

The field of sensing with fiber optics has grown rapidly over the past few years. Many technical papers on the subject have been published and many patents have issued.

If you have access to a good library, you may want to find the December 1981 issue of *IEEE Spectrum*. It contains an excellent article on fiber optic transducers by several Sperry Research Center engineers.

Some of the more interesting patents include Fiber Optic Transducers (US 4,408,829), Fiber Optic Musical Instruments (US 4,442,750), Fiber Optic Accelerometer (US 4,419,895), Fiber Optic Position Sensor (US 4,403,152), and Optical Fiber Tactile Sensor (US 4,405,197). Copies of U.S. patents are available for a nominal fee from the U.S. Department of Commerce, Patent and Trademark Office, Washington, DC 20231.

Manufacturers of fiber optic communications devices have published many fine booklets you may find helpful. You can find names of companies and their addresses in trade publications like *Laser Focus* and *Lasers and Applications*. Both these magazines publish annual trade directories loaded with useful information.

Several books for the experimenter interested in fiber optics are also available. Waldo Boyd, for example, has written for the Blacksburg Group *Fiber Optics* (Howard W. Sams & Co., 1982), an excellent book of fiber optic communications experiments. *A Practical Introduction to Lightwave Communications* (Howard W. Sams & Co., 1982), a book I developed for an IEEE course by the same name, has circuits and information about fiber optic communications that can be applied to fiber optic sensors. Finally, *The Forrest Mims Circuit Scrapbook* (McGraw-Hill, 1983) includes practical tips about working with fiber optics and building suitable transmitting and detecting circuits.

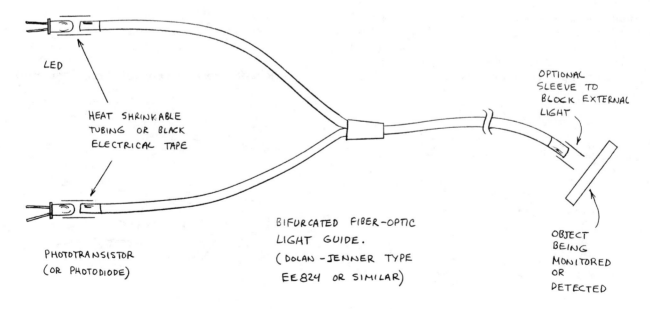

LED

HEAT SHRINKABLE TUBING OR BLACK ELECTRICAL TAPE

PHOTOTRANSISTOR (OR PHOTODIODE)

BIFURCATED FIBER-OPTIC LIGHT GUIDE. (DOLAN - JENNER TYPE EE 824 OR SIMILAR)

OPTIONAL SLEEVE TO BLOCK EXTERNAL LIGHT

OBJECT BEING MONITORED OR DETECTED

Fig. 7-7. Multipurpose fiber-optic sensing system.

Homemade Pressure-Sensitive Resistors

The conductive foam plastic in which MOS transistors and integrated circuits are often inserted to provide antistatic protection can be used to make pressure-sensitive resistors. The resistance of these do-it-yourself resistors can range from several tens of kilohms when no pressure is applied to a few hundred ohms or less at maximum pressure.

Figure 7-8 shows just one of many possible ways to assemble a conductive foam pressure-sensitive resistor. The basic resistor is simply a sandwich made by placing copper foil conductors on either end of a conductive foam cylinder or block. If you prefer, you can add embellishments (such as a plunger and return spring) to enhance the utility of the basic pressure-sensitive resistor.

The resistor illustrated in Fig. 7-8 can have a diameter ranging from that of a pencil eraser to a silver dollar. Copper foil for making the end contacts is available from hobby and craft shops. The foil is easily cut with ordinary scissors. If you cannot find copper foil in your area, an acceptable substitute is unetched copper cladded circuit board. In both cases, the copper should be buffed with a pencil eraser to prepare it for soldering. When the surface is shiny bright (*both* sides if you use foil), solder a length of wrapping or small diameter hookup wire to each end terminal.

Conductive plastic foam is available from many sources. If you don't happen to have any, try requesting a small piece from an electronics supplier or a firm or university that purchases integrated circuits in volume. Conductive foam can be cut with scissors or a hobby knife.

You can make a miniature pressure-sensitive resistor by using a ¼-inch mechanical paper punch to cut identical diameter circles of foil and a cylinder of conductive foam. After soldering leads to the foil disks, insert a copper-foam-copper-sandwich in a short section of miniature plastic tube like those in which points for lettering pens are sold. Two tiny apertures should be drilled in the side of the tube to provide exit ports for the leads. If you prefer a larger pressure-sensitive resistor, use a sawed off sec-

tion of a plastic pill bottle and proportionally larger sections of copper and plastic.

Applications for Pressure-Sensitive Resistors

Many applications exist for pressure-sensitive resistors. One possibility is a pressure-sensitive control that functions as a single axis joystick. Another is a programmable sensor for a weight sensitive scale. Still another is a simple accelerometer. In this role, a small weight such as a steel nut or lead fishing sinker attached to the upper, moving contact of the pressure-sensitive resistor would provide the necessary mass.

I've devised two simple circuits that illustrate how to use pressure-sensitive resistors in these and other applications. In Fig. 7-9, the pressure-sensitive resistor is connected as the var-

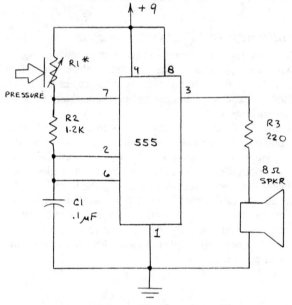

*RI: CONDUCTIVE FOAM PLASTIC PRESSURE SENSITIVE RESISTOR (SEE TEXT).

Fig. 7-9. Pressure-controlled tone generator.

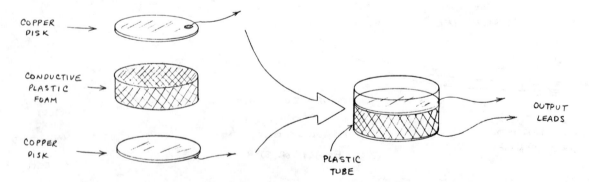

Fig. 7-8. Pressure sensitive resistor construction.

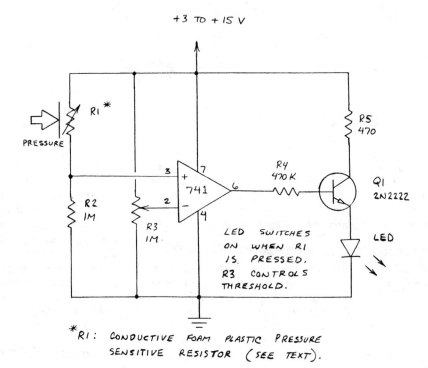

+3 TO +15 V

Fig. 7-10. Pressure-controlled comparator.

R1 *
PRESSURE

R2 1M

R3 1M

741

R4 470K

R5 470

Q1 2N2222

LED

LED SWITCHES ON WHEN R1 IS PRESSED. R3 CONTROLS THRESHOLD.

*R1: CONDUCTIVE FOAM PLASTIC PRESSURE SENSITIVE RESISTOR (SEE TEXT).

iable time constant component in a 555 astable oscillator audio tone generator. As the pressure on the resistor is increased, its resistance is decreased. This increases the circuit's frequency of oscillation. Although this circuit was devised merely to illustrate the use of a pressure-sensitive resistor in a straightforward analog or linear mode, it suggests possible applications in electronic music.

Figure 7-10 shows how a comparator can be connected to a pressure-sensitive resistor to provide a programmable two-state output. In operation, the switching threshold of the comparator is set by threshold adjust potentiometer R3. Pressure applied to R1 lowers its resistance, thus increasing the voltage applied to the comparator's noninverting input. When this voltage exceeds the reference voltage determined by R3, the comparator output swings to near the positive supply voltage. This turns on Q1 and illuminates the LED.

The circuit in Fig. 7-10 has practical applications as an input stage to a pressure sensing logic circuit or microcomputer. R3 permits the circuit to be adjusted over a range of sensitivities.

Going Further

Conductive foam pressure-sensitive resistors are not as sophisticated as commercial pressure sensing devices, but they are remarkably cheap and very easy to make. If you would like more information on the subject, Thomas Henry of Transonic Laboratories has published a brief article entitled "Conductive Foam Forms Reliable Pressure Sensor" in *Electronics* magazine (May 19, 1982, p. 161).

More About Pressure-Sensitive Resistors

There are many applications for pressure-sensitive resistors. For example, a pressure-sensitive resistor can serve as the transducer for an electronic scale or an accelerometer. When connected to an appropriate circuit, a pressure-sensitive resistor can provide a warning when an object which has been placed upon it is moved. Pressure-sensitive resistors can also be used in various kinds of keyboards and computer graphic input devices.

After the preceding discussion on pressure-sensitive resistors was published in *Computers & Electronics* magazine, that magazine published a letter from Scott Ellner who suggested yet another application for these versatile devices. Mr. Ellner wrote that he worked for an institution for severely mentally retarded and physically handicapped people, many of whom spend their days in wheelchairs. To protect these patients from receiving bed sores, Mr. Ellner wrote, specially designed cushions are necessary. Therefore, he was working on a special cushion fitted with an array of 260 pressure-sensitive resistors. Mr. Ellner's objective is to obtain a visual representation (LED array or computer screen) of the weight distribution of various patients seated upon the test cushion.

The sensors about which Mr. Ellner wrote were do-it-yourself devices fashioned from the conductive foam used to ship

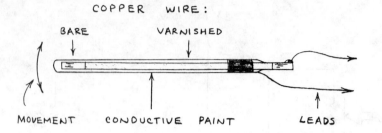

Fig. 7-11. Simple copper wire pressure sensor.

CMOS and other voltage sensitive semiconductors. Many other kinds of pressure-sensing resistors are also available. In this discussion I'll discuss several kinds of pressure-sensitive resistors, emphasizing an inexpensive commercial device. I'll also present some specific circuit and computer applications.

Do-It-Yourself Pressure-Sensitive Resistors

In 1969 I was attempting to measure the forces exerted on a small, homemade infrared-seeking guided rocket suspended in a wind tunnel. The wind tunnel was fashioned from a length of stove pipe fitted with air compression, straightener, and expansion stages. When suspended from the passenger side of my 1966 Chevy, the wind tunnel achieved an airspeed of 90 miles per hour when the car was driven at 70 mph, then the legal speed limit.

One of the force-measuring devices I devised was a short piece of copper wire coated with an insulating film. As shown in Fig. 7-11, the insulating material was removed from a short length of each end of the wire. The wire was then dipped into a commercially available conductive paint which was blended with minute particles of copper. After the paint dried, the coated wire formed a resistor whose resistance could be varied by bending the wire. Leads were attached to the assembly as shown in Fig. 7-11. One lead was attached directly to the exposed end of the copper wire. The other was attached to the conductive paint by means of a strip of tape or a small alligator clip.

I attempted to use the resistor shown in Fig. 7-11 to measure variations in the forces on a rocket in my wind tunnel. However, the oscillations of the rocket prevented accurate measurements. Nevertheless, the basic resistor in Fig. 7-11 is easy to make and may have other, more practical applications.

Do-it-yourself pressure-sensitive resistor can be made from a small disk of electrically conductive plastic foam of the kind in which CMOS ICs are sometimes shipped. This method was described in detail in the previous section.

Another type of simple pressure-sensitive resistor can be made by mounting a spring on the handle of a slide resistor. Though the sensitivity might not be as high as that of other methods, the results will be very repeatable. Back in 1958, one of the first radio transmitters launched in a model rocket used just such a device for an accelerometer. I remember watching that launch from a field near Colorado Springs along with a crowd of high school model rocket enthusiasts and our dads.

Commercial Pressure-Sensitive Resistors

Many different kinds of commercial pressure-sensitive resistors are available. For instance, Vernitech makes a potentiometer-type pressure-sensitive resistor that incorporates an infinite resolution potentiometer. This device offers a linearity of within 0.3 percent.

Also available are various kinds of electromagnetic and piezoelectric pressure-sensitive resistors. For information about manufacturers, see one of the electronics trade directories available at a good technical library or inquire at companies who represent various electronics manufacturers.

To my knowledge, the least expensive commercial pressure-sensitive resistors are manufactured by Interlink Electronics, Inc. Figure 7-12 is a drawing of one kind of resistor made by Interlink Electronics. The company labels this device a Force Sensing Resistor or FSR. Three FSRs can be purchased from the company by sending $5 plus $1 for postage and handling to the address given in the Appendix.

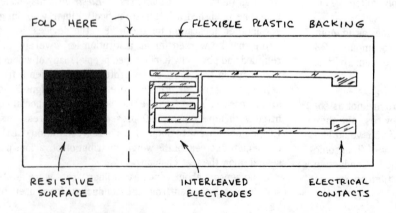

Fig. 7-12. Interlink Electronics pressure-sensitive resistor.

The FSR in Fig. 7-12 is printed on a thin sheet of clear, flexible plastic which can be easily cut with scissors. Referring to Fig. 7-12, on the left is a square-shaped deposit of material having a moderately high resistance. On the right is a pair of interleaved electrodes brought out to two terminals. In operation, the side of the FSR having the resistive coating is folded over the interleaved electrodes. When the resistive coating is squeezed against the electrodes, a variable resistance appears across the two terminals.

Figure 7-13 is a graph that plots (on a logarithmic scale) the resistance of an Interlink Electronics pressure-sensitive resistor versus an applied force. When the load applied to the FSR ranges from about 5 to 12 kilograms per square centimeter, the straight line log-log relationship plotted in Fig. 7-13 becomes a simple linear relationship as shown in Fig. 7-14. Note how, at least over this range, the change in resistance with respect to the applied load is very small.

Incidentally, both Figs. 7-13 and 7-14 are adapted from "Force Sensing Resistors," an application note published by Interlink Electronics. Among the applications for FSRs listed in this note are point-contact graphic tablets for computers, theft detectors, robot grip sensors, musical keyboards, musical drum pads, and theft detectors.

Application Circuits

It's quite easy to demonstrate the operation of a pressure-sensitive resistor with the help of a simple circuit. The two circuits that follow both use the symbol for the pressure-sensitive resistor shown in Fig. 7-15. This symbol is the one suggested by Interlink Electronics.

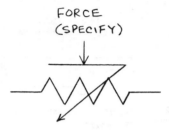

Fig. 7-15. Suggested FSR symbol.

Figure 7-16 is a simple tone generator made from a 555 timer chip configured as an astable oscillator. The frequency of oscillation of this circuit is governed by R1, R2 and C1. The approximate relationship is $1.44/(R1 + 2R2)C1$.

With the values given in Fig. 7-16 and when R1 is an Interlink Electronics FSR, the tone produced by the speaker ranges across the entire audio spectrum when the FSR is squeezed between thumb and forefinger. The tone range can be easily altered by changing C1. Increase C1 to reduce the frequency range.

Figure 7-17 is a straightforward comparator circuit that permits a pressure-sensitive resistor to switch an LED on or off as the pressure on the resistor is varied. The switching threshold of the circuit can be altered by changing the setting of R3. As

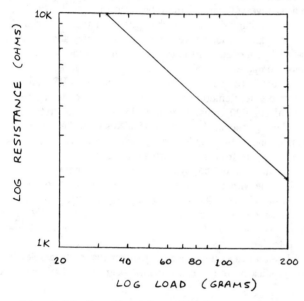

Fig. 7-13. Log/log plot of resistance versus force for FSR.

Fig. 7-14. Resistance of FSR in Fig. 7-13 at high load values.

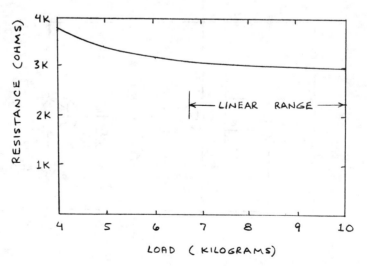

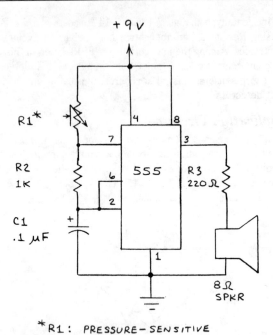

+9 V

R1*

R2
1K

C1
.1 μF

555

R3
220 Ω

8 Ω
SPKR

*R1: PRESSURE-SENSITIVE
RESISTOR

Fig. 7-16. Pressure-controlled tone generator.

Computer Applications

Some computer graphics tablets employ a surface coated with a resistive material. Applying pressure to the surface of the pad gives an output signal representative of the location of the touched region.

Simple pressure-sensitive resistors also have computer applications. For example, any of the homemade pressure-sensitive resistors described previously can be connected to the joystick input(s) of computers designed to accept variable resistance (potentiometer-style) joysticks. In this manner, joystick functions can be achieved simply by pressing on a pressure-sensitive resistor rather than by moving a joystick handle.

Several joystick circuit configurations are used by the various computer manufacturers. In the simplest configuration each pot in the joystick functions as a two terminal variable resistor. This is the approach used in IBM's PC*jr*. A somewhat more complicated approach is to connect one side of the pots in a joystick to a positive voltage and the other side to ground. This forms a pair of voltage dividers in which the rotor terminals supply a voltage which varies between the positive supply and ground as the stick is moved. This is the approach used in Radio Shack's Color Computer.

Figure 7-18 shows the internal circuitry of a PC*jr* joystick. The two potentiometers are linear taper devices with a resistance of 100,000 ohms. Two normally open push-button "fire" switches are included. Figure 7-18 also shows the pin connections of one of the PC*jr*'s two joystick connectors.

The leads from a pressure-sensitive resistor can be connected directly to the joystick connector of the PC*jr*. This is easily done with a wire-wrapping tool. Alternatively, one or two miniature phone jacks can be added to a joystick to permit the resistors to be connected to the joystick itself. I used the latter approach since the Berg type connectors used in the PC*jr* are hard to find.

In either case, it's important to know that noise coupled into the joystick ports can cause erratic operation. That's why the

shown, the circuit switches the LED on when the pressure on R1 is increased. If the input connections to the op amp (pins 2 and 3) are reversed, the LED will switch off as the pressure on R1 is increased.

The circuit in Fig. 7-17 can be easily modified. For instance, the LED can be replaced by a small relay such as Radio Shack's 275-004 if R5 is eliminated and the collector of Q1 is connected directly to the positive supply.

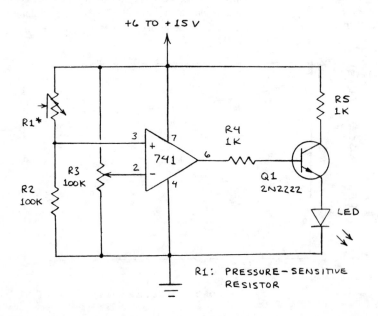

+6 TO +15 V

R1*

R3
100K

R2
100K

741

R4
1K

RS
1K

Q1
2N2222

LED

R1: PRESSURE-SENSITIVE
RESISTOR

Fig. 7-17. Pressure-controlled comparator.

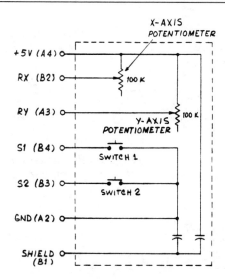

Fig. 7-18. Internal circuitry of IBM PCjr joystick.

joystick cables are shielded. For this reason, keep the leads to the pressure-sensitive resistor short or use two-conductor shielded cable and ground the shield as shown in Fig. 7-18.

This PCjr program will display the joystick value of a single pressure-sensitive resistor connected to the x-axis potentiometer:

```
10 CLS
20 X=STICK(0)
30 LOCATE 10,20
40 PRINT X
50 GOTO 20
```

When used with an Interlink Electronics FSR, this program emphasizes the low range of joystick values (from about 3 to 15). The following program permits an FSR to move a dot back and forth across the screen:

```
10 SCREEN (1)
20 CLS
30 X=STICK(0)
40 X=10*X
50 Y=20
60 PSET (X,Y)
70 PRESET (X,Y)
80 GOTO 30
```

Since the FSR works best with low joystick numbers, line 40 multiplies the retrieved value by ten. This provides the X coordinate for a horizontal line across the computer's display. As the FSR is alternately squeezed and released, a small dot moves back and forth along this line in 10-pixel increments.

If you have a Color Computer, you can connect one or more pressure-sensitive resistors to its joysticks if you first add a single fixed resistor in series with each sensing resistor. Connect the free end of one resistor to +5 volts (available at the joystick port) and the free end of the remaining resistor to ground (also available at the joystick port). The junction of the two resistors

then becomes the voltage divider output for the Color Computer joystick port.

The value of the fixed resistor depends on the resistance range of the pressure-sensitive resistor. Try values of from 1000 to 100,000 ohms. You can also experiment with which resistor is connected to +5 volts. For initial experiments, use a fixed resistor of 1000 ohms and connect its free end to ground. Connect the pressure-sensitive resistor's free lead to +5 volts.

Here's a listing that displays the Color Computer's joystick values:

```
10 CLS0
20 PRINT @ 133, JOYSTK(0);
30 PRINT @ 138, JOYSTK(1);
40 PRINT @ 148, JOYSTK(2);
50 PRINT @ 153, JOYSTK(3);
60 GOTO 20
```

CAUTION: Use care when attaching pressure-sensitive resistors or any other components to the joystick inputs of a computer. You may damage the computer, void its warranty, and disqualify it for repair by the manufacturer. Since digital computers use MOS integrated circuits that are susceptible to permanent damage caused by electrostatic discharge, remove any charge on your body by touching a grounded object. For best results follow the precautions recommended for handling and working with CMOS ICs. Finally, use caution to avoid exposing yourself to the possibility of electrical shock while working with a line powered computer.

Going Further

Pressure-sensitive resistors have numerous applications, many of which have yet to be fully developed. Experimenters can play an important role in developing new applications for these devices since they can be easily assembled from common materials or purchased at low cost.

For more information about pressure-sensitive resistors, see the FSR application note published by Interlink Electronics (address given in the Appendix). Thomas Henry of Transonic Laboratories wrote a brief application article on the subject entitled "Conductive Foam Forms Reliable Pressure Sensor" (*Electronics*, May 19, 1982, p. 161).

Adjustable Threshold Temperature and Light Alarms

If you enjoy gardening as much as I do, it's likely you could make good use of a circuit that sounds an alarm when the temperature falls to near freezing. Figure 7-19 shows a very simple circuit that signals a distinctive alarm tone when the temperature approaches 0°C. Since the temperature detection threshold is adjustable, the circuit can sound an alarm at any point over a wide temperature range.

The circuit makes use of a 741 operational amplifier operated as a voltage comparator. In operation, R2 functions as an adjustable voltage divider that provides a reference voltage to the

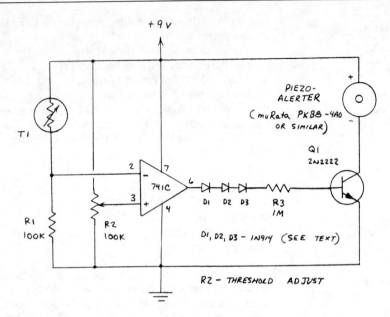

Fig. 7-19. Adjustable threshold temperature alarm.

noninverting input (pin 3) of the 741. Thermistor T1 and fixed resistor R1 form a second voltage divider that provides a temperature dependent voltage to the inverting input (pin 2) of the 741.

The resistance of T1 is inversely proportional to temperature. In other words, the resistance of T1 *increases* as temperature *decreases*. Therefore, the voltage output from the divider formed by T1 and R1 falls as temperature decreases. When this voltage reaches the reference voltage, the comparator output switches from low to high. This turns on Q1 which, in turn, actuates the piezoalerter.

The output from the 741 can be connected directly to R3, but the low-state output voltage will then turn Q1 partially on and cause the alerter to emit a low amplitude but audible tone. This off-state tone is eliminated by D1, D2, and D3. The total forward voltage drop of these diodes is about 1.8 volts, enough to cancel the effect of the low-state output voltage from the 741.

Thermistors can be purchased from Newark Electronics and other electronic parts distributors. For best results, use a thermistor having a room temperature resistance of from 25 kΩ to 50 kΩ. Glass bead and bulb thermistors are more expensive than other kinds, but since they can be safely immersed in water they are easily calibrated. Use crushed ice or snow to achieve an exact 0°C. calibration point if you plan to use the circuit as a freeze detection alarm.

The circuit in Fig. 7-19 can drive more powerful alarm devices such as sirens by substituting a relay for the piezoalerter. The following circuit shows how this is accomplished.

An Adjustable Light Detection Switch

Replacing the thermistor in Fig. 7-19 with a cadmium sulfide photoresistor allows the circuit to function as an adjustable threshold light/dark detection switch. Figure 7-20 shows one possible circuit configuration.

In operation, R2 sets the circuit's threshold. When the light intensity at PC1's surface is increased, the resistance of PC1 is decreased. This increases the voltage at the inverting input of the 741. When the reference voltage at the 741's noninverting input is properly adjusted via R2, the comparator will switch from low to high when PC1 is darkened. This turns on Q1 which, in turn, pulls in relay RY1.

The low-state output voltage from the 741 does not have sufficient amplitude to pull in the relay. Therefore, this circuit does not require the diodes used in Fig. 7-19.

Going Further

The sensors used in both these circuits are interchangeable. You can use a cadmium sulfide photoresistor in place of the thermistor in Fig. 7-19. And you can use a thermistor in place of the photoresistor in Fig. 7-20. Furthermore, both circuits are adjustable over a very wide temperature or light intensity range. Therefore, you should have little difficulty adapting one or both circuits to your specific application.

Finally, be aware that the reliability and accuracy of these circuits is determined both by your calibration procedure and the physical location of the finished circuit. For example, when the circuit in Fig. 7-19 is configured as a freeze detector, it may fail to operate consistently if the battery is exposed to temperature extremes. And it may sound false alarms or fail to operate if water bridges any of the leads from R1, R2, or the thermistor. Be sure to keep these caveats in mind when you play your application.

Detecting Sound

The recognition of speech by computers is a topic which has received considerable press over the past few years. Though machine speech recognition is a highly sophisticated technology the design and construction of electronic circuits that respond to the presence or absence of sounds is very straightforward.

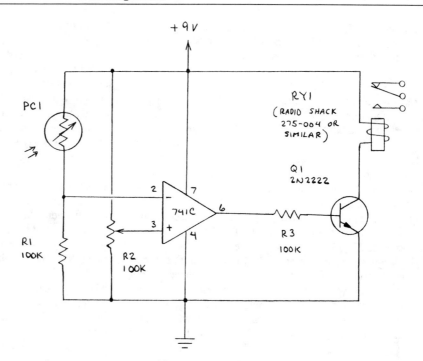

Fig. 7-20. Adjustable threshold light/dark detection switch.

I'm going to describe two such circuits and discuss some of their applications. First, let's review some basics of sound.

Sound

A vibratory movement of air that enters the human ear is perceived as a sound. The air movements that cause sounds are pressure waves that fluctuate around the normal atmospheric pressure of about 14.7 pounds per square inch (1013.25 millibars). Sound travels through air at a velocity of 1128 feet (344 meters) per second when the temperature is 68°F. This is much slower than the speed of light (984,000,000 feet or 299,800,000 meters per second). This, of course, is why the eye perceives a lightning bolt before the ear receives the sound.

The ear, like the eye, possesses an incredibly broad dynamic response. The typical human ear can detect sound pressures ranging from as little as 0.0001 microbar to 1000 microbars, a pressure range of 1 to 10,000,000.

Because the response of the human ear is so wide, sound is measured on a logarithmic scale. The unit of sound pressure is the decibel (dB). The faintest sound the human ear can discern, which has an intensity of 0.0000000001 microwatts per square centimeter, has been assigned a decibel level of 0. All other sound pressure levels are given relative to the 0-dB level. The intensity in dB of a particular sound is ten times the logarithm of the ratio of the sound intensity being specified (S1) to the 0-dB reference level (S2) or dB = 10 log (S1/S2).

A typical conversation yields a sound intensity of 0.001 microwatt per square centimeter at a distance of 3.25 feet. Inserting this value into the formula given before yields a decibel value of 70 [dB = 10 log (10 − 3/10 − 10)]. Table 7-1 lists the decibel levels for some other commonly encountered sounds.

Table 7-1. Levels of Commonly Encountered Sounds

Sound	dB
Jet aircraft at 20 feet	140
Threshold of pain	130
Propeller aircraft at 18 feet	121
Passing subway train	102
Niagara Falls	92
Passing truck or bus at 20 feet	80
Average automobile at 15 feet	70
Conversational speech at 3.25 feet	70
Average office	55
Average residence	40
Quiet whisper at 5 feet	18
Threshold of audible sound	0

A person's hearing ability is determined by age, sex, and previous exposure to loud sounds. It's well known, for instance, that the ears of older people have a diminished response to higher frequency sounds. The ears of young children are particularly sensitive, a fact which was clearly in evidence at a recent air show I attended. As the jet aircraft swooped by the spectators, virtually all the children placed their hands over their ears.

Teenagers are often warned about the hearing loss that will follow long term exposure to excessively loud music. Experimenters should heed the same warning and practice caution when experimenting with circuits that generate sounds. Years ago I learned to avoid placing an earphone in my ear until the circuit to which it was connected was switched on. On several occasions when I failed to follow this procedure, my reward was a painfully loud tone or noise. Thanks to these mishaps, there is a distinct hearing deficiency in one of my ears.

A Sound-Activated Relay

A sensitive sound-triggered relay can be designed around a single operational amplifier and a monostable multivibrator. One such circuit I've recently designed and tested is shown in Fig. 7-21.

The circuit in Fig. 7-21 uses a miniature electret condenser microphone that contains a built-in FET amplifier stage. The microphone I used, Radio Shack Cat. No. 270-092B, has a sensitivity of −65 dB relative to a reference of 0 dB of 1 volt per microbar at a frequency of 1 kHz. The frequency response of this microphone is nearly flat from 20 Hz to 10 kHz.

Referring to Fig. 7-21, the microphone's output is coupled through C1 into the inverting input of a 741 op amp. Feedback resistor R3 controls the gain of the op amp. The amplified signal is coupled through C2 to Q1, where it is inverted, and from there to the input of a 555 timer configured as a monostable multivibrator or "one-shot." The one-shot issues a pulse at pin 3 which has a duration controlled by R5 and C3. The values given in Fig. 7-21 give a duration of about 10 seconds. The armature of a relay connected to pin 3 of the 555 is pulled in during this time interval.

The values of nearly all the circuit components in Fig. 7-21 are noncritical. The sole function of R1 and R2 is to act as a voltage divider to bias the noninverting input of the op amp.

This permits the op amp to be powered by a single polarity supply. Though its value can be changed, C1 must be used for the circuit to function properly (because of the single polarity supply). C4 helps prevent the one-shot from being false triggered by stray electrical noise. The relay should be a low voltage, low-current unit such as Radio Shack's Cat. No. 275-004.

To test the circuit, rotate the shaft of sensitivity control R3 to its center position, apply power, and speak into the microphone. The relay should pull in and remain pulled in for about 10 seconds. If you continue speaking or if other sounds are present, the relay will remain pulled in. Adjust R3 to alter the sensitivity of the circuit.

The circuit is very sensitive when R3 is adjusted for peak amplification (R3 = 1 MΩ). While I was testing the prototype circuit, the relay remained pulled in continuously, even though my workshop was quiet. After spending some time checking and rechecking the circuit for wiring errors, I traced the problem to the air conditioner in the office that adjoins my workshop. The combined sounds of the air conditioner's fan, motor, and compressor were retriggering the circuit after each timing interval, thereby causing the relay to appear to be pulled in continually.

There are many possible applications for sound-activated relays. The circuit in Fig. 7-21, for instance, can be used to cause a toy car or robot to respond to a hand clap or even a spoken command. The circuit can also be used to switch on a light or other electrically operated device in response to a sound. Finally, the circuit can be used in a sound-activated intrusion alarm. If

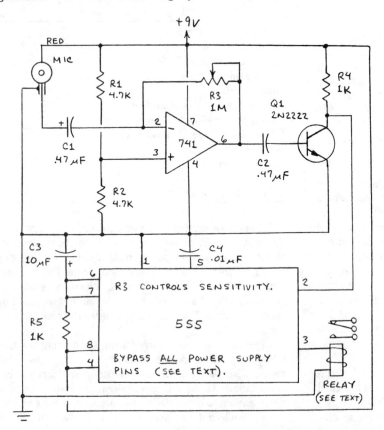

Fig. 7-21. Sound-activated relay.

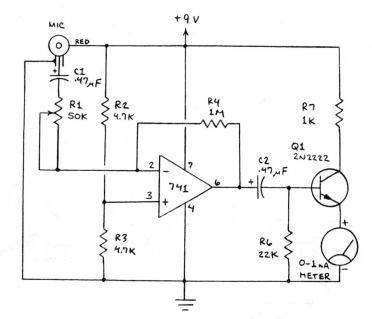

Fig. 7-22. Sound-level meter.

you try this application, however, it's particularly important to protect the system from both manmade and naturally occurring sounds that might cause false alarms.

A Sound-Level Meter

The first sound-level meters were installed in suitcase-size enclosures. However, the batteries required to power their vacuum tube amplifiers made them much heavier than a conventional suitcase. Despite their bulky size, those early meters were used to make many pioneering measurements of the intensity of sounds in offices, factories, street corners, train stations, and elsewhere.

Modern sound-level meters can weigh under a pound and fit in a pocket. One such meter sold by Direct Safety Company weighs 4.5 ounces and measures only 5.75 × 1.9 × 1.1 inches. This meter, which sells for $149, is powered by a single 2.7-volt battery and has a 40–120-dB range.

Figure 7-22 is the circuit for a simple uncalibrated sound-level meter I've recently designed. The circuit is similar to the input stage of the sound-activated relay in Fig. 7-21.

The microphone is an electret condenser unit such as Radio Shack's Cat. No. 270-092B. In operation, signals from the microphone are coupled through C1 and potentiometer R1 into the inverting input of a 741 op amp. The amplified signal is then coupled through C2 to Q1 which supplies current to a 0-1 milliampere meter. R1 and R4 control the gain, hence the sensitivity, of the circuit.

Since the op amp is operated from a single polarity supply, C1 is essential for proper operation. R2 and R3 form a divider that biases the noninverting input of the 741 at half the supply voltage. The meter can be a standard panel meter or a digital multimeter set to indicate current.

I tested the sound-level meter with a piezobuzzer that emitted a 6.5-kHz tone having a sound intensity of 90 dB (muRata PKB8-4A0 or Radio Shack 273-064). When the buzzer was placed

2 inches away from the microphone, the meter indicated an output current of 1 milliampere. Normal speech at a distance of 12 inches gave a current that fluctuated around 10 microamperes.

Adjustable Frequency Tone Source

While testing the circuits in Figs. 7-21 and 7-22, I found that a variable frequency tone source was a big help. Figure 7-23 is the circuit for a simple variable frequency tone source that drives a piezoelectric tone generator element. The circuit is designed around a 741 op amp configured as an astable multivibrator. R3 and C1 control the tone frequency.

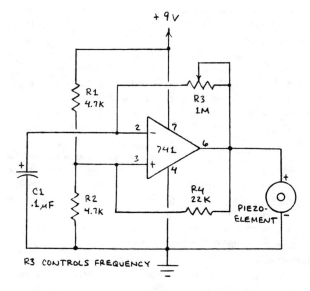

Fig. 7-23. Variable frequency tone generator.

Figure 7-24 shows how a 555 can be used as a variable frequency tone source. Here the tone frequency is determined by R1 and C1.

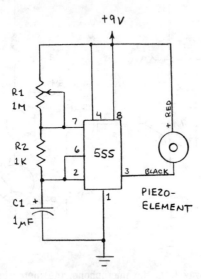

Fig. 7-24. 555 variable frequency tone generator.

Incidentally, piezoelectric tone generators produce a very penetrating, even uncomfortable sound. This is why they have become the sound source of choice in alarm devices like smoke detectors. Because their sound intensity can be so loud, use caution when working with them. I usually use ear protectors, especially when working with piezoelectric sound sources operated at high sound levels.

The Inverse Square Law

While experimenting with the sound-level meter in Fig. 7-22 and a piezoelectric sound source, I attempted to verify the inverse square law. Briefly, this law holds that the intensity of a sound wave, like a light wave, is inversely proportional to the square of the distance from the source. In other words, when a sound wave reaches a point 3 inches away from its source, its intensity is ⅑ the intensity when the wave was 1 inch from the source.

To perform the test I placed the sound source exactly 1 centimeter away from the microphone and adjusted the sound intensity until the meter indicated 250 microamperes. When the sound source was moved 2 centimeters away from the microphone, the meter indicated 60 microamperes. At 3 centimeters the meter indicated 12 microamperes, and at 4 centimeters the meter indicated 1 microampere.

These results differ from those predicted by the inverse square law. Table 7-2 compares my results with the predicted results.

The square law results are calculated relative to the initial reading of 250 microamperes. When the range was doubled from 1 to 2 centimeters, the test result was in close agreement with the predicted result. However, as the range was increased the

Table 7-2. Comparison of Test and Predicted Inverse Square Law Results

Range	Test Results	Square Law Result
1	250	250.00
2	60	62.50
3	12	27.78
4	1	15.63

test results were substantially less than those predicted by the inverse square law.

Several factors can account for the difference between the actual and expected results. For instance, the amplitude response of the microphone might not be linear. In any event, these results illustrate that the measurement of sound, like the measurement of light, requires careful attention to variables that might alter the expected results.

Going Further

If you are interested in the general topic of sound-activated devices, you will want to examine the field of machine speech recognition. One of the most challenging areas of electrical engineering today is machine speech recognition. It's been estimated that a digital computer will have to make some ten billion computations to reliably recognize a 10,000 word vocabulary spoken by a single person. Even then incredibly complex software will be required to enable such a machine to distinguish between such identical sounding words as pail/pale, piece/peace, soar/sower and so forth.

There now exist many computer-based systems that can reliably recognize from a few dozen to a few hundred words spoken by a single person after a training period in which the person pronounces each word for the benefit of the computer. But these systems are only reliable when the background noise level is low and the quality of the speaker's voice isn't altered by a cold or sore throat.

Contrast this capability with the incredibly sophisticated speech recognition system possessed by every human being. Most people can understand the same sentence spoken by thousands of persons using scores of dialects and accents. Moreover, people can discern conversations even when they are partially masked by ambient sounds. For example, when the level of the interfering noise matches that of the conversation (signal-to-noise level = 1:1) approximately half the words in the conversation can be understood. Many of the lost words can be filled in by assessing the context of the conversation. Even when the noise level is four times higher than that of the conversation, one-fourth of the words can still be understood.

Since the speech recognition capabilities of manmade systems are still rather primitive, there remain opportunities for experimenters to make important contributions. For further information on this fascinating topic, visit a good technical library and review some of the many articles and papers that have been published on speech recognition over the past decade. For starters, see the articles on speech recognition in the September 17, 1984 issue of *Electronic Products*. Also see "Voice Recognition

Systems and Strategies" (*Computer Design*, January 1983, pp. 67–70) and the special section on voice input and output in the April 21, 1983 issue of *Electronics* (pp. 126–143).

Audio Amplifier Experiments

In this era of digital electronics, it's easy to overlook the fact that one of the earliest electronic circuits is still the most widely used. That circuit is the ubiquitous audio-frequency amplifier.

Of the thousands of analog electronic circuits developed over the years, none is as versatile or possesses as much overall importance as the audio-frequency amplifier. Every public address system, radio, tape recorder, television, intercom, and transceiver contains at least one audio amplifier. Doctors use them to electronically monitor physiological functions such as respiration, pulse rate, and the electrical activity of the heart and brain. Scientists and engineers use audio amplifiers to boost the tiny signal levels from sensors such as accelerometers, strain gauges, photocells, and Geiger tubes. Experimenters and technicians use them to troubleshoot audio-frequency circuits (signal tracing) and for general purpose experimenting.

Since audio amplifiers are so commonplace, it's easy for those without electronic construction skills to perform experiments with them. Of course experienced experimenters may prefer to build their own custom amplifier, perhaps including features not ordinarily found in commercial amplifiers. In this discussion several kinds of commercial amplifiers and a do-it-yourself version you can easily assemble will be described. Then several experimental applications are described and others are suggested.

Suitable Amplifiers

The experiments which follow as well as many others can be easily performed with a low-cost commercial amplifier such as Radio Shack's Mini Amplifier-Speaker. This compact, battery-powered amplifier (Cat. No. 277-1008), which sells for about $12, has a frequency response of 100 Hz to 10 kHz. It has a sensitivity of 1 millivolt and an output power of 200 milliwatts. The amplifier is equipped with a volume control and two ⅛-inch phone jacks. One jack is the input port and the other is for an external speaker or earphone.

Some portable cassette tape recorders can also function independently as audio amplifiers. They offer the added benefit of permitting you to record unusual sounds or signals you happen to detect.

Many other commercial amplifiers can also be used. Especially well suited are public address systems and component amplifiers for home high fidelity systems. Keep in mind, however, that portable, battery-powered amplifiers provide much more flexibility. They also have no electrical shock hazard.

Some amplifiers are equipped with tone controls, but many are not. I've had excellent results by connecting a miniature, battery-powered equalizer between a battery-powered amplifier and an external earphone. The equalizer can substantially reduce the amplitude of both low-frequency noise (such as 60-Hz line noise) and high-frequency noise while emphasizing the audio frequencies to which the ear is most sensitive.

CAUTION: No matter what kind of amplifier you select, it is essential to use a great deal of caution if you elect to use any kind of earphone to monitor the amplified signal. The sound levels produced by an earphone connected to even a small battery-powered amplifier might be sufficiently loud to cause temporary or even permanent loss of hearing.

When using an earphone, *always* rotate the volume control to the quietest position before placing the phone in or against your ear. Then gradually increase the volume to an appropriate level. Always be careful to avoid high input signal levels when the earphone is in or near your ear.

A Do-It-Yourself Amplifier

Many readers may prefer to build their own amplifier to perform the experiments that follow. Therefore, before presenting the experiments, I'll begin by describing an easily assembled, basic do-it-yourself amplifier.

Figure 7-25 shows the circuit diagram for the amplifier. The input signal is coupled into the inverting input of a 741 operational amplifier. Other op amps can also be used. Gain control R1 doubles as both the input and feedback resistor. The amplified signal is fed into an LM386 power amplifier through R2, which serves as a volume control. The output from the LM386 drives an 8-ohm speaker through C3. An 8-ohm earphone can be substituted for the speaker. C2 sets the gain of the LM386 to 200. The gain can be reduced to 20 by omitting C2.

For best results, keep the connection leads of the circuit short and direct. This will reduce the amplifier's noise level and prevent the possibility of inductive feedback that might cause the amplifier to oscillate at high gain levels. To further reduce the possibility of oscillation, connect a 0.1-microfarad capacitor across the power supply pins of the two ICs.

An Underwater Hydrophone

Of the many experiments I've performed with audio amplifiers, listening to underwater sounds has been among the most interesting. When I was a high school student, some of my friends and I used a transistorized amplifier salvaged from a cheap, battery-powered public address system to establish a voice link between the deck and the bottom of a local swimming pool. One of us would swim down to the bottom with the microphone while the others remained on top to listen. Though, as you might expect, the words were bubbly sounding, it was actually possible to understand most of what was being said.

Later I used this same system to listen to the sounds produced by a captive porpoise. For both these experiments, the microphone was a low-cost crystal unit placed inside a mostly waterproof plastic bag.

Recently my interest in underwater sounds was revived while I was flipping through the pages of Edmund Scientific's catalog. There I spotted a surplus hydrophone (Cat. No. 41,759) for less than $10. I promptly ordered one of the units.

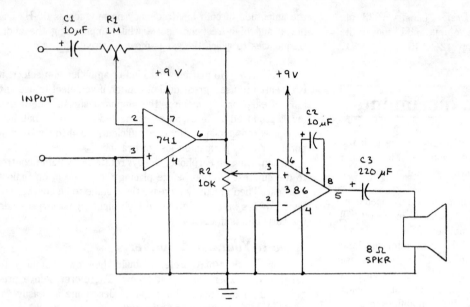

Fig. 7-25. A do-it-yourself audio amplifier.

As its name implies, a hydrophone is a microphone specifically designed for underwater use. The unit I purchased from Edmund Scientific is completely encapsulated in black rubber and plastic and can be used in either fresh or salt water to a depth of 300 feet. The unit has a low impedance and a capacitance of approximately 15,000 pF. The hydrophone's frequency response ranges from 10 to 6000 Hz, ± 4 dB (± 1 dB from 10 to 2000 Hz). Above 7000 Hz the hydrophone exhibits a rolloff greater than 12-dB per octave.

Figure 7-26 is a sketch of the hydrophone I purchased from Edmund Scientific. According to the specification sheet, this hydrophone may have one or two pairs of leads. The additional pair of leads, if present, facilitates connecting from two to four units in parallel. Indeed, you can buy a string of parallel-connected hydrophones from Edmund.

If the hydrophone you receive has an extra pair of leads, it's important to insulate the exposed end of the unused pair to prevent the possibility of inadvertently shorting the hydrophone's output. A small blob of silicone sealant like that used to assemble tropical fish aquariums works well.

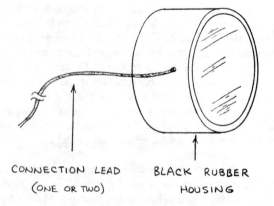

CONNECTION LEAD
(ONE OR TWO)

BLACK RUBBER
HOUSING

Fig. 7-26. Surplus hydrophone.

The specification sheet supplied with the hydrophone explains how to solder a shielded cable to the hydrophone's leads. To avoid using the tape suggested in the spec sheet, I used a slightly different method. First, I exposed the two strands of bare wire in the connection lead and folded one strand back along the insulated lead. I then placed the exposed conductors at the end of a length of shielded cable alongside the two strands as shown in Fig. 7-27.

After soldering the hydrophone's exposed leads to the shielded cable, I placed a short length of heat-shrinkable tubing over the connection and shrank it snugly over the soldered wires with the heat from a soldering iron. Next, I coated the entire connection with silicone sealant. Finally, I soldered a shielded ⅛-inch phone plug to the unused end of the shielded cable.

For best results, the hydrophone should be securely mounted to a suitable handle, rod, or pole. Otherwise, the strain placed on the connection lead may cause one or both the small wires to separate. Edmund Scientific's specification sheet recommends using a hose clamp to attach the body of the hydrophone to the flattened end of a length of electrical conduit (available from a hardware store). The connection leads should then be tied or taped at intervals along the length of the pole.

In some preliminary tests, I've used my hydrophone to listen to the sounds in the spring-fed creek that flows about 1000 feet from the old Texas farmhouse where these words are being entered into a word processor. This stream is populated by catfish, bass, blue gill, gar, and various kinds of turtles and frogs. Though these tests were conducted during January when the creek's inhabitants are much less in evidence, I could hear a variety of pops, crunches, gurgles, and the like. Even when the hydrophone was submerged in some of the deeper holes (8–10 feet), the sound of a twig tossed in the water could be clearly heard. Rain produced a static-like sound.

In the future, I plan to spend a good deal of time exploring the creek with my hydrophone. I also plan to take it out on some nearby lakes and to dip it in the waters of Corpus Christi bay

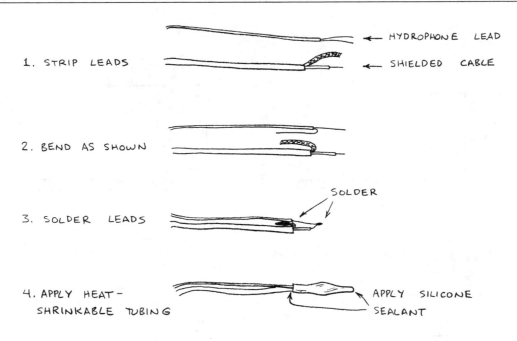

1. STRIP LEADS ← HYDROPHONE LEAD
← SHIELDED CABLE

2. BEND AS SHOWN

3. SOLDER LEADS SOLDER

4. APPLY HEAT-SHRINKABLE TUBING APPLY SILICONE SEALANT

Fig. 7-27. Connecting a shielded cable to the hydrophone.

on the Texas gulf coast. Finally, I plan to use the hydrophone in conjunction with a miniaturized electronic insect simulator. The latter device is a piezoelectric buzzer driven by an oscillator whose frequency rate can be altered.

CAUTION: For your own protection, *never* use a line-powered amplifier around water. To do so may result in a dangerous or even fatal electrical shock. The hydrophone experiments described here should be conducted with a battery-powered amplifier only.

Light and Lightning Detection

Another fascinating application for an audio amplifier is to monitor various light sources by means of a low-cost solar cell, photodiode, or phototransistor connected to the amplifier's input. Figure 7-28 shows various ways each of these devices can be connected to an amplifier.

You can use any of the various photodetector-amplifier combinations in Fig. 7-28 to transform into sound the pulsations or intensity fluctuations of any light source that is modulated at an audible frequency. For example, positioning the detector in the shadow of a hovering humming bird will cause the speaker connected to the amplifier to emit a tone that corresponds to the frequency of the bird's wing beats. A similar effect can be observed by positioning the detector so that it can receive flashes of sunlight reflected from the wings of flying insects.

You can also use any of the basic arrangements in Fig. 7-28 to monitor electrically or electronically modulated artificial light sources. All light sources directly powered by 60-Hz line current are modulated and will therefore cause an audio-frequency tone to be produced by the photodetector-amplifier com-

bination. Multiplexed light-emitting diode displays will elicit a similar effect.

If the light source emits invisible near-infrared, a lightwave receiver is particularly handy. It can determine whether or not the source is activated and the area it illuminates. I often use a photodiode-amplifier arrangement as an optical fiber continuity checker. The emission from a near-IR LED pulsed at an audible frequency is injected into one end of the fiber and the photodiode monitors the opposite end. Of course the continuity of short lengths of fiber can be visually checked with a visible source.

Some applications for a light-sensing amplifier are less obvious. For instance, the tungsten filaments in some car and truck headlights will vibrate when the vehicle rolls over a rough surface. This causes the filament to move in and out of the focal point of the lamp's parabolic reflector. The result is that the beam from the light is intensity modulated. The gong and bell-like sounds that result when these intensity fluctuations are detected by a light-sensitive amplifier are quite distinctive. You can produce a similar sound by striking the end of a flashlight with a pencil while illuminating a detector connected to an amplifier.

Of the many applications for a light-sensing amplifier, my favorite is lightning detection. An ordinary am radio indicates the presence of lightning by producing crackling and popping sounds. A light-sensing amplifier responds to lightning by producing similar crackles and pops. The major difference is that the light-sensing amplifier is directional. It can also detect lightning that is obscured by clouds and is therefore not ordinarily visible to the human eye. For best results, try detecting lightning at night. And be sure to avoid placing yourself in a location that might be vulnerable to a lightning stroke.

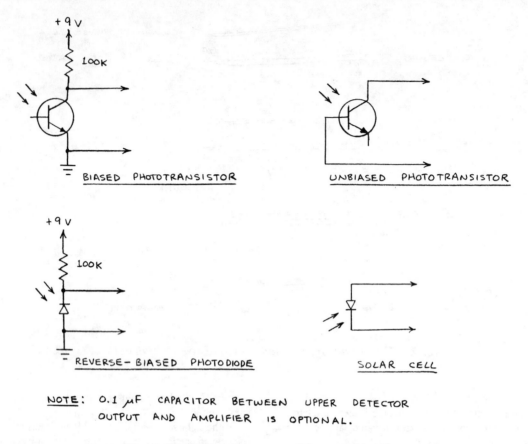

NOTE: 0.1 µF CAPACITOR BETWEEN UPPER DETECTOR
OUTPUT AND AMPLIFIER IS OPTIONAL.

Fig. 7-28. Light sensor-amplifier combinations.

A Simple Photophone

On February 19, 1880, Alexander Graham Bell and Sumner Tainter, Bell's lab assistant, became the first people to speak over a beam of light. For this historic demonstration they directed a beam of reflected sunlight through a pair of comblike grids made by scratching parallel lines in the silver coating of two small glass mirrors. One grid was mounted in a fixed position. The other was attached to the diaphragm of a modified telephone microphone. When voice was directed against the diaphragm, the grid to which it was attached moved back and forth in response to the voice signal. This caused a representative fluctuation in the sunlight passing through the two grids. The fluctuating light was detected by a homemade selenium light detector designed by Bell.

Professor Bell called his pioneering lightwave communicator the photophone. In the first version, the detector was separated from the grids by only a few centimeters. Soon afterward, Bell and Tainter were talking over beams of sunlight at distances of hundreds of feet. For many of these experiments they used a transmitter so simple it can be assembled by a child.

Figure 7-29 shows the construction details of a modern version of one of Bell's photophones. Bell's version used a very thin glass mirror attached to the end of a speaking tube. The mirror became alternately convex and concave in response to the un-

dulatory pressure of sound waves, thereby causing the divergence of the beam of sunlight reflected from the mirror to vary accordingly. Consequently the intensity of the beam at a distant point was amplitude or intensity modulated by the spoken words.

There are many ways to fashion a transmitter for a do-it-yourself photophone. The one shown in Fig. 7-29 is made by attaching a sheet of aluminum foil (shiny side out) to one end of a hollow tube. The tube can be cardboard, plastic, or metal. The foil can be attached with tape or a rubber band. Aluminized Mylar provides a smoother, flatter surface than aluminum foil. Therefore the beam it projects is narrower than that reflected from an aluminum foil transmitter. Unfortunately, aluminized Mylar is partially transparent and, therefore, has less reflectance than aluminum foil.

The receiver for a homemade photophone can use any of the light sensors of Fig. 7-28. For best results, however, use a silicon solar cell. The large surface area of this detector provides excellent collection efficiency without the need to use an external lens. Since the photophone is designed to be used in daylight, it's important to increase the signal-to-noise ratio by installing the cell in one end of a hollow tube. For best results, paint the inside of the tube with flat black paint or line the tube with black construction paper.

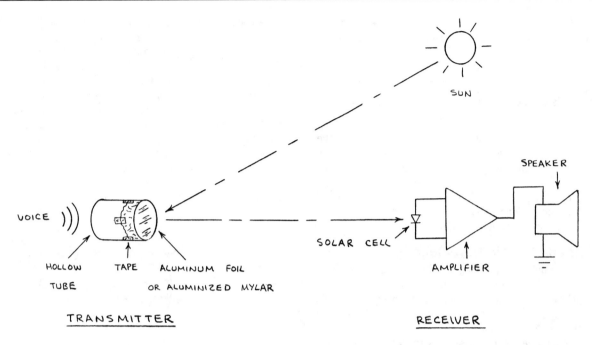

Fig. 7-29. A basic photophone system.

To use the photophone, simply reflect a spot of sunlight toward the detector. While holding the transmitter as still as possible, speak into its open end. The person at the receiver end should then hear every word you speak. The quality of voice reproduction will be influenced by the dimensions of the transmitter and the amount of ambient sunlight reaching the solar cell. For best results, the sun should be behind the receiver, and the receiver should be in a shady spot.

You can easily experiment with the basic photophone concept presented in Fig. 7-29. For example, add an equalizer to the output of the amplifier to enable the receiver to enhance the quality of the received signal. Or try different size transmitter tubes and reflective materials to see which combination works best.

I've made many kinds of photophone transmitters and receivers. The smallest transmitter is a 1-inch length of 1-inch diameter aluminum tubing. An ultrathin glass mirror is cemented to one end. I've even made transmitters by taping a flat sheet of aluminum foil or aluminized Mylar over a circular opening cut in a flat piece of cardboard.

As for receivers, I've used many different kinds of detectors and amplifiers. The most ambitious uses an 18-inch diameter glass parabolic reflector installed in a wood cabinet assembled for the purpose. A folding arm assembly permits a solar cell to be placed at the focal point of the mirror when the cabinet is opened. A transistor amplifier and speaker are installed inside the cabinet. For complete construction details, visit a library and find the February 1976 issue of *Popular Electronics* (pp. 54–61).

An experiment Otis Imboden and I once conducted will give you an idea of the versatility of the photophone concept. Otis is a photographer for *National Geographic* magazine who often photographs both gas and hot air passenger-carrying balloons. When the two of us were in Albuquerque, NM photographing a hot air balloon competition, I showed Otis some of my photophone equipment. Being the innovator he is, Otis suggested using as a photophone transmitter one of the large foil-covered reflectors he brought along to brighten up the passengers in the baskets of hot air balloons he was photographing. I gave a receiver to the pilot of a tethered balloon and Otis managed to speak to her by means of sunlight reflected from his giant photophone transmitter. Though I don't know if it was ever tried before, the combination of those two technologies could have been used to establish ground-to-air communications a century ago.

How much range will a photophone system give? A simple system using a solar cell receiver will easily give a range of up to a few hundred feet. For more range it's necessary to use a lens at the receiver or to do as Bell did and use a focused beam of sunlight to provide light for the transmitter.

In any case, it's very important that the person at the receiver end wear dark sunglasses and avoid staring at the bright reflection of sunlight from the transmitter. Staring at the beam for even a second or so will cause a scotoma, a temporary after image that leaves a blind spot in the field of vision. Staring at the reflected sunlight for more than a few seconds may cause temporary or even permanent damage to the retina of one or both eyes.

Going Further

Many electronics books describe applications for audio amplifiers. I've described several applications in *The Forrest Mims Circuit Scrapbook* (McGraw-Hill, 1983). A much better source is Calvin R. Graf's *Listen to Radio Energy, Light, and Sound*

(Howard W. Sams & Co., 1978). Cal's book describes dozens of experiments you can perform with a battery-powered amplifier.

Measuring the Flow of Air

There are various applications for devices that detect the movement of air. The design of such devices is usually very simple, and two simple devices you can build to measure the movement of air are described in the following. One device is a cup anemometer patterned after those meteorologists use to measure windspeed. The other is a hot-wire anemometer capable of detecting minute movements of air.

Applications for Air Movement Sensors

The most obvious application for devices that detect the movement of air is the measurement of windspeed. A closely related application is the measurement of the speed of a vehicle or aircraft. Airspeed indicators are also used to measure the velocity of air in a wind tunnel.

Air-movement detectors and sensors are sometimes used to monitor the blower in a heating or cooling system. The detector triggers a warning signal or shuts down the system when the air flow falls below a preset level. This same principle can be used to monitor the air flow in a clean room environment.

Air-flow detectors can be used to count objects on an assembly line or detect the edge of a nearby object. This application is accomplished by directing a jet of air toward the sensor. Objects passing between the jet and the sensor block the flow of air and actuate the sensor.

Finally, air-flow detectors have many uses in science and medicine. They can be used to monitor respiration and the flow of oxygen. Air-pressure switches can be used by disabled people to trigger electrical circuits and to operate computers. The operator simply puffs into (or sips from) a plastic tube to close the switch.

A Hot Wire Anemometer

The electrical resistance of a conductor changes with temperature. For example, a platinum wire that has a resistance of 2 ohms at 0°C has a resistance of 2.5 ohms at 100°C. In this case, the temperature coefficient of the wire is 0.0025/°C (0.5 ohm/2.0 ohm/100).

Air flowing past a heated wire tends to cool the wire, thereby lowering its resistance. By monitoring the resistance of the heated wire and taking the temperature of the surrounding air into account, it's possible to measure the speed of the air past the wire. A sensor designed for this specific purpose is called a hot wire anemometer.

Hot wire anemometers can be used to measure very small changes in air movement. Since the active surface area of the device can be quite small, hot wire anemometers are very useful for accurately portraying the flow of air and the turbulence around wind tunnel models. They can even be used to detect the movement of air created by the vibrating wings of a small insect.

Among the materials best suited for making hot wire anemometers are tungsten, platinum, and an alloy of platinum and iridium. Tungsten has a higher temperature coefficient of resistance than platinum (0.004/°C). When heated, however, tungsten tends to oxidize much more rapidly than platinum.

Figure 7-30 shows one kind of commercial hot wire anemometer. Notice that the active area of the probe is determined by the plating applied to either end of the tungsten wire. In recent years the sensing element of this basic probe has in many cases been replaced by a tiny quartz rod coated with a thin film of platinum. This sensor, which is called a hot film anemometer, responds more quickly to variations in air flow since a much smaller mass of metal is heated and cooled.

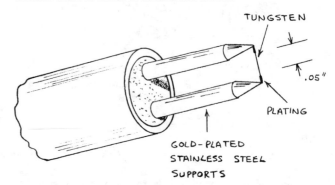

Fig. 7-30. A typical commercial hot-wire sensing element.

It's easy to experiment with hot wire anemometry since an ordinary flashlight bulb makes an effective hot wire sensor. All that's necessary is to remove the glass envelope from the bulb and monitor the current flowing through the filament. A change in the air flowing past the heated filament will change the resistance of the filament, hence the current flowing through it.

Of course the current flowing through the filament of the exposed light bulb must be much lower than that applied when the unbroken bulb functions as a light generator. Otherwise, the filament will be quickly consumed. Figure 7-31 shows a simple circuit I've devised that both applies a safe current and permits the monitoring of the current through the filament for this purpose.

In operation, a 7805 voltage regulator supplies a constant voltage that is applied to an incandescent flashlight lamp in series with R1, a 50-ohm, 5-watt resistor. Variations in the air flowing past the filament cause fluctuations in the resistance of the filament, hence the current through both the filament and R1.

R1 and the filament of L1 form a voltage divider. As the resistance of L1 changes in response to the air flow past its heated filament, the voltage applied to the inverting input of the 741 op amp varies accordingly. The 741 amplifies the voltage fluctuations and sends them to a voltmeter.

R5 controls the gain of the op amp [gain = −(R5/R2)]. R4 permits the output of the 741 to be zeroed when a measurement session is begun. Important: Unless the battery leads are short, it's important to bypass all the power supply connections with 0.1-μF capacitors. Connect the capacitors close to the two ICs.

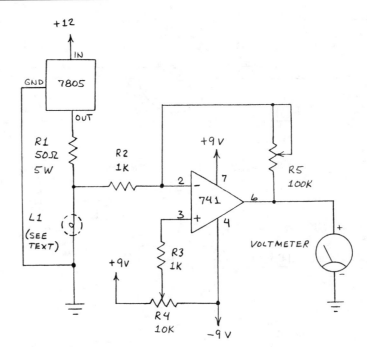

Fig. 7-31. Basic hot-wire anemometer circuit.

L1 is a PR13 flashlight bulb. Unless you want to use a lamp socket, solder a pair of insulated wire leads to L1. You must use care when removing the glass envelope from L1 for the glass is sharp and the filament is fragile. The method I use is to wrap several layers of masking tape around the envelope. I then squeeze the tape at the top of the bulb so that no glass is visible. Next, I place the taped envelope between the jaws of a C-clamp or a vise. I then slowly tighten the vise or clamp until the bulb pops. If the tape is pressed around the entire envelope, all the broken glass will usually lift away with the tape. Carefully use needle-nose pliers to remove any small shards of glass protruding from the metal base of the bulb.

CAUTION: A flashlight bulb may propel glass fragments a *considerable* distance if it is broken without appropriate protection. Wear safety glasses when breaking a flashlight bulb. Avoid using pliers or a hammer to break a bulb since the filament may be damaged.

The chief drawback of the circuit in Fig. 7-31 is the high current required by the 7805-R1-L1 combination (about 250 milliamperes). Even though this is only about half the current required by a 6-volt lantern light, it's much too high for powering the circuit with one of the 9-volt transistor radio batteries used to power the amplifier. Therefore, it's best to power this portion of the circuit with a pair of 6-volt lantern batteries connected in series.

The 7805-R1-L1 combination dissipates about 3 watts of power. Therefore, if the circuit is to be operated for more than a few tens of seconds, the 7805 should be fitted with a suitable heatsink. Likewise, R1 should be a 5- or 10-watt resistor. Too small a power rating will cause R1 to be destroyed.

The basic circuit in Fig. 7-31 is amazingly sensitive. While testing it, I found that slowly passing a hand by L1 caused an increase of about 1.5 volts in the output voltage when R5 was 100 kΩ. Gently exhaling and inhaling near the exposed filament of L1 caused the output voltage to swing even more.

It's important to note that when the gain of the circuit in Fig. 7-31 is made very high, setting the output to 0 by means of R4 can be very difficult since L1 is so sensitive. You might want to try placing a small cover or container over L1 while adjusting R4. Also, you can simplify the calibration procedure by operating the circuit at lower sensitivity levels.

L1's filament is very fragile. Therefore you may want to devise a protective housing to protect it. A length of plastic tubing placed over the metal socket is one possibility. Holes can be cut in the plastic to permit the flow of air. The filament should be kept dry since even a small droplet of water will cause the filament to be quickly destroyed if the circuit is activated.

The hot wire anemometer in Fig. 7-31 is much too sensitive to monitor more than the gentlest breeze. But it can be used to detect drafts sneaking into a house through doors, windows, and electrical outlets. It can also be used in experiments that detect respiration. A comparator can be connected to the output of the 741 so the circuit can indicate when the air flow exceeds a desired level.

A Cup Anemometer

A dc motor functions as a generator when its shaft is rotated by an external force. This principle can be used to make a very simple anemometer. I once applied this principle to make a miniature anemometer that measured the air speed of a wind tunnel. The wind tunnel, which was strapped onto the passenger side of a car, was used to test a miniature guided rocket. The anemometer was made by attaching a small balsa cone to the shaft

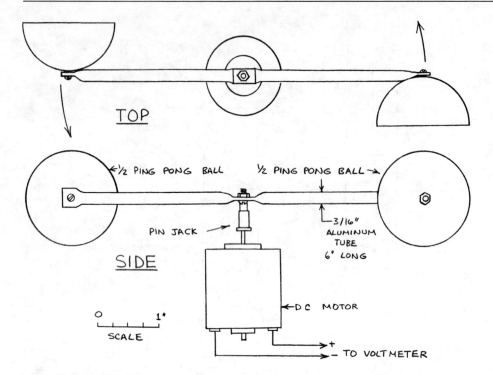

TOP

← ½ PING PONG BALL   ½ PING PONG BALL →

SIDE

PIN JACK →

└ 3/16"
ALUMINUM
TUBE
6" LONG

SCALE
O   1"

← DC MOTOR

→ +
→ − TO VOLTMETER

Fig. 7-32. Do-it-yourself cup anemometer.

of a dc motor. Four blades fashioned from the lid of a tin can were inserted into the balsa to form a propeller.

There are many ways to fashion a cup anemometer based on this principle. Figure 7-32 shows the details of the construction of a simple cup anemometer that I recently assembled. The two cups are halves from split ping-pong balls. They are attached with 4–40 hardware to the ends of a 6 × 3/16-inch hollow aluminum tube (available from hobby shops).

The ends of the tube are flattened with pliers and then drilled to receive the mounting screws. The center of the tube is flattened at a 90-degree angle to the flattened ends and drilled. The solder lug from a pin jack is bent at a right angle and secured to the center of the tube with 4–40 hardware. The receptacle end of the pin jack is then pressed onto the shaft of a small dc motor.

Use care when slicing the ping-pong balls in half. I used a sharp hobby knife and wore heavy gloves. Ping-pong balls are tough, so you must be careful.

To test the anemometer, I used a length of flexible metal strap (hardware stores) to secure the motor to one end of a sturdy aluminum rod. I then mounted the rod to the side mirror on the passenger side of a pickup truck and connected the motor's leads to a voltmeter. My son, Eric, then drove the truck at various speeds on a still day while I recorded the voltage readings.

Figure 7-33 is a graph that shows the ac and dc readings Eric and I obtained during the test session. The cups begin to rotate when the wind speed reaches 3–4 mph. The speed at which your anemometer begins to rotate and the slope of the calibration curve is dependent upon the motor you use. Note that the output from the motor is reasonably linear. This cor-

responds nicely with results I obtained with the wind tunnel anemometer described previously.

During the tests, I noted some fluctuations of the voltage output at certain speeds. When this occurred, the voltage reading would jump back and forth over a range of a few tens of millivolts. Therefore, I recorded what appeared to be the average voltage.

The anemometer in Fig. 7-32 can be improved by adding another pair of cups. In its present configuration, the cups don't always turn when the wind is below 10 mph unless they are perpendicular to the oncoming wind.

The motor must be protected from rain should this anemometer be installed outdoors. One possibility would be to install a split ping-pong ball over the top of the motor. It could be mounted to the 4–40 hardware that holds the pin jack in place. The split ball would rotate with the cups and keep rain from entering the top of the motor.

Going Further

You can find much more information about devices that detect and measure the flow of air at any good library. An excellent article on hot wire anemometry is "Hot Wire and Hot Film Anemometry" by Eric Nelson (*Sensors*, September 1984, pp. 17–22).

The TurboMeter is a compact anemometer with a shrouded fan and a digital readout. This unit measures winds up to 100 mph. The TurboMeter and other anemometers are sold by Edmund Scientific.

For more information about Honeywell's ultrasensitive air pressure switch, see *The Forrest Mims Circuit Scrapbook* (McGraw-Hill, 1983, pp. 138–140). Included in this reference

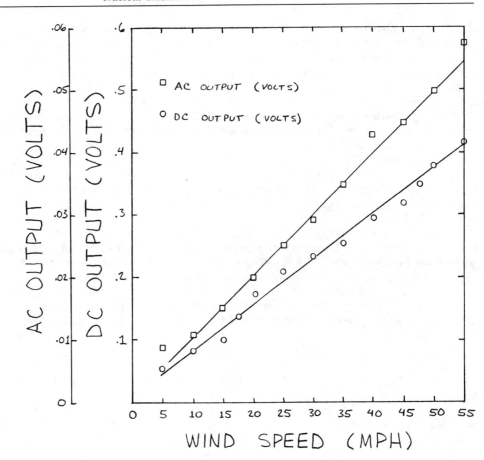

Fig. 7-33. Cup anemometer calibration curve.

are experimental circuits that permit disabled people to control external devices by puffing or sipping into a plastic tube. Also included is an experimental respiration detector circuit. The wind tunnel anemometer mentioned in this discussion is also described in this book (pp. 133–134).

Nuclear Radiation

The subject of nuclear radiation often generates concern. I was reminded about this recently when I took a portable Geiger counter into a department store while searching for radioactive materials. In the sporting goods section, I managed to thoroughly alarm a wide-eyed sales lady when the counter emitted a chorus of loud chirps after I pushed it into a display of a hundred or more Coleman lamp mantles hanging from a pegboard. As you may know (the sales lady didn't), lamp mantles contain radioactive thorium.

Radiation

Radiation is a very broad term for energy emitted by a source. Most forms of radiation are *nonionizing*. Although such radiation may heat or even physically or chemically alter a substance, it does not ordinarily cause atoms within the substance to gain or

lose electrons and thus become *ionized*. Nonionizing radiation includes visible light, infrared radiation, and radio waves.

There are, of course, exceptions to most rules. For example, focused optical radiation from a laser having a power density of at least from 1 to 10 megawatts per square centimeter (depending upon the wavelength) will create a spark-like miniature plasma of ionized air molecules. Ionizing radiation, however, is a descriptive phrase almost always reserved for very short wavelength electromagnetic radiation such as gamma rays and both alpha particles (the nuclei of helium atoms) and beta particles (fast moving electrons). These particles ionize atoms when they collide with them at very high speeds.

Substances which emit ionizing rays or particles are collectively termed *radioactive*. The effect of ionizing rays and particles upon most forms of life can be profound. For this reason alone, considerable effort has been devoted to the development of sensitive and reliable detectors of radioactivity.

One of the most important early detectors of radioactivity was invented by Hans Geiger. A sealed metal tube filled with air, argon, or other gases and containing a wire electrode, Geiger's detector is operated by placing a high voltage across the wire electrode and the metal tube. The potential is set below the breakdown point of the gas.

A gamma ray or alpha or beta particle entering a thin glass or mica window in the side of the tube will cause the molecules of gas along its path to become ionized. The resulting low impedance path allows a brief flow of current through the tube. Additives called *quenching agents* cause the gas to quickly deionize following passage of the ray or particle. This greatly increases the life of the tube and enhances its detection capability in high radiation fields.

The current pulse through a Geiger tube can be easily amplified and caused to generate an audible click. In high radiation fields, the clicks will merge into a continuous rash or even a buzz. A simple integration circuit can convert the noisy clicks into a series of chirps that indicate by their frequency of occurrence relative radiation intensity. An analog meter, digital readout, or simply a flashing LED may also be used to indicate relative radiation levels.

Geiger tubes are still very commonly used to detect the presence of radioactivity. They are relatively expensive and fragile, however, and they require a high operating voltage.

The *scintillation effect* radiation detector utilizes a fluorescent mineral or crystal such as thallium or cesium-activated sodium iodide or bismuth germanate. Ionizing radiation entering the crystal causes the emission of light which is detected by a sensitive photomultiplier tube.

Though a scintillation crystal alone is a solid-state device, the requirement of a fragile photomultiplier tube with an accompanying high voltage power supply gives scintillation detectors the same drawbacks as Geiger tubes. Recently, however, new kinds of scintillation crystals have been developed that emit at slightly longer wavelengths which can be detected by silicon and gallium arsenide phosphide photodiodes.

One such crystal is cadmium tungstate. When stimulated by ionizing radiation, it emits green light having a peak wavelength of about 530 nanometers. This light can be detected by a suitable photodiode connected to a low-noise, high-gain amplifier.

A scintillation crystal paired with a photodiode forms a rugged solid-state detector of radioactivity. A simpler approach, however, is to detect radioactivity *directly* with a semiconductor. Silicon pn junction diodes can detect gamma rays and both alpha and beta particles. Although not as sensitive as other detectors, they are mechanically sturdy and require no high voltage power supply. Even commercially available silicon diodes, photodiodes, solar cells, and zener diodes have been used to detect ionizing radiation.

Cadmium telluride is a more sensitive detector of ionizing radiation than silicon. Cadmium telluride detectors are used in nuclear medical research, space research, brain scanners, and in compact personal radiation monitors.

Direct Reading Dosimeters

The direct reading dosimeter is a modern form of the venerable electroscope, one of the first devices capable of indicating the presence of static electricity. The electroscope is based upon the well known principle that unlike charges attract and like charges repel.

Traditional electroscopes are made by folding a rectangular piece of gold leaf into two equal halves and hanging the gold leaf from a conducting support in a glass bottle. The electroscope is "charged" by touching a positively or negatively charged object or electrode to a metal sphere attached to the conductor emerging from the bottle. This applies an equal charge to both leaves of gold foil. Since like charges repel, the leaves will then fly apart and defy gravity until their charge gradually leaks away into the surrounding air or is intentionally shorted to ground.

So how does the electroscope detect radiation? The charged leaves of the electroscope will gradually give up their charge by transferring electrons to or receiving electrons from the surrounding air. Unless it contains moisture, however, air is a poor conductor of electrons. Radioactive particles and rays have the ability to strip electrons from atoms that form air. These ionized atoms provide a conductive path for the charged leaves of an electroscope. Therefore, assuming no other leakage paths are present, the distance between the leaves of a charged electroscope is proportional to the cumulative radiation which has entered the space around the electroscope's leaves.

A typical pencil dosimeter made by Dosimeter Corporation is smaller than a penlight. The device is prepared for use by first inserting the end containing the contact pin into the socket of a charger. The contact pin is mounted within a flexible metal bellows. When the dosimeter is pressed down into the charger's socket, the opposite end of the contact pin makes contact with a metal member (the frame) inside the dosimeter. A charge of about 170 volts is then transferred to the frame via the contact pin.

The charge is also applied simultaneously to a metal-plated quartz fiber suspended from and in electrical contact with a hinge on the frame. Since the frame and the fiber are given equal and like charges, the fiber swings away from the frame.

The sealed portion of the dosimeter that houses the quartz fiber and its frame is filled with dry air and is called an *ion chamber*. Radiation entering the chamber ionizes atoms of air, thereby providing a conductive path for some of the electrons on the fiber. As the charge on the fiber is gradually diminished by a radiation field, the quartz fiber moves toward the frame.

The position of the fiber can be viewed against a scale inside the dosimeter by peering through the instrument's eyelens while pointing the contact end toward a light source. Light enters the instrument through a glass collar around the contact pin.

A key component of the dosimeter is the contact pin-bellows assembly. Since the bellows pulls the contact pin away from the frame when the instrument is pulled from the charging socket, the charged quartz fiber is electrically isolated from the outside world. Consequently, a typical dosimeter typically loses only about 0.25 percent of its charge per day.

Pencil dosimeters are a very important radiation monitoring device. And unlike most electronic monitors, they are truly pocket sized. Although they don't provide an indication of the *rate* of incoming radiation, they do give an accurate measure of *cumulative* exposure. And though they require an external charger, one charger can be used to service dozens of instruments.

A Homemade Electroscope

Figure 7-34 shows a simple electroscope you can make from homemade materials. For best results use a heavy gauge copper wire for the conducting support. Round off both ends of the wire

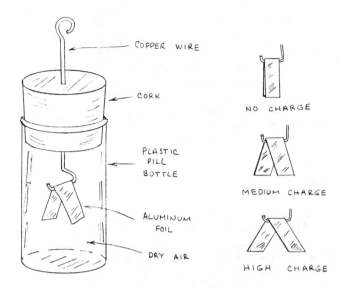

Fig. 7-34. Simple do-it-yourself pill bottle electroscope.

with a small file and fine sanding paper. This eliminates sharp edges and burrs that would otherwise serve as discharge points for the charged leaves.

Some hobby and craft stores sell gold leaf which you can use to make a traditional electroscope. I've used both standard and heavy-duty grocery store brands of aluminum foil with good results. The standard gauge works best.

Prepare the foil by cutting it with sharp scissors to the size you plan to use. The finished size isn't critical so long as the leaves don't touch the sides of the bottle. Next, smooth the foil by placing it on a flat surface and stroking it a few times with a smooth pencil or your thumb. For best results, the foil should be as flat as possible. But this can be tricky to achieve since the foil tends to curl when it's stroked.

After you smooth the foil, fold it in two equal sections over the edge of a paper card and again smooth both sides. Then remove the card and hang the foil leaves over the support wire

which has previously been inserted through a cork and formed as shown in Fig. 7-34. The leaves should be closely spaced and parallel to one another.

You can charge this homemade electroscope simply by touching the exposed end of the support wire with a comb you've stroked a few times through your hair. If both your hair and the comb are dry, the leaves will fly apart and remain erect for at least a few seconds. On a very dry day, the leaves will remain erect for quite some time.

If the leaves appear to stick to the sides of the bottle, use a larger container or smaller leaves. If the leaves fail to diverge when the electroscope is charged, make sure they have not become attached to one another at any slight nicks or frayed edges along the edge of the foil.

Failure of the leaves to diverge may indicate moist air in the electroscope's bottle (its ion chamber). Assuming everything else checks out but the leaves still fail to repel one another, it may be necessary to replace the air in the bottle with dry air.

Several companies sell cans of pressurized, filtered, dry air for blowing dust from photographic film and mechanical devices. I've had good results using such air in my homemade electroscopes. Just remove the cork and leaves, squirt a dose of dry air into the bottle, and quickly replace the cork and leaves.

When you place a radioactive sample (more on this later) near the exposed electroscope wire, the leaves will begin to collapse at much faster than their normal rate. Move the sample away and the leaves will stop falling. This proves that a charged electroscope, even one made with household materials, can indeed detect ionizing radiation.

A Better Electroscope

Figure 7-35 shows another homemade electroscope with which I've experimented. This electroscope differs from the first in that it has but one moving leaf. The second leaf has been replaced by a rigid rectangle of two-sided, copper-plated, printed circuit board.

Referring to Fig. 7-35, the support member is a copper wire soldered to the back side of the PC board and bent around to the front side to form a support for the movable leaf. The leaf

Fig. 7-35. Better quality homemade miniature electroscope.

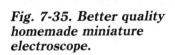

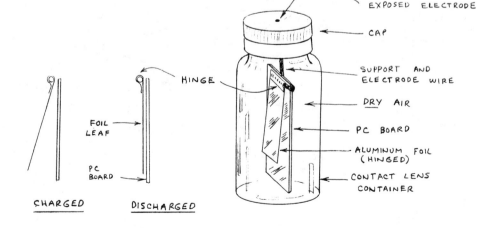

is a strip of smooth, flat aluminum foil. Wrap one end of the foil strip around a wire having a diameter larger than that of the support wire. Then slip the resulting foil tube over the support wire to form a hinged leaf.

This electroscope can be installed in a plastic pill bottle much like the one in Fig. 7-35. I used, however, a small glass bottle used to ship soft contact lenses. The bottle has a soft plastic cap into which you can easily insert the support wire. The cap provides a secure seal and is easily removed.

Note that this electroscope doesn't have a protruding wire contact. This greatly reduces the possible leakage paths, thus preserving the electroscope's charge for a longer time.

A Nonelectronic Electroscope Amplifier

You can amplify the motion of the leaves of an electroscope by reflecting a narrow beam of light from one of the leaves. If the reflected spot of light is directed toward a white card marked with an appropriate scale, you can easily measure movements of the leaf.

Figure 7-36 shows how you can use a helium-neon laser for this purpose. Note that one side of a sheet of aluminum foil is shinier than the other. For best results, make sure the shiny side of the electroscope leaf faces *outward*.

If you place the white card a few feet away from the electroscope, the reflected spot of light will move as much as 6 inches or even more as the leaf falls from its fully charged, erect position. Since the foil surface is not perfectly flat, the reflected spot of light will be fairly large and blurred around the edges. Nevertheless, by arranging the electroscope, laser, and card in suitable positions, you should have little difficulty detecting very tiny movements of the electroscope's leaf when you place a sample of radioactive material near its external electrode.

Radiation Source

By now you may be wondering where to obtain a radioactive source to check out a homemade electroscope. One possibility is to purchase a source from a manufacturer or distributor of radiation monitors. For instance, Dosimeter Corporation sells for about $35 a source containing 5 microcuries of cesium-137.

You can obtain other sources for considerably less. The cheapest source is probably a thorium impregnated lamp mantle. Such mantles, which are made by Coleman, Aladdin, and Gaz, contain thorium-232, an isotope that can be bred into uranium-233, the fissionable isotope used in nuclear reactors and weapons. Thorium-232 emits alpha particles and decays through a series of ten *radiodaughters* (subsequent elements), the first being a beta emitter, radium-228.

A second, reasonably low-priced radioactive source is the polonium-210 used in Staticmaster dust removers. A product of Nuclear Products Company, Staticmasters are soft brushes equipped with a replaceable cartridge containing a strip of metal coated with polonium-210 permanently encapsulated in tiny ceramic beads. The polonium-210 emits a potent spray of alpha particles that ionizes the air near the brush. This provides a conductive path for the static charge that attracts dust particles to glass lenses, phonograph records, and photographic film.

Staticmaster brushes (they come in two sizes) and replacement cartridges are available from some audio equipment and camera stores. If you just want a radioactive source, save your money and buy only a replacement cartridge. Be sure to comply with the safety precautions supplied with the cartridge!

Experimenting with a Geiger Counter

S.E. International's Monitor 4 is a compact Geiger counter which indicates radiation levels by means of a moving coil meter,

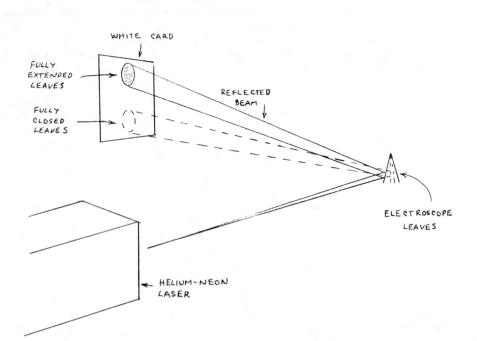

Fig. 7-36. Amplifying the movements of an electroscope's leaves with a laser.

flashing LED and an audible chirp generator. The instrument incorporates CMOS circuitry and consumes only about 3 milliwatts in a low radiation field. This provides a life of up to 2000 hours for the 9-volt battery that powers the instrument so long as the radiation level doesn't exceed a nominal background count. The Monitor 4 uses a Geiger tube having a mica end-window to permit the detection of alpha particles. The tube also detects beta particles and gamma and x-radiation. For more information, write S.E. International.

I've used the Monitor 4 to detect radiation from several sources. For example, a Coleman lamp mantle lying on a flat surface produces a reading of from 0.1 to 0.2 milliroentgens per hour (mR/h) when the Geiger tube's port is placed directly over the mantle. Rolling the mantle into a tight bundle and placing it next to the port gave a reading of from 0.4 to 0.5 mR/h.

When I placed a sheet of paper between the mantle and the Geiger tube, the radiation level was only slightly lowered. Since alpha particles are stopped by paper, the bulk of the radiation appears to be beta particles emitted by the radium-228 by-product of the thorium in the mantle.

I've also used the Monitor 4 to detect alpha emission from the polonium-210 microbeads in a Staticmaster IC200 static eliminating brush. When the Geiger tube port is placed directly against the grid over the polonium-210 microbeads, the radiation level exceeds the Monitor 4's maximum detection level of 50 mR/h. Alpha particles are blocked by only a few centimeters of air. When the grid over the polonium-210 is placed exactly 1 centimeter from the Geiger tube port, the radiation level is 10 mR/h. At 2 centimeters, the level is too low to measure.

Polonium-210 decays with time, and Staticmaster cartridges are stamped with an expiration date. Therefore, the measurements you obtain may differ from those I obtained.

Incidentally, another proof that the emissions from the polonium-210 are indeed alpha particles is to place a sheet of paper between the Geiger tube port and the Staticmaster cartridge. When I tried this, the radiation level was too low for the meter to indicate even when the alpha source was placed directly over the Geiger tube port.

You can use the Monitor 4 and similar monitors to detect very low levels of radiation if you first determine the natural background count. The background count can range from several to more than a hundred counts per minute, although the typical count is usually between 10 and 25 counts per minute.

The origin of the background count is cosmic rays and the natural radioactivity present in soil and perhaps the building materials of your home or office. The background count can vary with atmospheric conditions. For instance, a tightly sealed home can collect a higher than usual accumulation of radioactive radon gas emitted by construction materials such as brick and tile. Likewise, inversion layers in the atmosphere can trap larger than usual quantities of naturally occurring radon, thereby increasing the background count.

After you measure the background count, you can then proceed to check suspected low-level radiation sources. For instance, on a day when the background count was 16 counts per minute (averaged over 5 minutes), the center of the screen of a color television set I use as a computer monitor gave a count rate of 28 per minute. Although this figure is nearly double the background rate, it's still very low.

Other radioactive sources around my home, checked on a day when the background count averaged 11 counts per minute, include a glazed brick used as a step for a storage building (40 counts/minute) and a ceramic tile entryway (16 counts/minute). An ionization-type smoke detector with an internal radiation source, presumably polonium, produced no detectable radiation above the background count.

Other household radiation sources include older pieces of earthenware glazed with orange or red pigment containing uranium oxide. And though they are no longer manufactured due to their hazardous properties, watches and clocks with hands and numbers coated with radium impregnated luminescent paint are still around.

Mr. Milo Voss, Manager of Safety, Health and Plant Protection at the Ames Laboratory of the U.S. Department of Energy, is one of many health physicists who have studied low-level background radiation. In a recent telephone conversation, Mr. Voss recounted how aerial surveys have spotted higher than usual radiation levels over some cemeteries and golf courses. Apparently the source of this radiation is the natural radioactivity of the granite headstones in the cemeteries and the phosphate fertilizer spread over the greens of the golf courses. Mr. Voss also observed that pilots and passengers of high flying aircraft are exposed to higher than usual levels of radiation.

Summing Up

The subject of nuclear radiation always generates considerable controversy, particularly when so-called minimum acceptable exposure levels are discussed. Some health physicists believe no level of exposure can be incurred without some risk to the population. Others feel this view is far too extreme, particularly in view of the naturally occurring background radiation to which we are all subjected.

Milo Voss, for instance, has studied thorium in some detail and concluded in a 1979 report that the material is relatively safe unless it is inhaled or ingested. On the other hand, Walter Wagner, a Veterans Administration health physicist, has filed a $300 million class action lawsuit against the Coleman Company and other manufacturers of lantern mantles. Mr. Wagner is convinced that the thorium in the mantles constitutes a public health hazard.

Despite their divergent views regarding exposure levels, both sides agree that, for better or worse, naturally occurring radioactive sources abound. They are found in granite, bricks, grains, soils, and even in our bodies. The ceramic housing of a DIP integrated circuit can be slightly radioactive!

Another area of agreement is that radiation, no matter its source, is very difficult to accurately measure. I've had considerable experience measuring light, but measuring radiation is far trickier. That's because there are various kinds of radiation, each with distinctive energy levels.

You can find out much more about the many aspects of nuclear radiation at any good library. You can also obtain helpful advice about measuring radiation from the various companies that make radiation monitoring instruments.

Experimenting with an Ultrasonic Rangefinder

Several years ago I walked the grounds of the Arkansas Enterprises for the Blind while blindfolded. Upon my head I wore a unique pair of spectacles whose bridge contained a triangular array of three ultrasonic transducers. A cable connected the spectacles to a pocket-sized electronics package.

As I scanned my surroundings, a sequence of strange and exotic swept frequency tone bursts was fed into my ears by plastic tubes. Depending upon the target's distance and surface texture, the sounds ranged from *drzzz . . . drzzz . . . drzzz . . .* to *whoosh . . . whoosh . . . whoosh.*

Having long been interested in the design of electronic travel aids for the blind, I was fascinated by the engineering of the eyeglass device I tried. On the other hand, I found the sounds it produced much more confusing and difficult to understand than the simple go-no go tones emitted by the much smaller and lighter eyeglass-mounted infrared travel aids I had previously developed.

Both infrared and ultrasonic travel aids for the blind, a topic I'll cover in more detail in future columns, are still undergoing development. One of the latest uses the unique ultrasonic ranging system Polaroid developed for its SX-70 camera. This system has many useful applications, and its design and operation are the subject of this discussion. First, however, let's find out more about ultrasonic sound.

Ultrasonic Sound

Sound waves having frequencies below and above the limits of normal human hearing are termed, respectively, *infrasonic* and *ultrasonic* sound. Infrasonic sound has a frequency less than a few hertz. It is generated by earthquakes, machinery, and moving air. Ultrasonic sound has a frequency greater than 20 kHz. It is generated by mechanical and electronic sources, jingling keys, rustling leaves, and animals such as bats.

Applications for low intensity ultrasonic sound include measurement of mechanical stress, flaw detection, and nonoptical imaging devices such as those used to view the fetus of a pregnant woman. High intensity ultrasonic sound is used for soldering, surgery, mixing liquids such as water and oil, and forcing dirt and oil from objects immersed in an ultrasonically agitated liquid.

Although all these applications are important, none has generated as much interest as two of the very first applications for ultrasonic sound, distance measuring and object detection. As early as 1918, scientists had developed practical systems for detecting submarines by reflected waves of ultrasonic sound. In World War Two, this technology became widely known as *sonar*, an acronym for *so*und *na*vigation *a*nd *r*anging. Today military applications for sonar range from detection of submarines over a range of 10 kilometers or more, mine detection, and guidance and control of various homing weapons.

The best known civilian applications for sonar include the detection of fish and depth sounders. Detection applications in which the sound waves are propagated through air include ultrasonic intrusion alarms and, as described before, travel aids for use by the blind and automatic focusing systems for cameras. This brings us to Polaroid's SX-70 ultrasonic ranging system, the subject of the remainder of this discussion.

Polaroid's Ultrasonic Ranging System

After Polaroid introduced its automatically focused SX-70 instant picture camera, many inventors, firms, and experimenters expressed more interest in the product's ultrasonic rangefinding system than in the camera itself. A few years ago Polaroid responded to this interest by introducing its Ultrasonic Ranging System Designer's Kit. The kit contains two preassembled circuit boards, two instrument grade electrostatic transducers, two 6-volt Polapulse batteries, a battery holder, and an instruction manual. The kit is available for $150 from the Polaroid Corporation.

This remarkable kit requires only that the battery holder leads be clipped or soldered to one of the two circuit boards. It can then be used as an ultrasonic rangefinder complete with a 3-digit LED display. It can detect and indicate the distance to objects within a range of 0.9 to 35.2 feet. It has a resolution of ± 0.12 inch at distances out to 10 feet and ± 1 percent over the entire detection range.

Having had considerable experience designing and testing infrared travel aids for the blind, I have particularly enjoyed experimenting with Polaroid's ultrasonic ranging system. The unit is much more sensitive than I had expected. For example, it will reliably detect a 1-inch diameter unfinished wood pole at 18 feet. It will detect a 9-inch diameter utility pole at 28 feet. And it will detect power lines 19 feet overhead.

The only major drawback is the inability of the system to reliably detect targets having a surface which is smooth with respect to the wavelength of the ultrasonic sound. Borrowing from optical terminology, such targets can be termed *specular reflectors*. If the surface texture of the target is rough with respect to the wavelength of the ultrasonic waves, the target can be termed a *diffuse reflector*.

Flat, specular targets whose surface is normal to the oncoming sound waves are readily detected. When the target is off-axis to the beam of sound, however, the ultrasonic waves are reflected at an angle away from the source.

This can give rise to anomalous readings if the diverted off-axis beam eventually strikes a second surface having acceptable target characteristics. The sound will reflect back to the intended target and from there to the ranging system. The range measurement displayed by the readout will therefore be the two-path distance to the second, unintended target. The intended target will be ignored. This phenomenon is illustrated in Fig. 7-37.

Infrared rangefinders that use LEDs and diode lasers exhibit this same problem when used to detect specular reflectors such as glass, polished marble, and automobiles with a high gloss surface. Therefore I was not surprised when the ultrasonic system could not detect these targets at an off-axis angle.

What was surprising was the wide range of targets having relatively rough surfaces which in fact appear specular to an

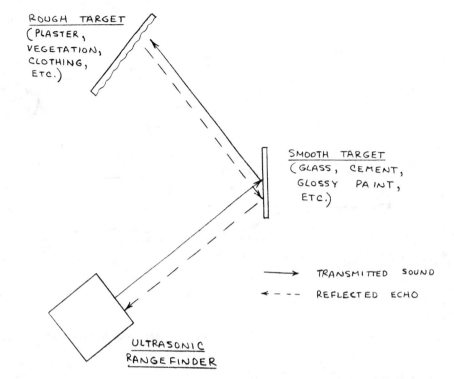

Fig. 7-37. How a target with a smooth surface can cause false distance reading.

ultrasonic beam. Typical targets that are difficult or impossible to detect at other than the normal angle include smooth cement walls and driveways, plywood, painted surfaces, automobiles, flat metal plates, and many others.

Fortunately, most targets include diffusely reflecting features that make possible detection even at off-axis angles. For example, the curved surfaces of most cars usually cause some of the ultrasonic beam to be reflected back toward the source. Overall, there are more diffusely reflecting targets (fabric, vegetation, shingles, people, posts, carpets, etc.) than specular targets. Nevertheless, it's prudent to be aware of the difficulties posed by specular reflectors when using an ultrasonic ranging system, particularly since Polaroid's otherwise excellent rangefinder manual fails to discuss the subject.

The Transducer

The key component of the Polaroid system is an instrument grade electrostatic transducer that doubles as both an ultrasonic speaker and microphone. In operation, a foil diaphragm stretched tightly over a concentrically grooved metallic backplate form a capacitor. In the receive mode, the capacitance of the transducer is altered by incoming sound waves. In the transmit mode, the electrostatic force of a charge placed across the capacitor causes the foil to move.

The transducer emits a relatively narrow sound cone (nominally 20 degrees in divergence at the −20-dB points). The best operating frequency for the unit falls between 50 and 60 kHz. These parameters are clearly summarized in the plots, adapted with permission from Polaroid's manual, which are shown in Fig. 7-38.

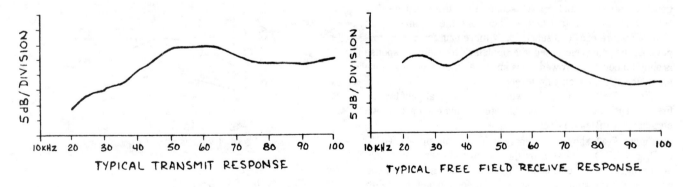

Fig. 7-38. Transmit and receive response of Polaroid's transducer.

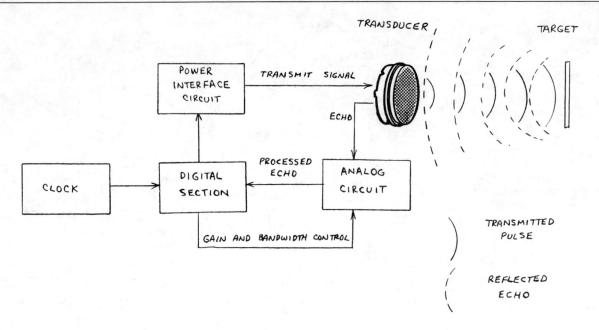

Fig. 7-39. Block diagram of ultrasonic circuit board.

The Ultrasonic Circuit Board

The transducer is connected directly to the ultrasonic circuit board, a slightly modified version of the board found in the SX-70 camera. Figure 7-39 is a block diagram showing the major sections of this board. The functions shown in Fig. 7-39 are implemented by three custom chips.

When activated, the ultrasonic circuit board applies to the transducer at intervals of about 200 milliseconds a one-millisecond chirped sequence of 14 pulses at 60, 57, 53, and 50 kHz. The pulses have an amplitude of 300 volts. The four different frequencies are used to minimize the possibility that the reflection characteristics of the target and, perhaps, its surroundings might cause destructive interference, thus cancelling the reflected sound wave.

The received signal, which may represent a single reflected echo or a series of echoes from various targets, is boosted by a cleverly designed amplifier which incorporates 16 levels of time-dependent gain control. The echo from a nearby target is generally stronger and arrives sooner than the much weaker echo from a more distant target. Therefore, the automatic gain control feature provides substantially more gain for distant targets. As the gain is increased, the amplifier's frequency response is simultaneously narrowed. This improves the receiver's noise immunity at very high gain levels.

Figure 7-40 is a graph showing the theoretical gain for the first 8 steps. Steps 9–16 resemble Step 8, and each successive step has a 4-dB gain increase.

Figure 7-41 shows the binary gain control signals (GCA, GCB, and GCC) generated by the timing and control circuitry on the ultrasonic circuit board. Also shown in Fig. 7-41 is the narrowing of the amplifier's bandwidth at and beyond the eighth gain step. Note the parallel listing of echo times and distances

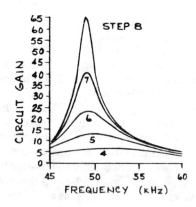

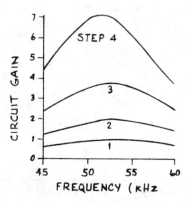

Fig. 7-40. The first eight gain steps of the receiver circuitry.

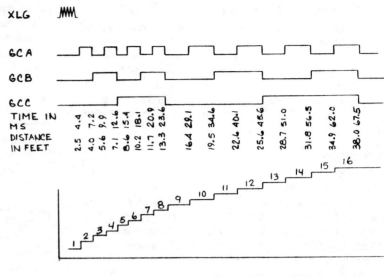

Fig. 7-41. Operation of the ultrasonic circuit board's gain control logic.

to the target. (Changes in air temperature will cause slight changes in these readings.)

Figure 7-42 summarizes the entire transmission and reception sequence for a single chirp directed against three targets within the detection field. Note how the circuit shapes each echo and then generates a clean, square pulse representing only the first echo. This signal is designated MFLOG and is available at pin 15 of the ultrasonic circuit board. The transmitted chirp signal is available at pin 16 of the board. The time duration between these signals represents twice the distance to the target.

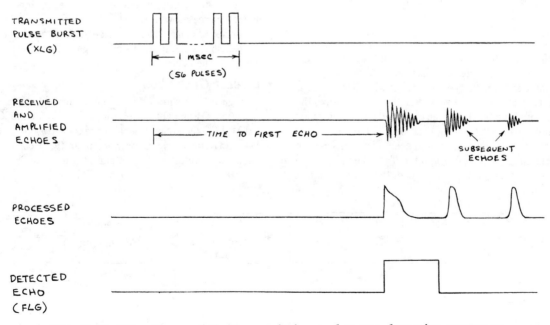

Fig. 7-42. Waveforms of the transmission and target detection sequence.

The Experimental Demonstration Board

The utility of Polaroid's ultrasonic ranging system kit is greatly enhanced by the experimental demonstration board. This board is essentially a custom 3-digit counter that converts the echo time signals from the ultrasonic circuit board into distance to the target to the nearest tenth of a foot. CMOS ICs are used throughout, and the range is displayed on a 3-digit LED readout. A red filter is included to increase the visibility of the display in the presence of bright ambient light.

Testing the System

The kit is supplied with the two boards connected by a 6-conductor ribbon cable. Therefore, all that's necessary to use the kit as a 0.9- to 35.2-foot rangefinder is to solder the two battery leads to the experimental demonstration board.

For best results, install the boards, transducer, and battery holder in a suitable enclosure. Make sure the transducer leads do *not* become disconnected and that exposed metal parts of the transducer, including its connections, do *not* touch either board.

I have temporarily used a compact 7 × 6 × 1-inch wood cigar box. Holes for the display and power switch can be cut through the box top, and a circular aperture for the transducer can be made in the bottom of the box. This arrangement will allow you to view range measurements while pointing the unit away from your face and toward targets of interest. For best results, of course, install the boards in a permanent metal or plastic enclosure.

Incidentally, Polaroid warns that the transducer *must* be properly connected to the ultrasonic circuit board before power is applied. Otherwise, the high voltage chirps may damage the board. One of the two transducers supplied with the kit is connected to the ultrasonic circuit board by a short, shielded cable. If the cable connections are flexed frequently, be sure to inspect the solder connections at regular intervals to make sure they are secure.

Applications for the System

One of the most interesting applications for Polaroid's rangefinder technology is Tailmate, a sophisticated detection system for trucks and trailers developed by Gregson Holdings Ltd.

When a Tailmate-equipped vehicle is placed in reverse, the system is actuated and the driver can then read on a cab installed readout the distance to a loading dock or other object. If the driver wants to stop the vehicle a specific distance from an object, he can enter the distance into the Tailmate unit. Three feet before the preset distance is reached, the system emits a pulsed warning tone. When the desired distance is reached, a continuous tone is sounded. The system can also measure the distance to overhead objects.

Another application is the Sona Switch, an indoor-outdoor object detection system designed for automatic door openers, vehicle detection, and security systems. The Sona Switch is a product of Electronic Design and Packaging Company.

Other applications include robotics, vehicle height detectors, automatic controls for agricultural equipment, and measurement of the product level in silos and storage tanks. A particularly interesting application is a wheelchair guidance system that helps quadriplegics maneuver a motor-driven chair through narrow passages like doorways and halls.

Experimenting with the Rangefinder

Polaroid's rangefinder can be used with little or no modification for many applications. If you want to interface the device to a computer bus, you may find helpful an article on the subject by Steve Ciarcia ("Home In on the Range!," *Byte*, Nov. 1980, p. 32).

An Audible Output for the Rangefinder

Some of the many useful applications for Polaroid's ultrasonic ranging system can be enhanced by the addition of an output tone whose frequency varies with the range to the detected target. Figure 7-43 shows in block diagram form an experimental circuit I've designed that converts the echo time of this rangefinder into a chirped tone. A single chirp is produced for each rangefinding cycle.

Briefly, an ultrasonic burst from the ranging system triggers a one-shot which, in turn, causes the capacitor in a sample-and-hold circuit to begin charging. When the echo is received, the capacitor stops charging and is immediately discharged by an analog switch. During the time before the echo is received, a voltage-to-frequency converter produces a tone whose frequency increases with the charge on the capacitor. The result is a series of chirps. Nearby targets (short echo time) give low pitched chirps and more distant targets (longer echo time) give higher pitched chirps.

Circuit Operation

Figure 7-44 shows the complete circuit for the chirper. In operation, Q1 and Q2 serve as input buffers for the XLG (start) and MFLOG (echo time) signals from the ultrasonic circuit board. The XLG signal triggers a 555 timer (IC1) configured as an astable multivibrator. The output from IC1 (pin 3) remains high for a time determined by R7 and C2.

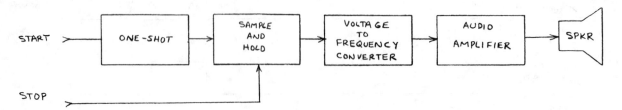

Fig. 7-43. Block diagram of audible output circuit for Polaroid's ultrasonic rangefinder.

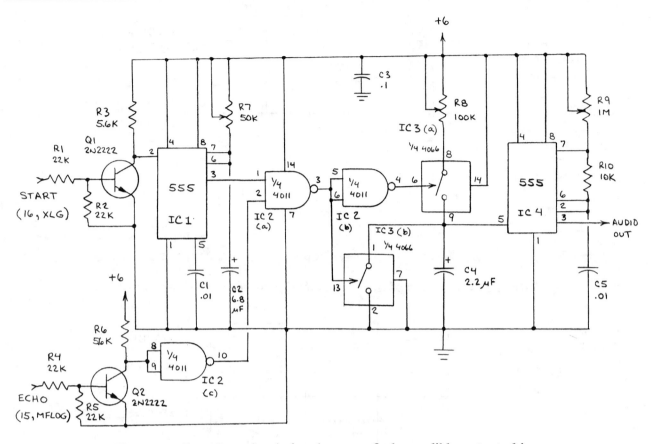

Fig. 7-44. Complete circuit for the rangefinder audible output chirper.

The MFLOG signal is ANDed by IC2A and IC2B, and, when both signals are present, analog switch IC3A is closed. This allows C4 to be charged through IC3A and R8.

IC4 is a 555 configured as a voltage-to-frequency converter tone generator. Its control input is connected directly to C4. Therefore, as C4 begins to charge, the oscillation frequency of IC4 begins to rise.

When the echo arrives, the MFLOG signal goes low and the AND gate output also goes low. This opens IC3A so that C4 is no longer charged. Simultaneously, IC3B is closed by IC2A and the voltage on C4 is dumped to ground. The voltage-to-frequency converter immediately ceases to oscillate. Depending upon the adjustments of R7, R8, and R9, the audio output from the circuit ranges from low pitched thumps (nearby objects) to high pitched chirps (distant objects).

You can better understand the circuit's operation by referring to the oscilloscope waveforms in Figs. 7-45 and 7-46. The upper trace in Fig. 7-45 shows the 1-millisecond tone burst that is applied to the ultrasonic transducer by pin 16 of the ultrasonic circuit board. The lower trace shows the square wave which begins with the transmitted tone burst and ends when the first echo is received. It appears at pin 15 of the ultrasonic circuit board.

Figure 7-46 shows the critical waveforms in the chirp generator circuit for three different target ranges (2.4, 6.3, and 9.5

feet). The upper trace shows how the echo time is increased for each of these ranges. In a millisecond sound travels approximately 0.89 feet. Therefore, the round trip time for the 9.5 feet range should be 9.5 × 2 × 0.89, or 16.9 milliseconds. This is in close agreement with the indication of about 16.75 milliseconds on the lower trace in Fig. 7-46.

Using the Circuit

The operation of this circuit is very much dependent upon the settings of potentiometers R7, R8, and R9. R7 controls the one-shot's pulse duration. If it is set to produce a pulse that exceeds in duration the time between range cycles (about 200 milliseconds), the output sound will be a series of double or triple thumps or beeps.

R8 controls the charging time of C4. If its resistance is too low, the voltage-to-frequency converter will produce a steady tone. If its resistance is too high, the voltage on C4 may take too long to reach the levels required to alter the tone from IC4.

Finally, R9 controls the pitch of the chirps. For best results, set R9 to produce very high pitched chirps when the system is detecting very distant targets. Nearby targets will then give a characteristic *thump . . . thump . . . thump.*

Going Further

The circuit in Fig. 7-44 is not optimized, and many variations

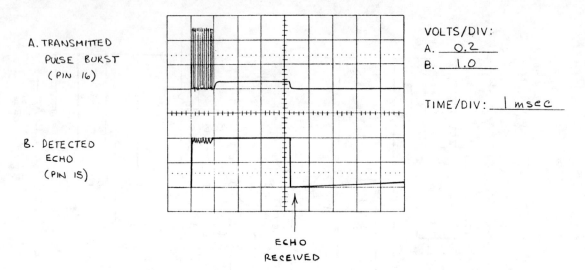

A. TRANSMITTED
   PULSE BURST
   (PIN 16)

B. DETECTED
   ECHO
   (PIN 15)

VOLTS/DIV:
A. 0.2
B. 1.0

TIME/DIV: 1 msec

ECHO
RECEIVED

Fig. 7-45. Key waveforms of the ultrasonic circuit board at target detection.

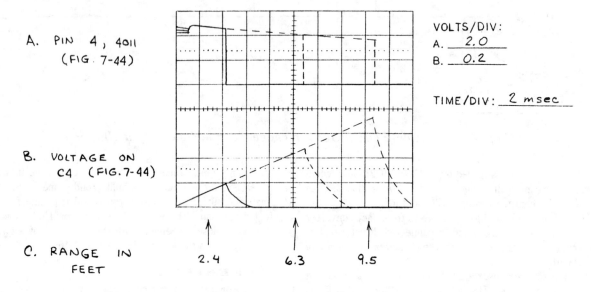

A. PIN 4, 4011
   (FIG. 7-44)

B. VOLTAGE ON
   C4 (FIG. 7-44)

C. RANGE IN
   FEET

2.4   6.3   9.5

VOLTS/DIV:
A. 2.0
B. 0.2

TIME/DIV: 2 msec

Fig. 7-46. Key waveforms of audible output circuit (Fig. 7-45).

are possible. For example, you can remove C4 and connect in its place a piezoelectric alerter such as Radio Shack's 273-065. The alerter will emit tone bursts whose *duration* is perceptibly longer when distant targets are being detected. Since all the bursts are brief, no matter what the range, longer pulses from the alerter seem louder than very short pulses. The overall effect provides an interesting alternative to the voltage-to-frequency converter chirped tone output.

More About Ultrasonics

This material appeared originally in *Computers & Electronics* (June 1983). One of the applications briefly mentioned was an automatic wheelchair guidance system for quadriplegics. The system helps these people perform tasks like guiding a motor driven wheelchair through narrow passages like doorways and halls.

After reading the column, David L. Jaffe, the wheelchair's developer, sent along additional details about this fascinating device. Mr. Jaffe, who is with the Veterans Administration Medical Center in Palo Alto, CA, designed the prototype chair as a student project at Stanford University.

Stanford has a well deserved reputation for encouraging its engineering students to tackle challenging sophisticated design projects. Mr. Jaffe's project, which was sponsored by the Veterans Administration, was carried out at the Smart Product Design Laboratory of Stanford's Mechanical Engineering Department.

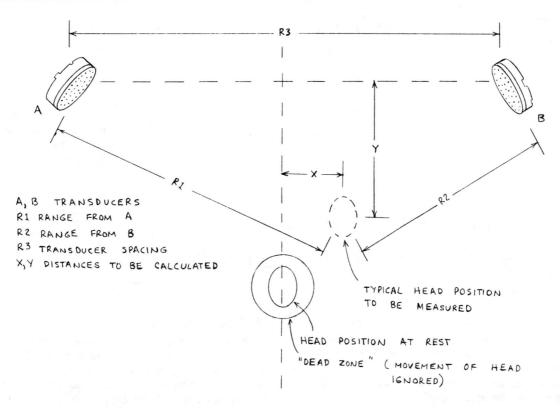

A, B TRANSDUCERS
R1 RANGE FROM A
R2 RANGE FROM B
R3 TRANSDUCER SPACING
X,Y DISTANCES TO BE CALCULATED

Fig. 7-47. David Jaffe's ultrasonic head detection system for motorized wheelchairs.

Mr. Jaffe's use of Polaroid's rangefinders is more sophisticated than implied in the literature I received from Polaroid. According to his letter and a diagram I've reproduced as Fig. 7-47, "Two sensors were aimed at the user's head. Then, by computer-aided triangulation, the user's head position in rectangular or cylindrical coordinates was determined."

"In operation," Jaffe wrote, "the user would tilt his/her head in the direction that he/she wished the chair to travel. The magnitude of tilt controlled the speed. The user's head became a proportional 'joystick.' "

The chair was also equipped with forward-facing ultrasonic sensors to detect potential obstacles. The chair was automatically stopped when an obstacle was detected within a preprogrammed distance.

The chair had still other ultrasonic sensors. Jaffe writes, "Side facing sensors allowed the vehicle to travel automatically at a fixed distance from a wall on either side of the chair. The cruise-control mode enabled the vehicle to navigate in a straight line at a constant speed. The net result was a vehicle that could be used by individuals who did not have use of their hands. The Polaroid sensor system provided a man/machine interface that did not require physical contact."

Recently Jaffe has developed a much improved model of his prototype wheelchair. The new chair uses a single board STD computer and is programmed entirely in FORTH. Also, the head detection sensors have been remounted *behind* the user's head to simplify the process of entering and leaving the wheelchair.

Already one version of the new chair has been delivered to a user in France.

Jaffe's work with the Polaroid ultrasonic sensors has stimulated several other clever ideas. If you wish additional information about the automated wheelchair, write David L. Jaffe at the VA Medical Center (3801 Miranda Avenue, Palo Alto, CA 94304).

A Cassette Recorder Analog Data Logger

The portable cassette tape recorder was invented to provide a convenient means for both recording and playing back speech and music. Computer hobbyists in the mid-1970s found an entirely unexpected role for cassette recorders when they discovered they can be used to save computer program listings and data.

The cassette recorder can also be used to store nonspeech or music analog data such as light intensity, temperature, velocity, revolution rate, and many other parameters. I've spent a good deal of time experimenting with these applications.

It's surprisingly easy to store analog data on tape. All that's necessary is an appropriate sensor and a circuit that transforms the signal from the sensor into a variable audio-frequency signal.

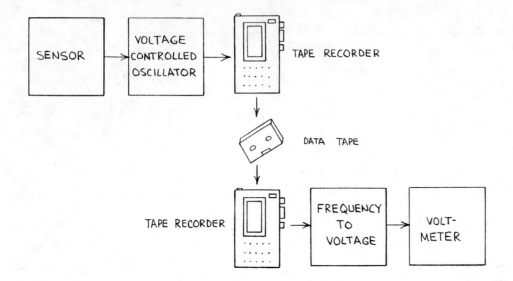

Fig. 7-48. Cassette tape storage and retrieval system.

The variable frequency signal can be connected directly to a cassette recorder's microphone jack or it can be transformed into an audio frequency tone and transmitted to the recorder's microphone.

Several methods are available for extracting and decoding a signal that has been saved on a cassette tape. The simplest is to connect a digital frequency counter to a cassette recorder's phone jack. The signal can also be decoded by means of an oscilloscope or frequency-to-voltage (F/V) converter circuit. Figure 7-48 is a block diagram of a complete analog data storage and retrieval system that employs an F/V converter circuit.

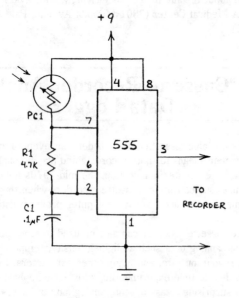

Fig. 7-49. Typical light-sensitive oscillator circuit.

A Typical Sensor Circuit

Figure 7-49 shows a typical sensor circuit that can be connected to the microphone jack of many cassette recorders. Though the circuit shown is designed as a light-sensitive oscillator, the cadmium-sulfide photoresistor can be replaced by a temperature sensing thermistor or other variable resistance sensor.

In operation, the 555 oscillates at a frequency determined by the resistance of PC1. As the light level on PC1's sensitive surface increases, the resistance of PC1 decreases, thereby increasing the circuit's oscillation frequency. C1 has been selected to keep the maximum frequency within the frequency response range of typical battery-powered cassette recorders.

The circuit in Fig. 7-49 is merely one of many suitable variable frequency oscillators. For example, simple bipolar and unijunction transistor oscillators and various kinds of op-amp and timer integrated circuit oscillators can also be used. In some cases, an oscillator circuit is not even necessary.

For example, a magnet sensor switch connected in series with a resistor and a flashlight cell will generate a series of pulses when the sensor switch is placed near a rotating object to which a small magnet has been affixed. This method can be used to transform to an audio tone the rotation of many kinds of objects including bicycle wheels, anemometers and engine shafts.

A Frequency-to-Voltage Converter

There are several basic circuits and chips that transform to a representative voltage a variable frequency signal. I've experimented with a circuit that uses a 555 timer IC and several circuits using the 9400 and the LM331, two chips designed specifically as F/V converters. Though all these circuits work well, I prefer the LM331.

Figure 7-50 shows a basic F/V converter designed around the LM331. Since the phone output of most cassette recorders has a very low impedance (typically 8 ohms), it's necessary to

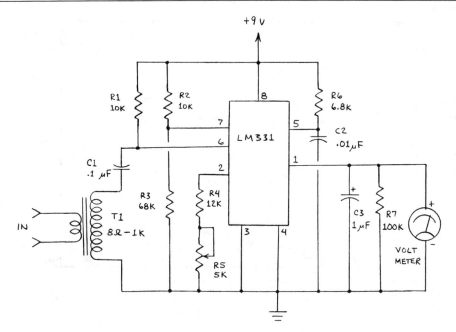

Fig. 7-50. Frequency-to-voltage (F/V) converter circuit.

couple the output signal from the recorder to the F/V circuit by means of a small audio transformer.

In operation, one of the inputs of a comparator in the LM331 is biased at a voltage determined by R2 and R3. The signal from the recorder is inductively coupled through T1 to capacitor C1 and from there to the second input of the comparator. When the amplitude of the incoming signal exceeds that of the reference voltage provided by R2 and R3, the output of the comparator changes state until the input signal level falls below the reference voltage.

The output from the comparator is connected to a monostable multivibrator (one-shot) in the LM331. Each time the comparator switches, the one-shot triggers and closes a current switch. This allows output filter capacitor C3 to charge for a time determined by the time constant of R6 and C2. R7 functions as a bleeder resistor that continually discharges the charge on C3, thereby causing the voltage on C3 at any given instant to correspond to the average charge available from the current source. Therefore, the voltage across C3 is directly proportional to the frequency of the incoming signal.

The output from the LM331 should be applied to a high-input impedance voltmeter. R5 allows you to calibrate the circuit for a particular output voltage at a specific input signal frequency. Depending on your particular application, you may wish to use either a conventional analog meter or a digital meter. The pointer of an analog meter permits trends of a changing frequency to be easily observed, a capability not possible by watching the flickering digits of a digital meter. On the other hand, the digital readout of a DVM or DMM permits a more accurate measurement of output voltages.

Figure 7-51 is a graph that shows the linear response of the LM331 to a variable frequency ranging from 1 to 10 kHz. The graph was made by recording the voltage output from the LM331 as the frequency of the signal applied to its input was altered.

Table 7-3 gives the actual measurements used to plot the graph for a sine wave signal:

Table 7-3. Output Voltages Obtained for Various Frequencies

Input Frequency (kHz)	Output (V)
1	1.35
2	1.67
3	2.48
4	3.22
5	3.98
6	4.76
7	5.53
8	6.25
9	6.95
10	7.13

These results were obtained when the circuit in Fig. 7-50 was powered by a 9-volt battery. When the circuit is powered by a supply delivering from 10 to 12 volts, the output voltages can be easily adjusted by means of R5 to be $1/100$ the signal frequency. This is accomplished by applying a 5-kHz input signal and adjusting R5 for an output of 5 volts. The output voltages will then range from almost exactly 0.1 volt at 100 Hz to 10 volts at 10 kHz.

Incidentally, as you can see by referring to Fig. 7-51, when the incoming signal was a square wave, a nonlinear knee occurred when the signal frequency was 3 kHz. Other anomalies will occur if the amplitude of the incoming voltage is too high or low. In any case, the voltage of the input signal should range between −0.2 volt and the supply voltage.

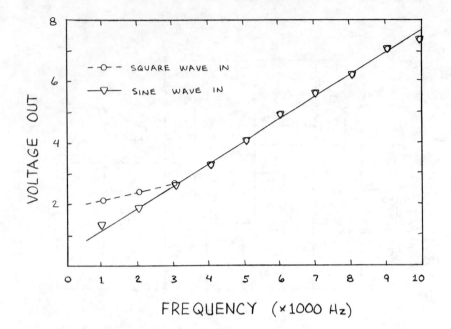

Fig. 7-51. Typical LM331 F/V calibration curve.

Selecting a Recorder

One of my prized possessions is an Olympus L200 micro-cassette recorder. This tiny, voice-actuated machine weighs only 4.4 ounces and measures 4.2 × 2.0 × 0.5 inches. Miniature recorders like the L200 are ideal for recording data from a shirt pocket. Unfortunately, however, microcassette recorders lack the frequency range of recorders that use standard cassettes.

For this discussion I spent a good deal of time testing the low frequency response of the L200 and Radio Shack's CCR-82 computer cassette recorder. Over the frequency range between 100 and 1000 Hz, the CCR-82 worked exceptionally well. The L200, however, has poor frequency response below about 400

Hz. Though a F/V circuit connected to the L200 can be adjusted to decode frequencies below 400 Hz, this hampers the decoding of higher frequency signals.

Figure 7-52 is a graph comparing the performance of the CCR-82 and the L200 over the 100- to 1000-Hz range. As you can readily see, the performance of the CCR-82 is very linear over this entire frequency range, thereby making this recorder an excellent choice for analog data recording applications. Unfortunately, the poor low-frequency response of the L200 causes a sharp knee in its response. When used to record higher frequency signals, the L200 should work nearly as well as the CCR-82 since the L200 has a specified frequency response of 400 to

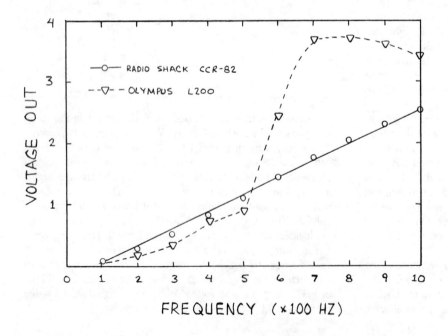

Fig. 7-52. Typical cassette recorder calibration curves.

6000 Hz. Unfortunately, despite a series of tests, I was unable to obtain consistent results with the L200 at higher frequencies.

It's important to understand how to make graphs like the one in Fig. 7-52 if you wish to store analog data on cassette tape. Such graphs are essential for properly calibrating an analog data logging system.

First, it's necessary to store on tape a known sequence of audio frequency tones spaced uniformly across the desired signal range. In the case of Fig. 7-52, I connected a signal generator directly to the microphone input of both the L200 and the CCR-82 and stored 10-second tone bursts at 100 Hz intervals between 100 and 1000 Hz. I then connected the input of the F/V converter circuit in Fig. 7-50 to the output of each recorder and played back the tapes. While each tape was played, I recorded the voltage for each signal frequency indicated on a digital multimeter connected to the output of the F/V converter. Finally, I plotted the data from each tape on the graph in Fig. 7-52.

Graphs like the one in Fig. 7-52 are often called calibration curves. If the curve is in fact a straight line, as is the case with the CCR-82 in Fig. 7-52, the response of the system is linear and it's very easy to correlate an output voltage from the F/V converter with its respective signal frequency. On the other hand, a nonlinear response, such as the one given by the L200, limits the usefulness of an analog data recording system.

In any case, it's important to note that the readout procedure requires that the recorder's volume control be properly adjusted. During my tests, I found that the CCR-82 worked best when the volume control was set to just above the 3 point on its dial. The L200 worked best when the volume control was set to between 5 and 6. Other recorders may require very different settings.

Using the System

Before using the analog data recording system, it's necessary to adjust the sensor circuit so that the frequency of the signal it generates falls within the proper range. Likewise, it's necessary to make sure the F/V converter is properly adjusted.

If you plan to take down the decoded voltages by hand, it's important to make sure each data sample is recorded for at least five seconds. This will allow the DVM readout to settle long enough for you to record the reading. If you have access to a chart recorder, the samples can be much briefer since the pen will plot as a continuous readout the voltage output.

Many modifications can be made to a cassette recorder analog data logging system. The simplest is to change the input sensor circuit to permit the storage of various kinds of information. If your recorder is equipped with a REMOTE jack, you can turn the drive motor on and off by an external timing circuit. This will allow you to place up to 180 10-second data samples on a thirty minute tape. If the samples are recorded at intervals of one hour, this arrangement would allow you to take data for seven and a half days on a single tape.

Finally, if you have a personal computer you may be able to interface the LM331 F/V decoder to one of its joystick ports. For example, Radio Shack's Color Computer has voltage dependent joystick ports which can be directly connected to the LM331 F/V converter so long as the output voltage from the converter doesn't exceed 5 volts. Use caution when making direct connections to a computer's joystick ports for you may damage the machine if you exceed the permissible voltage levels. You may also void the computer's warranty.

Electronic Aids for the Handicapped

One of the most rewarding pastimes for electronics experimenters is the design and construction of electronic aids for the handicapped. The Telephone Pioneers of America know this as well as anyone. For many years members of this organization have built beeping balls for blind children and other electronic aids for handicapped people.

Many companies and a few foundations make and sell numerous kinds of electronic aids for the handicapped. Some of these aids, such as the reading machines for the blind made by Telesensory Systems, Inc., are amazingly sophisticated. They are also expensive.

Prosthesis (pros' thē sis) is the technical term for an electronic or mechanical substitute or supplement for a missing or defective part of the body. I've found that many people in need of a prosthetic device are not aware that such devices exist. For instance, I recently observed an elderly blind man attempting to cross a busy street. While helping him across, I informed him that his cane was far too short since he could extend it less than a foot in front of his extended foot. He was delighted to learn that *long* canes are readily available.

Many people, perhaps some reading this very column, have a hearing deficiency that can be readily alleviated with the help of a hearing aid. But people tend to be more receptive to wearing eyeglasses, a vision prosthesis, than hearing aids.

In a moment we'll look at several kinds of aids you can make. First, let's investigate a valuable service you might be able to render hearing aid users.

Repairing Hearing Aids

Much of my early solid-state education was gained by repairing defective hearing aids obtained from hearing aid dealers for a few dollars or less. It soon became apparent that most of these aids would work properly when the power switch, earphone jack (if present), battery contacts or volume control was cleaned with a tiny spurt of spray solvent. Sometimes a frayed phone cord was the culprit, but this only occurred with so-called body aids. Most aids today are worn in or behind the ear and include an integral phone connected to the ear via a plastic tube.

I originally used the repaired aids to make miniature transistor radios. This is easily accomplished by removing the microphone and installing in its place a tuned circuit made from a miniature ferrite-core coil and disk capacitor. A germanium diode is used to detect the signal.

While a college student, I put this experience to work repairing damaged and defective hearing aids for a local dealer for $5.00 each. Considering the very high replacement cost of such devices (often several hundreds of dollars and sometimes much more), the fee was a bargain.

You can easily put together a hearing aid repair kit which will solve most common hearing aid problems. You'll need some facial tissue, cotton swabs, and alcohol to clean accumulated debris from exposed portions of the aid. A pencil eraser or other abrasive surface (e.g., very fine grit sanding paper) is handy for cleaning battery contacts. A small can of contact cleaner with a thin spray tube or nozzle is essential for cleaning switch contacts, phone jacks, and volume controls.

Sometimes switch contacts will wear or no longer engage completely. A good pair of midget needle-nose pliers or a pair of pointed tweezers can remedy this problem.

You'll also need a small hand lens and a portable desk lamp to find problems and see what you're doing. And you'll need a multimeter to check batteries and make continuity checks of switch contacts and phone wires.

Your kit will be complete with the addition of a set of jeweler's screwdrivers and a miniature, low-wattage soldering iron. You can put everything in a small fishing tackle box or tool box. You'll soon find the kit can be used for much of your experimental work also.

You must gain experience troubleshooting and repairing salvaged hearing aids before attempting to repair a person's only audio link with the outside world. Hearing aids are very fragile and usually very expensive so you must exercise *considerable* care when attempting to repair them. Other than minor cleaning, leave the repair of units under warranty to the dealer or manufacturer.

Spare parts can be expensive and difficult to find. You can sometimes salvage parts from discarded aids. Or you can order them from manufacturers or hearing aid supply firms. A friendly hearing aid dealer may provide you with addresses or let you look through his parts catalogs. Be aware, however, that many dealers do not repair aids and may not have such information.

You can give salvaged aids you have repaired to a hospital, charity, or nursing home. This might provide you with a tax deduction (charitable gift) which will help finance your project. After you've become competent repairing salvaged aids, you can try your hand repairing aids in everyday use. Call a local nursing home or hospital to find out about aids which need service.

Do-It-Yourself Prosthetic Devices

It's relatively easy for experimenters to design and assemble prosthetic devices for blind and deaf people. You might also be able to build various kinds of control devices which will enable immobile or paralyzed people to turn on various combinations of lights, a radio, television, and other devices from a fixed location. Let's examine the possibilities.

Aids for the Blind

Probably the simplest gadget you can build for a blind friend or relative is a light probe which produces a sound when a photodetector is illuminated. Light probes have been made since 1912 when British scientist Fournier d'Albe built a device he called the "Exploring Octophone." D'Albe's device employed a selenium cell in a Wheatstone bridge and signaled variations in light intensity with a musical tone.

You can duplicate d'Albe's light probe with a simple circuit such as the one in Fig 7-53. This circuit can be easily installed in an aluminum cigar tube. The circuit is a two-transistor regenerative amplifier. The oscillation frequency of the circuit is dependent upon the light level at the sensitive surface of the cadmium sulfide photocell. The audio output is supplied by a small earphone which serves as a miniature speaker. The volume is sufficiently high so that the user can hear the tone but not so loud as to distract nearby persons.

The probe can be made very small. Several years ago I built a probe, complete with mercury button cell and hearing aid receiver, into a small plastic tube measuring about $3/8 \times 1$ inch. This small size was made possible by the tiny but surprisingly effective hearing aid receiver unit. I intended to attach the probe to a key chain but, because of its small size, promptly misplaced it.

There are many ways to make circuits which produce a tone whose frequency is dependent upon light. Figure 7-54, for example, is an integrated circuit version designed around the readily available and inexpensive 555 timer chip. Although this circuit is very sensitive and can be adjusted to provide various frequency ranges and threshold levels, it consumes more current than the transistor version in Fig. 7-53.

An important advantage of the circuit in Fig. 7-54 is that it can drive a small speaker or other transducer with more current

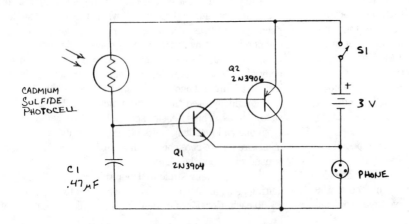

Fig. 7-53. Simple transistorized light probe for the blind.

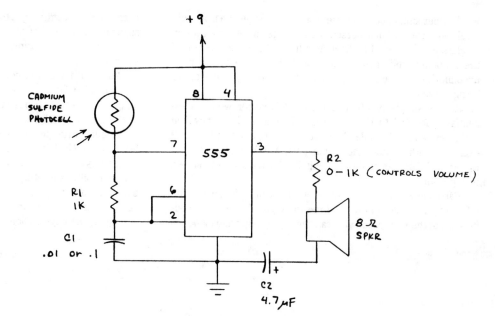

Fig. 7-54. 555 light probe with high volume output.

than the transistor circuit. This may prove a significant advantage to blind people with a hearing problem, a not too uncommon combination among older individuals.

Still another light probe circuit is shown in Fig. 7-55. This circuit uses the amazingly versatile LM3909 LED flasher chip as a light controlled audio oscillator.

Light probes have many practical applications. Blind telephone operators have used them to find illuminated indicator lights on a switchboard. They can also be used to determine if room lights are on (or off), whether or not appliances with pilot lights are on and to find color changes in clothing and wall surfaces.

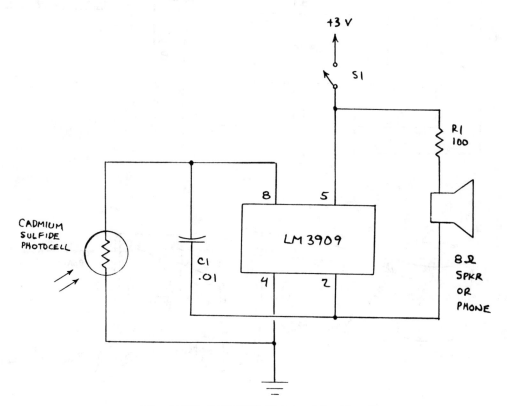

Fig. 7-55. LM3909 light probe circuit.

Another simple but very useful aid for blind people is a liquid level indicator. A blind person usually determines when a cup or glass has been filed by keeping a finger tip inside the rim of the container while pouring the liquid. This procedure can be uncomfortable if the beverage is hot and unseemly if guests are being served.

A very simple two-tone liquid level indicator which is essentially identical to the light probe in Fig. 7-55 is shown in Figs. 7-56 and 7-57. Indeed, you may wish to build an LM3909 oscillator with a phone jack input port. You can then connect either the liquid probes shown in Fig. 7-57 of the photocell shown in Fig. 7-55.

Partial or even total loss of vision can be an unfortunate side effect of advanced *diabetes mellitus* (or *sugar diabetes*). A modification of the light probe idea can be used to help blind diabetics

perform their own tests of the sugar level in their urine. The sugar content is ordinarily determined by dipping a strip of test paper into a urine specimen. Color changes in the paper denote the sugar level.

You can detect the color of a test strip by making a light tight chamber fitted with a cadmium sulfide photocell and a small lamp. When a test strip is placed in the chamber, the resistance of the photocell will be directly related to the color of the strip.

I built a urine monitoring device similar to this in 1966. At least one such device is commercially available today. If you want to assemble one for a blind diabetic, you'll need to spend time carefully calibrating the unit. You *must* also make absolutely sure the unit can be reliably operated by a blind person. Erroneous readings could prove hazardous to the person you wish to assist. The device should be used under the supervision of a

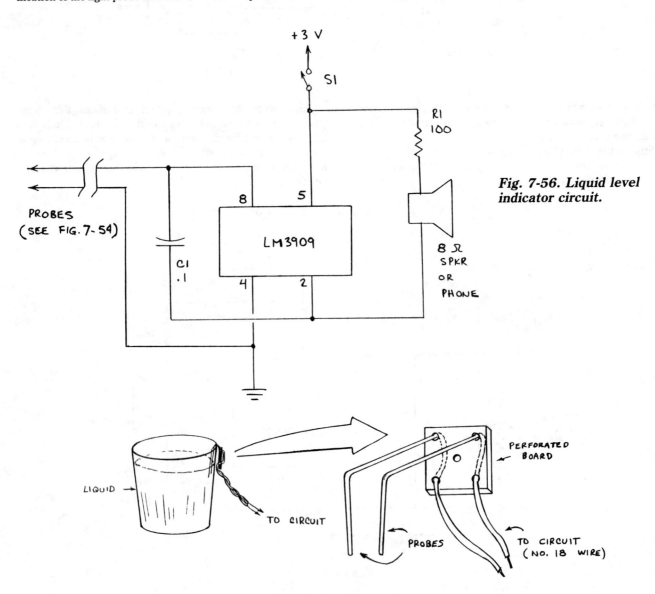

Fig. 7-56. Liquid level indicator circuit.

Fig. 7-57. Liquid level indicator probe assembly.

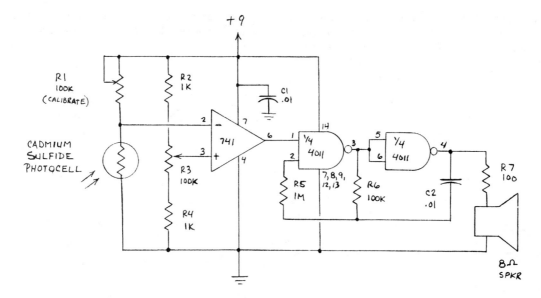

Fig. 7-58. Color detection circuit for urinalysis.

physician. Various state and federal agencies regulate the manufacture of prosthetic devices and medical aids.

If you would like to experiment with this application, which has several nonprosthetic uses also, Fig. 7-58 is an elementary circuit which will get you started. The 741 operational amplifier is operated without a feedback resistor so that is functions as a comparator. R3 sets the switching threshold for the comparator. The photocell is connected as one arm of a voltage divider. When the divider output reaches the switching threshold set by R3, the comparator switches on and actuates the tone generator made from two gates in a 4011.

In a practical version of this circuit, the photocell would be pointed at an illuminated test strip, which you can obtain at a pharmacy, and the output from the divider would therefore be fixed at a voltage representative of the reflectance of the strip. The shaft of R3 would then be rotated until the tone is heard. Raised markings at predetermined points would form a scale for R3's knob (which should have a stubby pointer).

Aids for the Deaf

We've already discussed the repair of hearing aids. In a pinch, you can actually *make* a usable hearing aid from a miniature audio amplifier circuit or a low-cost portable commercial amplifier. You'll also need a microphone and earphone.

A combination like this will be much bulkier than even an ordinary hearing aid, and it will be subject to clothing noises unless the microphone is attached in an unobstructed location. It will also be much more subject to feedback oscillation. Furthermore, it will not necessarily be as optimized for the amplification of audio frequencies as are the specialized components of most hearing aids. Nevertheless, it may prove very helpful in an emergency or when a person's hearing aid is being repaired.

The totally deaf cannot be helped by conventional hearing aids. Ways to signal totally deaf people include a light panel,

cathode ray display, or printer. These methods require that frequent attention be given the output device.

A simple yes-no signalling system can be made with a tactile stimulator. Such devices can be made from piezoelectric materials which, when electrically stimulated, vibrate or poke against the surface of the skin.

A more homely but simpler approach which rarely fails to get a person's attention is an eccentric weight attached to the shaft of a small dc motor. When the motor is spinning, the vibration produced by the offset weight is easily felt even if the packaged unit is carried in a pocket.

Several years ago I read that this form of tactile stimulator was being used to signal deaf workers at an unnamed location. This kind of stimulator does not require continual monitoring as is the case with warning or indicator lamps. And it can be particularly important to persons who are both blind and deaf.

When used as a simple alerting device, a tactile stimulator must usually be connected to an external control circuit. Wire connections are awkward but simple. Radio or infrared links are best but costly.

The control device can be as simple as a door bell or as vital as a fire alarm. For deaf people with teleprinters it may be a ring detector which signals an incoming call.

Additional Information

Articles about electronic aids for the handicapped usually stimulate many questions. If you want additional information about aids for the blind, write the American Foundation for the Blind (15 West 16th Street, New York, NY 10011) and request information about services provided by the foundation. One such service is the sale of aids and appliances for blind and visually impaired persons. A catalog listing such aids is published by the foundation, but you should request a copy only if you wish to make a purchase or are seriously interested in working with blind people.

For many years the Veterans Administration's Department of Medicine and Surgery (Washington, DC 20420) has sponsored research into problems of the blind and deaf. This organization has also sponsored the exploratory development of various kinds of aids for persons with partial and total hearing and vision impairment.

You can also find information on this subject by spending some time at your local library. Medical and technical libraries are best.

EIGHT

Piezoelectronics, Thermoelectronics, and Experimental Circuits

- The New Piezoelectronics
- Experimenting with Piezoelectronics
- Experimenting with a Piezoelectronic Speaker
- The Solid-State Heat Pump
- An Experimental Security Alarm
- Experimenting with Small DC Motors

Piezoelectronics, Thermoelectronics, and Experimental Circuits

The New Piezoelectronics

The piezoelectric effect was discovered in 1880 by Pierre and Paul Curie. They found that a voltage is produced when pressure is applied to crystals of quartz, tourmaline, and Rochelle salt. A year later they discovered the converse effect, that when an electric field is applied to these same crystals they will expand or contract. The name applied to the wonderful effect discovered by the Curie brothers comes from the Greek *piezien*, meaning *to press*. It is pronounced with the accent on the first syllable (pi-e-zo).

We are surrounded by modern devices and gadgets that exploit the century-old piezoelectric effect. The quartz crystals found in CB radios and scanners, digital watches and clocks, and computers and TV games provide an electromechanical resonance so precise that oscillators designed around them may have a stability that varies only a few cycles in a million.

Some phonograph pick-up heads use a piezoelectric crystal or ceramic element to transform the movements of a needle sliding in the groove of a record into an analog of the sound recorded therein. There are also piezoelectric microphones. Conversely, high-frequency tweeters and some miniature earphones use a piezoelectric element to convert an audio signal back to sound. Consider, for example, the piezoelectric wafer that directly drives a tweeter cone. The wafer is a sandwich of two piezoelectric disks separated by a corrugated centervane. When a signal is applied, one disk expands while the other contracts, thus moving the cone to produce a sound.

The ultrasonic sound emitted by electrically stimulated piezoelectric transducers is used to find fish, measure water depth, agitate solvents that clean jewelry and miniature mechanical assemblies, repel certain insects, focus cameras, and guide blind people around obstacles. Sparks produced by mechanically stimulated piezoelectric elements are used to light furnaces, hot air balloon burners, and the fuel in some cigarette lighters. The ionization produced by high-voltage piezoelectricity is used to neutralize the static charge on phonograph records.

Low-intensity ultrasonic sound emitted and detected by piezoelectric transducers can safely enable the outline of a fetus in its mother's womb to be viewed on a CRT. On the other hand, high intensity ultrasonic waves from similar transducers can scramble an egg without breaking its shell, set cotton ablaze, shatter gallstones, weld, and even kill, small animals such as fish, frogs, and mice.

Not all crystals exhibit the piezoelectric effect. Normally a crystal contains an equal number of positive and negative charges, and is therefore in a state of equilibrium. In other words, it is electrically neutral. If the shape of a crystal is changed by external mechanical pressure, then both the positive and negative charges move, and the conditions for the flow of an electrical current are present.

If, however, the charges move in the same direction, they neutralize or cancel one another and there is no current flow. For a current flow to exist, the charges must move in *opposite* directions, and that is what happens in crystals that lack a center of symmetry and, therefore, exhibit the piezoelectric effect.

It is important to realize that the piezoelectric effect is always accompanied by the physical movement via compression or expansion of the atoms in a crystal. Therefore, a piezoelectric crystal produces a current only when its atoms are in motion. Applying a constant pressure to the crystal will not squeeze out a continuous current!

Thus when a piezoelectric element is struck by a small hammer, a pulse of current is produced while the element is being compressed. When the element springs back to its former shape, a second burst of current having a polarity opposite that of the first is produced. An ac voltage can be elicited by applying a fluctuating pressure to a piezoelectric element.

The dimensions of a piezoelectric crystal change only when the applied voltage changes. The resulting contraction and expansion have dimensions of very small magnitude, often on the order of a few mils or even much less. Yet the velocity at the surface of a piezoelectric crystal vibrating hundreds of thousands of times per second can be substantial. If, for instance, a crystal surface expands 0.01 inch in 0.00001 second (100 kHz), then the surface of the crystal must move at a velocity of no less than 83.3 feet per second or 56.8 mph. It may, in fact, move considerably faster.

Though the current produced when a piezoelectric crystal is squeezed is very small, the electromotive force can be on the order of a few hundred or even a few thousand volts. This is sufficient to create a visible arc several millimeters in length.

Though the converse effect, the physical movement of an electrically stimulated piezoelectric crystal, is not nearly so apparent as a highly visible spark, its existence can be demonstrated by an appropriate position-sensing transducer, perhaps one which itself employs the piezoelectric effect to generate an output voltage proportional to an applied movement. Its existence may also be made apparent by secondary effects. For example, a rapidly vibrating piezoelectric wafer produces an audible tone.

Piezoelectric Materials

During the past century many materials that exhibit the piezoelectric effect have been identified. Only a few, however, have been found suitable for practical applications, and of these quartz is the unquestioned leader. Though the piezoelectric properties of quartz are not nearly so pronounced as those of Rochelle salt, quartz has excellent mechanical properties and good temperature stability.

Lithium niobate, a crystal with unique optical properties, is used in some piezoelectric applications. However, it is more temperature sensitive than quartz.

Aside from quartz, the most important piezoelectric materials are manmade ferroelectric ceramics such as lead zirconate titanate. These ceramics, unlike quartz and other piezoelectric crystals, are polycrystalline and would, therefore, seem unsuitable for piezoelectric applications. They are given piezoelectric properties during their manufacture when the application of a strong electric field polarizes the material.

Applications for Piezoelectricity

We've already mentioned some of the better known applications for piezoelectricity, some of which have been in use for more than fifty years. For instance, many of the applications that depend upon the electrically triggered motion of a piezoelectric element can be traced to the *bimorph*, the first piezoelectric device invented by the Curie brothers.

The basic bimorph shown in Fig. 8-1 is a sandwich formed by attaching thin piezoelectric bars to either side of a metal strip. Electrically conductive coatings may be applied to either side of one end of the bimorph to form terminals. A voltage applied to the terminals causes one bar to contract in one dimension and expand in the other while the opposite movement occurs to the second bar. The resulting forces cause the bimorph to bend. By changing the polarity of the applied voltage, the bimorph will bend in the opposite direction.

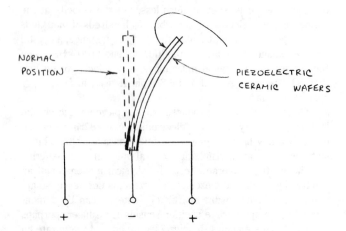

NORMAL POSITION

PIEZOELECTRIC CERAMIC WAFERS

+   −   +

Fig. 8-1. The piezoelectric bimorph or bender.

Recently an entirely new array of applications for the piezoelectric effect was announced by Piezo Electric Products, Inc. Formed in 1980 to develop new piezoelectric devices, the firm acquired the piezoelectric product manufacturing facilities of Gulton Industries.

So far Piezo Electric Products has announced several new devices developed by Eric and Henry H. Kolm, latter day versions of the Curie brothers and vice presidents of the new firm. The piezoelectric relay or actuator is a miniature solid-state relay which is faster, longer lived, and less noisy than conventional electromechanical relays. Figure 8-2 is a drawing of how a piezoelectric relay is constructed. The chief drawback of present devices is the requirement of about 40 volts of actuation.

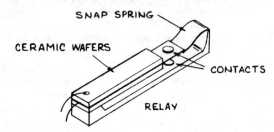

SNAP SPRING

CERAMIC WAFERS

CONTACTS

RELAY

Fig. 8-2. The piezoelectric relay developed by Piezo Electric Product, Inc.

The piezoelectric relay technology developed at Piezo Electric Products has been applied to the design of a new dot matrix printhead. The new printhead uses piezoelectric bending elements instead of solenoids to drive small pins against a carbon ribbon. If successful, a piezoelectric dot matrix printer should operate at a higher speed with considerably less heat production than a conventional dot matrix printer.

A third application involves piezoelectric quadrature motors and fans. The fan operates on the principle of the insect wing and uses a hundredth of the power of a conventional blower of similar output. One application is a miniature cooling fan for electronic components.

The inverse of the piezoelectric vibrating fan is a solid state generator that produces power from moving air or gas. According to Piezo Electric Products, "We have proof-of-concept models of electrical generators which operate off the acoustic energy in the exhaust of internal combustion engines. We anticipate that automobile mufflers of this type would replace the alternator and noticeably improve the fuel economy of the engine."

The firm has also developed a piezoelectric windmill that generates electricity when piezoelectric elements are vibrated by moving air. The company claims such windmills ". . . can be effectively made in the form of small units resembling snow fences, highway barriers or other structures capable of supporting a number of small vanes resembling the leaves of a tree. Such piezoelectric windmills would also operate over a wider range of wind velocities than rotary windmills, and hopefully cost less per watt of installed power."

Another idea from Piezo Electric Products is a bicycle generator. It has much less drag than conventional generators and directly produces the high voltage required for high brightness gas discharge lamps.

Other companies are also developing exotic new applications for piezoelectric technology. Watson Industries sells for $295 a piezoelectric gyroscope which it claims is superior to laser ring gyros. The piezoelectric gyro is compact and weighs only about 10 oz. Its sensitivity is sufficient to detect a rotation rate as little as 0.04 degree per second.

National Semiconductor has developed a CMOS chip that is powered by an accompanying piezoelectric element. Developed by Gould Inc., the chip derives its power from a rotating tire, and is designed to transmit a coded signal when tire pressure falls below a certain point.

Many kinds of piezoelectric accelerometers and force transducers have been developed. There is even a piezoelectric micrometer system manufactured by Polytec Optronics, Inc. This system includes a power supply that provides an adjustable voltage to miniature piezoelectric elements mounted upon precision micrometers. Model P-252 provides a total excursion of 25 mm with a piezoelectric fine tuning of up to 20 microns (at 0.02 micron per volt).

Several years ago a Japanese television company invented a high voltage piezoelectric transformer for use in color TVs. West Germany's Siemens AG has taken this principle in another direction by developing a miniature piezoelectric isolation transformer. As shown in Fig. 8-3, a voltage applied to one side of a piezoelectric ceramic wafer induces an acoustic wave that propagates across the wafer to a second pair of electrodes. The voltage produced by the acoustic wave is transferred to the second electrode pair and used to control an external circuit such as an SCR or TRIAC.

One such device developed by Siemens is available for only $.70 in large quantities. It is the PZK 20 Piezo Ignition Coupler, and it can be used in applications that would instantly destroy the LED in an optocoupler. It's very fast (the acoustic wave travels 2 km/s), it provides very high isolation (the piezoceramic element is an insulator) and it produces its own output voltage.

Although the recent emphasis in piezoelectrical technology has been upon ceramic elements, quartz is still of major importance. The Statek Corporation has made notable advances at both ends of the quartz frequency spectrum.

On the low end, Statek has developed a quartz crystal oscillator that fits within a miniature TO-5 transistor can. A quartz crystal smaller than the point of a sharp pencil provides an oscillation frequency as low as 10 kHz. The crystal and its miniature hybridized circuit can withstand a shock of 1000 Gs.

On the high end, Statek has developed what it believes is the smallest 1-MHz microprocessor crystal. The new crystal is about one-fourth the size of an 8-pin miniDIP. Its calibration accuracy of ±0.05 percent is achieved by a process in which the miniature crystal is etched from a quartz wafer and fine-tuned to the correct dimensions by laser trimming.

Learning More

Many texts about electronics cover piezoelectricity in much more detail than I have in this brief introduction. Some encyclopedias, particularly those dedicated to science topics, cover the subject quite well.

If you want to experiment with piezoelectric crystals of your own making, *Crystals and Crystal Growing* (Alan Holding and Phylis Singer, Anchor Books, Doubleday & Co., 1960) has a chapter containing the recipe for making Rochelle salt single crystals larger than a sugar cube. A diagram and photo show how to attach with mineral oil aluminum foil electrodes to each side of the crystal. When given a tap with a small hammer, the crystal will light up a small neon lamp. Though this book is dated, it might be available in libraries or stores that deal in paperback books.

Experimenting with Piezoelectronics

The voltage produced when certain crystals and ceramics are mechanically flexed is called *piezoelectricity*. The effect is reversible, and piezoelectric crystals and ceramics will contract or expand when a voltage is applied across them. Many modern electronic devices and circuits rely upon the piezoelectric effect, and the best way to learn how piezoelectric devices operate is to try them in actual circuits and experiments.

The Piezoelectric Microphone

The so-called *crystal* microphone is a piezoelectric acoustic transducer. Early crystal microphones used a Rochelle salt crystal element. Today the piezoelectric element of many such microphones is a polarized ceramic wafer about the size of a fingernail. The ceramic is easy to mass produce. It is also stronger and more moisture resistant than Rochelle salt.

Thanks to the reversible nature of the piezoelectric effect, many high impedance record player cartridges and earphones

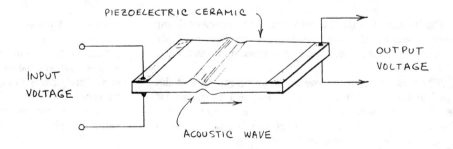

Fig. 8-3. Operation of the piezoelectric isolation transformer developed by Siemens AG.

use piezoelectric elements. Indeed, a piezoelectric microphone can function as a small speaker, and a piezoelectric earphone can be used as a microphone.

You can learn much about the operation of a piezoelectric microphone with the help of an oscilloscope. Connect the leads of the microphone directly to the scope's probe. Set the vertical sensitivity to about 0.1-volt/division. Adjust the sweep speed to about 1-millisecond/division.

First, speak or whistle into the microphone. The scope's CRT will display a visual analog of the sound, and, depending upon the proximity of the microphone to your mouth, the amplitude will range from about 0.1 to 0.5 volt. The waveform overlaps the no-signal centerline and is therefore ac in nature.

Next, rap the microphone with a pencil or thump it with a finger. If the microphone is an economy version, the scope's CRT will display a ringing pulse with an initial peak of perhaps 40 or 50 volts. The duration of the initial pulse will be about 0.1 millisecond.

Better designed, highly damped piezoelectric microphones will produce only a very low voltage when tapped or thumped since their element is designed to prevent inadvertent high voltage spikes that might damage the input stage of a preamplifier. This could happen if a microphone were dropped or otherwise given a strong blow.

You can perform a dramatic experiment to demonstrate the high voltage output of a highly stressed piezoelectric microphone element by connecting the leads from the microphone to a neon glow lamp as shown in Fig. 8-4. Select a very cheap or discarded microphone, perhaps one with a damaged foil diaphragm, since it is necessary to remove the microphone's cover.

If you remove the foil, do not directly strike the element to light the neon lamp. Instead, strike the metal support bar that bridges two opposite corners of the element and provides a mounting point for the diaphragm. Be careful; the sole support for the piezoelectric element is probably a pair of rubber vibration-damping bumpers on two opposite corners of the element. The element is easily detached from these supports. Also, the two leads emerging from one side of the element are very fragile.

The Piezoelectric Push Button

Piezoelectric push buttons are used to ignite the fuel of some cigarette lighters, laboratory burners, and home furnaces. They produce a brief spike of up to 18,000 volts and can make an arc up to 3/16 inch in length.

For several years I've experimented with a Model 3652 high voltage push button made by Vernitron Corporation. Similar devices made by Vernitron are used as solid-state igniters for outdoor cooking grills. The company has also manufactured hundreds of thousands of 0.1 inch piezoelectric ceramic cubes used to power the flash in compact cameras.

A pictorial view of the 3652 high voltage push button is shown in Fig. 8-5. The piezoelectric element is a compact slug about 5/8 inch by 3/16 inch. Most of the unit's size is taken by the spring-loaded trip hammer that strikes the piezoelectric element.

To operate the unit the push button is pressed downward with a force of a few pounds. This compresses the upper spring and moves the cam toward the pin on the trip hammer. When the cam pushes the pin attached to the hammer toward the drive slot, the hammer is triggered and slammed with a good deal of

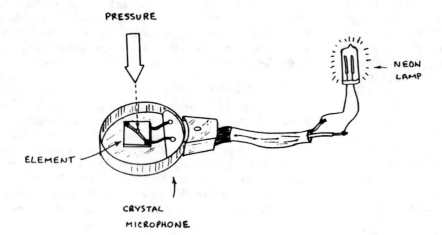

Fig. 8-4. Flashing a neon lamp with a crystal microphone element.

Tap the center of the foil diaphragm with a pencil or thump it with a finger and the lamp should flash. The voltage pulse will be up to a millisecond wide and its amplitude may reach a few hundred volts!

It's not necessary to remove the diaphragm to conduct this experiment. If you wish to remove the foil, peel it from around the edge of the microphone case first. Then *carefully* pull it away from the central metal support that is attached to the piezoelectric element. Small scissors may help.

force against the piezoelectric element by the energy stored in the upper spring. When the push button is released, the lower spring, which was compressed by the downward motion of the hammer, returns the hammer to its resting position where it is immediately ready to again be driven against the piezoelectric element.

The arc produced by the piezoelectric push button can be viewed by placing the output electrode near the unit's metal frame. For best results, the arc should be viewed in subdued

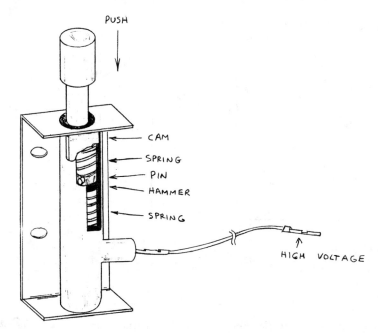

Fig. 8-5. Vernitron 3652 piezoelectric high voltage push button.

light or against a dark background. Unless you want to feel a potent tingle, keep your fingers away from the output electrode when the button is pressed!

An interesting experiment is to connect a piezoelectric push button to a long xenon flash tube. When the button is pressed, a thin violet arc will immediately appear between the tube's electrodes.

With suitable rectification it should be possible to use a piezoelectric push button to charge a capacitor to a very high voltage. Of course the capacitor would have to be rated for the expected voltage. I plan to explore this subject farther since it might make possible a very simple power supply for Geiger counters and infrared image converter tubes.

An 18,000-volt piezoelectric push button is available from Edmund Scientific.

The Ceramic Filter

The ceramic filter, a most unusual piezoelectric device, is dependent upon the mechanical resonance of a piezoelectric ceramic wafer. When a signal is applied to its input, a surface wave is induced in the ceramic. If the frequency of the wave matches the resonant frequency of the ceramic, the wave will travel along the surface of the ceramic where it induces a piezoelectric voltage at a second pair of electrodes. Otherwise no signal is passed through the filter. In effect, then, the ceramic filter functions like a frequency selective isolation transformer.

Ceramic filters are widely used as 455-kHz intermediate range filters in am radio receivers. They are also used as 10.7-MHz filters in fm receivers and television sets. At these frequencies the size of the filter is much smaller than an equivalent electronic filter. For example, a typical 455-kHz ceramic filter is a disk 0.2 inch across and from 0.1- to 0.4-inch thick. If the signal applied to a center electrode and a common electrode on the back side of the disk is at or very near 455 kHz, then the

disk will vibrate and induce an electrical signal at a third electrode around the upper edge of the disk.

Figure 8-6 is a circuit that demonstrates the operation of a 10.7-MHz ceramic filter such as the SFE 10.7MA5-A made by muRata Corporation of America and available from Radio Shack.

In operation, two inverters in a 7404 or 74LS04 hex inverter form a high frequency oscillator whose output signal is buffered by a third inverter and fed into a ceramic filter. The frequency of the oscillator, which is determined by R1, can be adjusted from about 9 to 19 MHz with the component values shown. Much lower frequencies can be produced by increasing the value of C1.

Figure 8-7 shows the signal from the oscillator before and after its passage through the filter when R1 is adjusted to provide the filter's peak frequency response. Note that the output signal appears to have about twice the amplitude of the input signal. Actually, the output signal is an ac version of the single polarity input signal, hence the apparent doubling of its amplitude.

Also note that the output signal is phase delayed and is a much smoother, cleaner version of the input signal. The input signal might be cleaned up somewhat by optimizing component placement and using direct, point-to-point wiring instead of the solderless breadboard I used. Why is the output better shaped than the input? The acoustic wave that travels across the surface of the ceramic serves to dampen imperfections in the input signal.

How effective is a ceramic filter? Figure 8-8 is a frequency response plot I made with the help of the circuit in Fig. 8-6 and an oscilloscope. Note that the filter has a double peak with a nearly −3-dB valley or ripple at the specified peak response region. Of more significance is the rapid decrease in response beyond the −6-dB points. The −3-dB bandwidth is about 390 kHz. At −10 dB the acceptance window is about 500 kHz.

My measurements do not agree as closely as I would have liked or expected with those given in muRata's published spec-

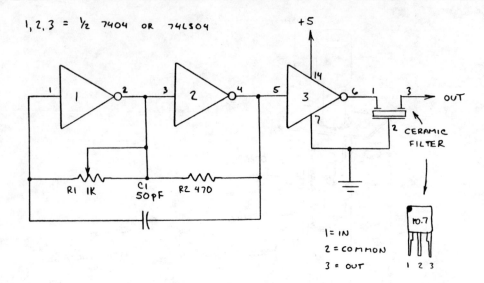

1,2,3 = ½ 7404 OR 74LS04

Fig. 8-6. Ceramic filter test circuit.

ifications for the SFE 10.7MA5-A. The -3-dB bandwidth, for example, is given as 280 ± 50 kHz. Though the ripple for this filter is not given, a graph published in muRata's literature suggests a ripple considerably less pronounced than the -3 dB I measured.

A 10.7-MHz Ceramic Oscillator

Quartz crystals are normally used to regulate precision oscillators. I've found that a ceramic filter will also work, but without the precision quartz provides.

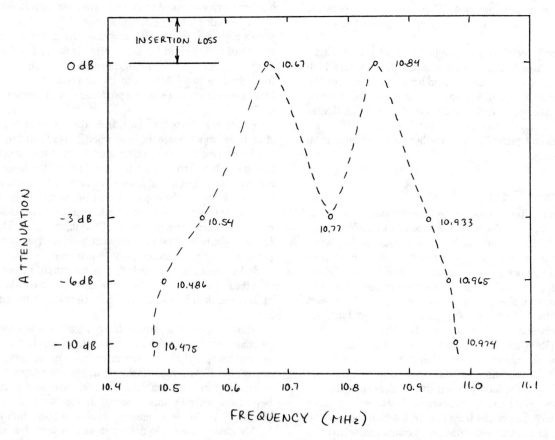

Fig. 8-7. 10.7-MHz signal at either end of a ceramic filter.

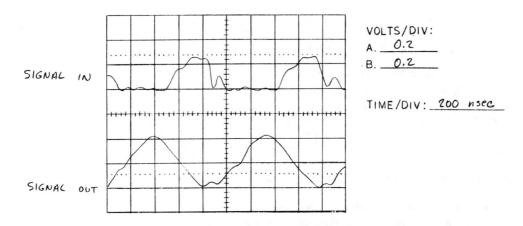

VOLTS/DIV:
A. _0.2_
B. _0.2_

TIME/DIV: _200 nsec_

Fig. 8-8. Measured bandpass of 10.7-MHz ceramic filter.

Figure 8-9 shows a 10.7-MHz ceramic oscillator. The circuit is virtually identical to the one in Fig. 8-6. The only exception is that C1 in Fig. 8-6 has been replaced by the ceramic filter.

The variety of frequencies available with ceramic filters is much less than the vast number of quartz crystal frequencies. Nevertheless, the circuits in Figs. 8-6 and 8-9 suggest an interesting application: a matched radio-frequency (rf) oscillator and frequency-sensitive detector. The signal from the transmitter can be coupled directly or through the air via a fast risetime LED or by radio waves. Additional circuitry will be required to implement this application. And you will have to observe FCC regulations that apply to rf emissions. If you use a line-powered supply to operate the circuit, a nearby TV set may be subjected to severe video interference.

The chief advantage of this application is the very low cost and compact size of the ceramic filter used in both the oscillator and receiver. The ceramic filter used to produce the plot in Fig. 8-8 oscillated at 10.71950 MHz when R1 was 500 ohms. A second filter gave a frequency of 10.72105 MHz. This relatively minor difference is of little consequence since the oscillator can

be tuned a few tens of kilohertz in either direction by changing the setting of R1.

Piezoelectric Alerters

Crystal microphones and speakers are designed to operate across a wide band of audio frequencies. Piezoelectric alerters, however, are generally designed to operate at a fixed or relatively narrow audio frequency band. They are true solid-state sound sources.

The first commercial piezoalerter of which I am aware was the Mallory Sonalert. (Mallory is now a division of Emhart Industries.) Sonalerts are available in various kinds of housings having a range of audio outputs. Most include self-contained drive circuitry.

I first purchased a Sonalert in 1966 and a few years later used it to measure the velocity of a model rocket in flight. The Sonalert, a Model SC628 emitting a tone of 2.9 kHz, and its battery were installed in the base of a model rocket by the young members of a model rocketry club I then sponsored. The rocket's engines were installed in pods attached to the rocket's cen-

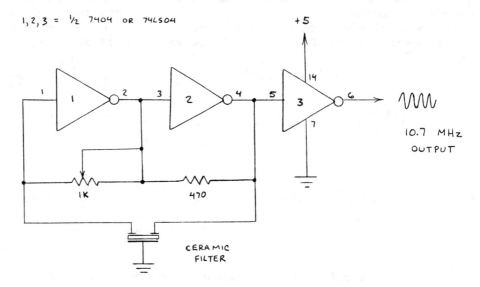

Fig. 8-9. 10.7-MHz ceramic oscillator.

ter tube. The sound from the Sonalert was tape recorded from the ground during the rocket's flight. By measuring the doppler shift, it was possible to measure the rocket's velocity.

Alerter Construction and Operation

Thanks to their miniature size, low current consumption, and penetrating sound, piezoelectric alerters are commonly used in digital watches, clocks, smoke alarms, pagers, appliances, calculators, and games. A typical alerter is a metal disk from 25 to 40 mm in diameter upon which is bonded a smaller disk of piezoceramic material. A conductive film is deposited over the ceramic layer, and electrodes are attached to it and the metal disk.

Often alerter disks include a *feedback electrode* made by isolating a small section of the metal film on the back of the piezoceramic material. The feedback electrode, which is shown in Fig. 8-10, simplifies the design of driver circuits and stabilizes the alerter's oscillation frequency. Piezoalerter disks can be purchased alone or installed in plastic holders complete with connection leads. Versions with self-contained driver circuits much like the Mallory Sonalert are now available from several companies.

It is essential to properly mount an alerter disk for maximum sound output. If the vibrating portion of the disk is cemented or otherwise attached to a mount, severe attenuation of the device's sound output will occur.

Figure 8-11 shows three acceptable ways to mount an alerter disk. The *center mount* permits the outer rim of the disk to vibrate; the *edge mount* permits the entire disk to vibrate. Both these methods permit the disk to vibrate across a range of audio frequencies.

The *nodal mount*, also shown in Fig. 8-11, is best for a very loud, single-frequency tone. The node of a piezoalerter disk is a concentric ring around the center of the disk at which vibration at a fixed frequency is at a minimum or even nonexistent. Ideally, the diameter of the nodal ring is 0.55 times the diameter of the metal disk. The actual diameter, however, varies from the predicted value due to the presence of the piezoceramic disk and nonuniformities in the metal disk.

One way to find the actual location of the nodal ring is to sprinkle fine sand or powder on a piezoalerter disk being driven

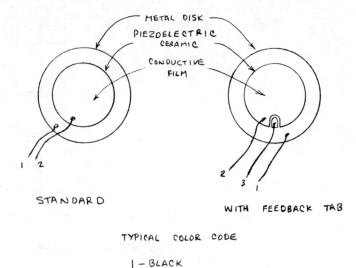

TYPICAL COLOR CODE

1 — BLACK
2 — RED
3 — BLUE

Fig. 8-10. Piezoelectric alerter elements.

at a desired frequency by a suitable oscillator. The powder particles will gradually bounce into the nodal region and form a thin, circular ring around the center of the disk.

Piezoalerter Driver Circuits

A piezoalerter can be driven directly by a variable frequency signal generator. Even alerters having nodal mounted disks can be operated across the audio spectrum, although edge and center mounted disks work best across a wide band of audio frequencies.

Figure 8-12 shows a simple single transistor driver for a piezoalerter having a feedback terminal such as the Model PKM11-6A0 from muRata Corporation of America.

The PKM11-6A0 can be operated over a specified range of from 3 to 15 volts (mine works down to 1 volt) and has a current consumption over this range of from 2 to 12 milliamperes. Its output sound pressure level ranges from more than 80 dB at 3 volts to more than 90 dB at 15 volts. Its resonant frequency is

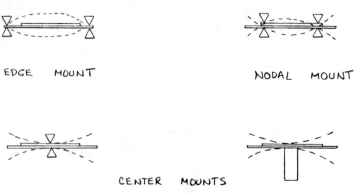

Fig. 8-11. Three mounting arrangements for piezoelectric alerter circuits.

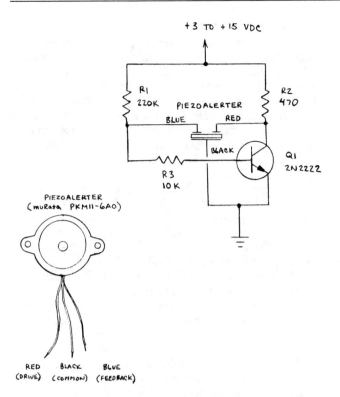

Fig. 8-12. Single transistor piezoelectric alerter driver circuit.

is controlled by the dimensions of the feedback tab on the alerter disk and not the components of the oscillator. For example, changing R1 over a range of from 100 kΩ to 330 kΩ altered the shape of the waveform but not the frequency. The frequency is nearly independent of changes in V_{cc}.

Figure 8-13 shows a simple single chip CMOS oscillator suitable for driving a piezoalerter. This circuit is adapted from one in a Gulton Industries application note. Notice how the 4049 gates are connected in parallel to permit higher drive current.

The circuit in Fig. 8-13 has the advantage of having an adjustable frequency. A breadboard version I built operated over a range of from about 185 Hz to 7 kHz. The frequency change, however, was not gradual but occurred in steps. When the piezoalerter reached its resonant frequency of around 7 kHz, changing R2's resistance had no effect.

The circuit in Fig. 8-14 will drive at a variable frequency both piezoalerters with and without feedback terminals. Unlike the circuit in Fig. 8-13, this circuit provides a gradual, non-stepped output tone. A slow *pock . . . pock . . . pock* sequence can be produced by using a 0.47-μF capacitor for C1.

The operation of a piezoalerter's feedback electrode can be graphically demonstrated by connecting the anode of a red LED to the blue lead of the alerter in Fig. 8-14. Connect the LED's cathode to ground. The output from the blue lead easily exceeds a few volts, more than enough to forward bias the LED and cause it to emit a dim glow.

Keep in mind there is no electrical connection between the feedback electrode and the main electrode on the piezoelectric ceramic disk. The voltage at the feedback terminal is true piezoelectricity generated in response to the pressure wave that appears in the piezoelectric ceramic disk in response to the drive signal. The LED demonstration shows how a piezoelectric device can function as a true solid-state transformer or isolator.

within 700 Hz of 6.5 kHz. It has an operating temperature range of from −20° to +60°C. And it weighs only 1.5 grams.

A test version of the circuit in Fig. 8-12 drove the alerter at a frequency of 6772 Hz when V_{cc} was 3 volts. This frequency

Fig. 8-13. Single IC piezoelectric alerter driver circuit.

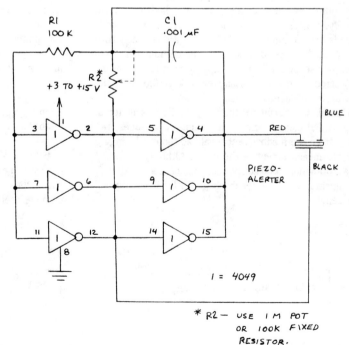

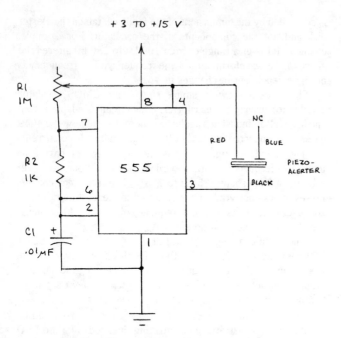

Fig. 8-14. Adjustable frequency piezoelectric alerter driver.

Using an Alerter as a Filter

Figure 8-15 shows how to demonstrate the use of a piezoalerter as a ceramic filter. The Model PKM11-6AO I used exhibited frequency response peaks at 2.3, 7.0, 18, 27, and 45 kHz. Although a scope is helpful, it's possible to monitor the filter's operation by simply listening to the change in amplitude of the filter's sound output as the signal generator's frequency is varied. Of course this method only works at audio frequencies.

Incidentally, I attempted to measure the delay introduced by the piezoelectric ceramic with the help of a dual-trace 100-MHz oscilloscope. The speed of sound in the ceramic is around 5000 meters per second. Since the gap between the main and feedback electrodes on the piezoalerter disk is 0.5 mm, the expected delay is 100 nanoseconds.

Though the driver circuit for the test, the 555 oscillator in Fig. 8-14, provided clean leading and trailing pulse edges, the signal elicited from the feedback terminal had too much ringing for an accurate measurement of the delay. Although I *think* I monitored a 100-nanosecond delay, I cannot be certain due to the sloppy appearance of the feedback pulse. Perhaps you will have better results.

Other Alerter Ideas

The very narrow audio spectrum produced by piezoalerters makes them ideal for use in experiments with sound. With the help of a microphone and oscilloscope, you can easily demonstrate constructive and destructive interference of sound waves. Try pointing the microphone at the alerter while moving the microphone back and forth. Or point both the alerter and the microphone at a flat metal or plastic panel which you can move back and forth. The proper arrangement will reveal on the oscilloscope screen a periodic amplitude fluctuation in the received signal.

Note that in an enclosed room the sound of an alerter can vary dramatically in intensity. For example, as I type this a 2.9-kHz alerter is sounding off in the electronics shop several paces from my desk. Moving my head a fraction of an inch in a certain direction changes the perceived sound from an annoyingly loud tone to a barely perceptible sound.

This very dramatic amplitude fluctuation is a result of the single-frequency acoustical waves from the alerter forming complex interference patterns throughout my office. Negative interference causes the formation of *dead spots* where the sound is virtually imperceptible. Constructive interference forms regions where the sound is uncomfortably shrill.

Sounds from radios, televisions, phonographs, and people span a wide range of audio frequencies. Therefore, the effects of interference are not nearly as noticeable.

The effects of the acoustical interference caused by the pure tone emitted by an alerter may or may not be desirable. It is certainly attention getting to walk by an alerter and notice the changes in sound intensity. But it can also be confusing, particularly if you are trying to find the source of the sound in an enclosed room having many flat, hard reflecting surfaces! The resultant interference problems can be avoided by using multiple or swept-tone alerters.

If you enjoy experimenting, try using a piezoalerter as a microphone. You'll find that alerters with nodal mounting function as *frequency selective* sound detectors. Or try adding to an alerter a tube or reflector to form a directional sound source. You can try operation at resonant ultrasonic frequencies. And you can develop various kinds of sonic radar circuits. You can even try operating an alerter under water. If you do, however, keep your circuits battery powered to avoid a shock hazard.

Alerter Precautions

Data sheets for piezoalerters note that mechanical shock can cause these devices to generate high voltage spikes that can damage their drive circuitry and perhaps other associated cir-

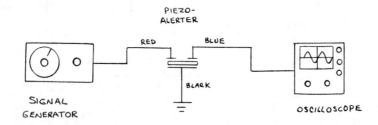

Fig. 8-15. Using a piezoelectric alerter as a signal filter.

cuits. This problem can be alleviated by installing an appropriately rated protection diode directly across the alerter.

Another precaution concerns the placement of an alerter on a circuit board. Be sure to mount the alerter on a rigid, fixed portion of the board. If the alerter is mounted on a cantilevered portion of a circuit board, it may set up vibrations in the board which may substantially reduce its sound output.

Finally, a precaution I've *not* seen in the data sheets concerns the shrill sound which can be produced by some alerters. I've found that the sound can easily produce a piercing headache. While experimenting with the circuits described here, I eventually resorted to covering the aperture of the alerter with clay or tape to muffle somewhat the sound output. Another option is to wear ear protectors.

Quartz Crystal Oscillators

The final piezoelectronic component we will consider is the quartz crystal. Precision cut wafers of quartz are used to make piezoelectric resonators having exceptional frequency stability.

Figure 8-16 shows an ultrasimple crystal-controlled unijunction transistor oscillator that uses only four components. The quartz crystal replaces a capacitor normally used in this circuit. The oscillation frequency can be tuned from about 50 kHz to exactly 1 MHz when the crystal has a resonant frequency of 1 MHz. Tuning is accomplished by altering the resistance of R1.

If you monitor the output of the oscillator in Fig. 8-16 with an oscilloscope, you will notice that the oscillation frequency tends to change in jumps as R1 is adjusted. This is a result of the crystal oscillating at various harmonics of its 1-MHz resonant frequency. Near 1 MHz, the oscillator quickly locks onto the crystal's resonant frequency.

The circuit in Fig. 8-16 is useful for understanding the operation of a simple quartz crystal-controlled oscillator. It can

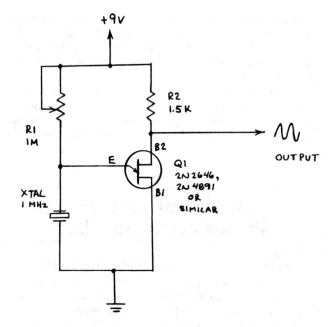

Fig. 8-16. Ultrasimple crystal controlled UJT oscillator.

also be used to supply a marker frequency to calibrate oscilloscopes, signal generators, and shortwave receivers.

Figure 8-17 shows a very useful crystal-controlled clock-pulse generator. The circuit is designed around Intersil's ICM7209, a CMOS general purpose timer chip. The crystal can be any quartz crystal having a resonant frequency of from 10 kHz to 10 MHz. The circuit consumes only about 11 milliamperes when powered by a 5-volt supply and requires only four external components.

Fig. 8-17. Crystal controlled clock generator.

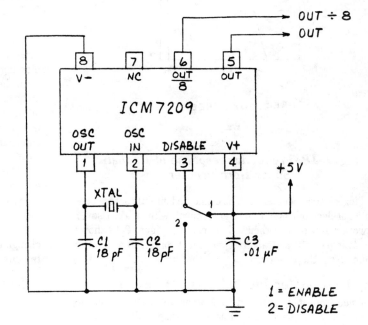

For more information about the circuit in Fig. 8-17, see the "Project of the Month" in the September 1981 issue of *Popular Electronics*. Back issues of this magazine are available at most libraries. Or, see *The Forrest Mims Circuit Scrapbook* (McGraw-Hill, 1983, p. 114).

If you would like to know more about quartz crystal oscillators, invest an hour or so at a good library. Many excellent quartz oscillator circuits have been published over the years. Particularly popular are quartz crystal-controlled rf oscillator circuits.

Experimenting with a Piezoelectric Speaker

Piezoelectric speakers are commonly used as high frequency tweeters in high fidelity speaker systems. Since the moving surface of such speakers is comparatively small, sounds having a frequency approaching 40 kilohertz can be generated.

Figure 8-18, for example, is the frequency response curve for a typical piezoelectric tweeter, Radio Shack's Realistic Super Horn (Cat. No. 40-1381). This tweeter has an impedance greater than 1000 ohms at 1 kHz and more than 20 ohms at 40 kHz. It has good transient response and faithfully reproduces single frequency tones. It's also very lightweight, sturdy, and vibration resistant.

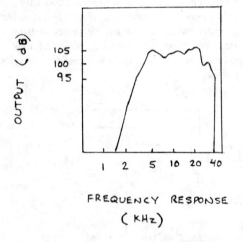

Fig. 8-18. Frequency response of Realistic Super Horn tweeter.

Piezoelectric tweeters can be connected directly across conventional electromagnetic speaker terminals without a crossover network and with or without a level control. Figure 8-19 shows how to implement a straightforward level control to adjust the volume from the tweeter.

Applications for Piezoelectric Tweeters

Aside from their intended function as hi-fi tweeters, piezoelectric speakers have several additional applications. For in-

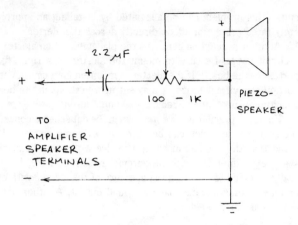

Fig. 8-19. Piezoelectric tweeter volume (level) control.

stance, a piezoelectric tweeter can be used as a high-impedance microphone.

More significant are applications in which the speaker serves as a sound emitting source for a warning device or ultrasonic generator. I've experimented with several circuits that implement these functions. They're easy to assemble, and you may wish to try the two described in the following.

A Multioutput Alarm

Figure 8-20 is a simple but very effective multiple output alarm circuit made from both halves of a 556 dual timer chip. When connected to a piezoelectric tweeter, this circuit emits three different attention-getting alarm signals.

In operation, the left timer is connected as a multivibrator that switches on and off at a rate of about 0.5 Hz. Let's refer to this circuit as the *autoswitcher*. The right timer is connected as a multivibrator that oscillates at a frequency of a few kilohertz. We'll refer to this portion of the circuit as the *audio oscillator*.

Switch S1 permits three operating modes to be selected. When S1 is switched to position 1, the audio oscillator operates independently of the autoswitcher and the speaker emits a steady, piercing tone. When S1 is switched to position 2, the output from the autoswitcher is connected to the trigger input of the audio oscillator through R3. The speaker then emits a distinctive alternating tone that resembles a "twee-dell" siren. When S1 is switched to position 3, the output from the autoswitcher is connected directly to the audio oscillator's trigger input and the speaker emits a series of penetrating tone bursts.

The circuit in Fig. 8-20 can be used to drive a small 8-ohm electromagnetic speaker. But a piezoelectric tweeter provides much better results. The sound is very penetrating *and* very loud. The volume can be increased by raising the supply voltage to 15 volts. The various sound outputs can be altered by simply changing the various RC components of one or both halves of the circuit.

A Sonic/Ultrasonic Siren

A piezoelectric tweeter makes a very effective sound source for a sonic/ultrasonic siren. I've designed and assembled an ex-

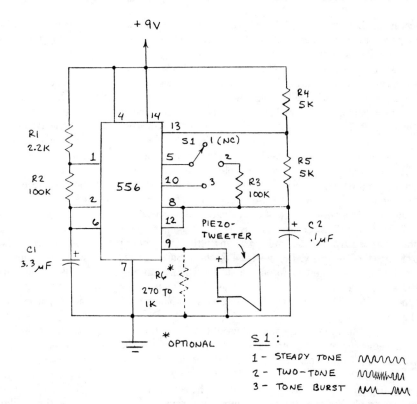

Fig. 8-20. Multiple outlet piezoelectric speaker driver.

perimental version of such a siren to test its effectiveness as a warning or intruder alarm. I've also experimented with the effect upon barking dogs of the siren in its ultrasonic mode. More about this later.

Figure 8-21 is the circuit diagram for the siren. In operation, the SN76488 complex sound generator operates as a self-recycling, swept-frequency, up-down oscillator. The frequency range of the oscillator is determined by R2 and C2. The sweep rate is determined by R1 and C1.

With the values shown in Fig. 8-21 the oscillator can be adjusted via R2 and C2 to sweep from a lower range of a few hundred hertz to a few kilohertz and an upper range of from about 15 to 50 kHz. The cycle rate can be varied from a few tenths of a hertz to several hertz by altering the setting of R1.

Testing the Siren

There's really no need to explain how to test the siren circuit when it's operated in the audible frequency mode. You simply

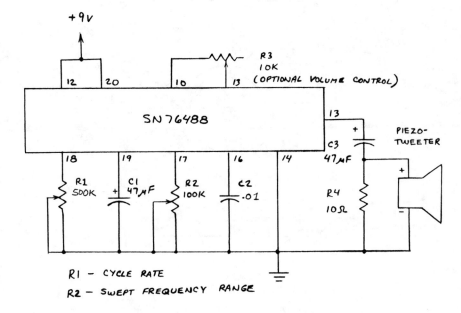

Fig. 8-21. Sonic/ultrasonic siren circuit.

adjust it for the sound you prefer. CAUTION: The sound can be very penetrating. You should cover the speaker while testing the siren indoors to protect your ears.

Operating the siren in its ultrasonic mode is trickier since you cannot hear the output. This is the operating mode I've used in an experiment to silence barking dogs.

A young person with normal hearing can detect sound having frequencies approaching 24 kHz. Animals such as dogs and, especially, bats can hear sounds having even higher frequencies. The upper limit for older people is usually much lower, perhaps 10 kHz or even considerably less. People whose hearing has been impaired by exposure to excessive sound or disease may have even lower frequency response levels. The siren in Fig. 8-21 can cover this entire range of frequencies and beyond.

While testing the siren circuit in my office, a neighbor's pack of four dogs began a chorus of yelping when a fifth dog wandered too close to their territory. I connected a Realistic Super Tweeter to the siren, adjusted R2 for ultrasonic operation, went outdoors and pointed the tweeter at the dogs. They immediately stopped barking! The worst offender repeatedly shook his head as he walked away.

Although this brief experiment achieved the desired result, later the dogs completely ignored a second barrage of ultrasonic waves during another barking episode. I'm not sure why the second experiment was not successful. Perhaps the wind was blowing from the wrong direction. Or maybe the sound of their own barking blocked the effect of the ultrasonic waves.

In any event, I quickly stopped a third round of barking the next day with a few quick bursts of ultrasonic waves. Unfortunately, all subsequent attempts to discourage the dogs from barking have been unsuccessful. Indeed, the siren now causes the dogs to bark even more vigorously!

Though I've concluded my brief experiment aimed at curing barking dogs, I would be interested to hear from any readers who obtain better results. Of course you should be careful that your experiments not offend the owners of barking dogs or that, as in my case, they stimulate rather than eliminate barking spells.

Assembling a Permanent Siren

The circuit in Fig. 8-21 uses so few parts you can assemble it *inside* the empty space in the plastic enclosure of a Realistic Super Tweeter (Cat. No. 40-1380) or similar enclosed piezoelectric tweeter. This tweeter's enclosure can be opened by removing the screws inside the three recessed holes on the back panel of the enclosure. When the screws have been removed, place the enclosure on its back panel and remove the front portion of the enclosure. This will keep the fragile piezoelectric element and its paper cone from falling out of its receptacle and possibly being damaged.

There's ample room in the enclosure for the necessary components, controls, and a 9-volt battery. Figure 8-22 shows one way the components can be installed.

Begin assembly by *carefully* removing the piezoelectric speaker element from its plastic mounting cylinder. Drill in the side of the cylinder a hole for mounting a 9-volt battery holding clip. Then drill holes for R1 and S1 as shown in the view of the back panel in Fig. 8-23.

I assembled the siren circuit on a 1.5 × 2-inch piece of perforated board with preetched copper foil patterns suitable for DIP ICs. Figure 8-24 shows the parts layout I used.

You can mount the completed board directly to the speaker connection solder tabs inside the enclosure (back side of the back panel). First, unsolder the piezoelectric element's leads (careful, they're fragile). Then place melt some solder blobs on the backside of the board. Melt some more solder on the solder tabs. Then hold the board against the tabs (component side facing the speaker cylinder) while remelting the solder on one tab. When the solder hardens, heat the second tab. The board will then be anchored securely in place.

Next, *carefully* solder the speaker leads to the circuit board.

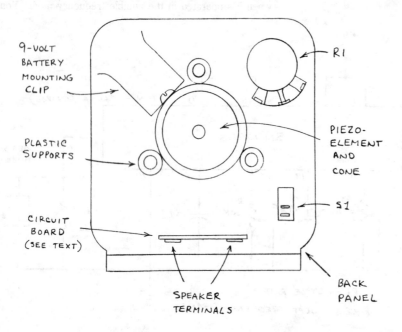

Fig. 8-22. Assembly details for permanent version of the siren.

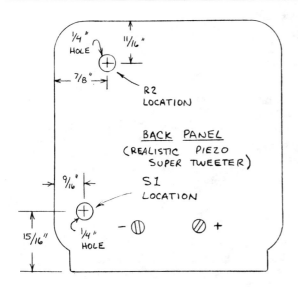

Fig. 8-23. Location of controls on back panel of tweeter.

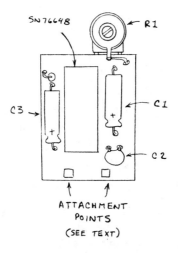

Fig. 8-24. Parts layout of siren circuit board.

Then solder the leads from a 9-volt battery clip to the board and one pole of S1. Finally, connect leads between the board and the second pole of S1 and the terminals of R1.

You'll have to adjust the swept frequency range before closing the enclosure since R2 is attached to the circuit board. If you want to be able to make quick frequency changes, you can mount R2 in the side of the enclosure. There isn't room to mount it on the back panel of the speaker I used. Of course, you can exchange R1 for R2 if you don't need to change the cycle rate but do want to change the swept frequency range.

Going Further

There are many ways to make sonic and ultrasonic sirens. Over the years many different transistorized and integrated cir-

cuit versions have been published, and you may want to experiment with some of them.

Of course sound generator chips greatly simplify the design and construction of such circuits. In addition to the SN76488 I used for this circuit, you can use the SN76495, SN94281, and various other sound synthesizing chips.

If you're troubled by barking dogs, you may want to consider adding a sound activated trigger to the siren. Adjust the circuit to deliver a burst of ultrasonic waves each time it detects a bark. This approach may very well solve your problem.

Finally, you can boost the output from the siren circuit by adding an external amplifier. The amplifier can be as simple as a single transistor or a powerful high fidelity unit. A good choice is the LM386 amplifier chip.

The Solid-State Heat Pump

On my desk as I write is a cream colored wafer a bit larger than a postage stamp and about 4-millimeters thick. From it emerges a pair of wires, one red and the other black. When these wires are connected to a 6-volt battery, one side of the wafer almost immediately becomes very warm, even hot. Remarkably, the opposite side becomes very cold. If the air is sufficiently humid, frost will appear on the cold surface within seconds.

This extraordinary wafer is known as a *thermoelectric module*. However, it might more appropriately be termed a *solid-state heat pump*. Its operation is completely reversible for when the connections to the battery are switched the hot side of the wafer becomes cold and the cold side becomes hot. Moreover, the module will generate an electrical current when its two opposing surfaces are maintained at different temperatures.

The phenomena I've described are collectively known as the *thermoelectric effect*. Although most people knowledgeable about electronics know about the heat and power generation aspects of the effect, surprisingly few are aware of the cooling phenomenon. One reason may be that engineering schools and books about electronics often give little or no emphasis to this very remarkable effect.

The cooling ability of a thermoelectric module like the one on my desk can be easily demonstrated with the help of some commonly available hardware. Figure 8-25 is a photograph of a module made by the Cambion Division of the Midland-Ross Corporation. An extruded aluminum heatsink and a small aluminum box are attached to opposite sides of the module with a rubber band. Heatsink compound insures a good thermal bond between the module and the two attachments.

The heatsink, which should be attached to the hot side of the module, is placed in a shallow pan of water as shown in Fig. 8-26. A teaspoon or so of water is then placed in the aluminum box. Within minutes after the module is connected to a 6-volt battery, the water will be frozen solid and the metal box will be coated with a furry layer of frost. Add more water and the module will produce a cube of ice. The efficiency of this miniature freezer can be increased by making an insulating chamber for

***Fig. 8-25. A thermoelectric module.** (Courtesy Electronic Connecter Division of Midland-Ross Corp.)*

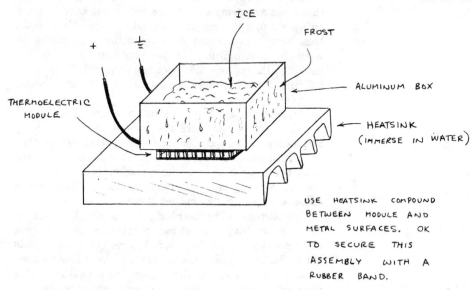

Fig. 8-26. A simple thermoelectric ice cube maker.

USE HEATSINK COMPOUND BETWEEN MODULE AND METAL SURFACES. OK TO SECURE THIS ASSEMBLY WITH A RUBBER BAND.

the box from foamed plastic panels held together by tape or a rubber band.

Discovery of the Thermoelectric Effect

The discovery of the thermoelectric effect can be traced to 1821 when the German physicist Thomas J. Seebeck found that when two conductors of different materials are connected to form two junctions as shown in Fig. 8-27, a voltage will appear across the free ends of the conductors if the two junctions are maintained at different temperatures. Seebeck's discovery formed the genesis of temperature sensing thermocouples.

The Seebeck effect can be easily demonstrated with ordinary hardware or even pocket change. For example, a homemade thermocouple made by wrapping a few turns of copper wire around one end of a steel nail will generate 6 or 7 microamperes

at a few millivolts when heated by the flame of a match. Even more power can be obtained by overlapping the edges of a penny and a nickel and securing the coins together with an alligator clip. After being heated by a succession of several matches, the coins I tried produced 60 microamperes at 6 millivolts.

In 1834, Jean-Charles-Athanase Peltier, a French watchmaker, passed a current through a junction of two dissimilar metals. He found that, depending upon the direction of the current, the junction would become warm or cool.

Metals remained the exclusive thermoelectric material for many years. Some of the better thermoelectric junctions are copper-constantan, iron-constantan, and chromel-alumel. Constantan is an alloy of copper and nickel. Chromel is an alloy of nickel and chromium. Alumel is an alloy of nickel and aluminum. These and other junctions are widely used today in the manufacture of temperature-sensing thermocouples.

230

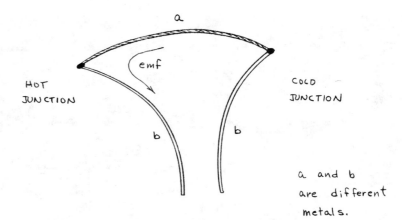

Fig. 8-27. A basic Seebeck junction.

In the Seebeck mode, metal junction thermocouples generate only a few microvolts per degree Celsius. Semiconductor couples may exhibit Seebeck coefficients of hundreds or even thousands of times greater. In the early 1950s, Abram F. Ioffe in the Soviet Union and H. J. Goldsmid in England independently found that semiconductors such as bismuth telluride make excellent thermoelectric materials. Ioffe's group made demonstration power generators and refrigerators. Goldsmid's group made junctions that exhibited a drop of as much as 65°C below room temperature. Scientists in the United States later discovered that lead telluride is also an excellent thermoelectric material.

Semiconductor Thermoelectric Devices

Today most semiconductor thermoelectric devices are based upon lead telluride or bismuth telluride. The selection of the alloy is based largely upon the preferred operating temperature of the module. For example, one firm employs a quaternary alloy of bismuth, tellurium, selenium, and antimony. The alloy is appropriately doped to provide an n-type or p-type semiconductor.

Figure 8-28 shows the construction of a simple single-junction semiconductor thermoelectric device. The upper ends of the two semiconductor bars are soldered to a common copper header, and their opposite ends are soldered to separate copper headers. Electrical connections are made to separate headers. Since practical thermoelectric modules are usually made from arrays of many such junctions or couples, thin plates of ceramic

are attached to both sides of the module to electrically isolate the individual junctions. The ceramic permits reasonably good heat transfer and, at the same time, prevents electrical shorts between adjacent modules.

Referring back to Fig. 8-28, when a direct current is passed first through the n-type semiconductor bar and then through the p-type bar, heat is pumped from the upper side of the module to the lower side. Conversely, when the polarity of bias is reversed, heat is pumped from the lower side to the upper side.

In either case, the side from which heat is removed rapidly becomes very cool while the opposite side becomes very warm. If the heat isn't removed from the warm side, some of it will be radiated and conducted back to the cold side. Eventually the module will reach a point of equilibrium and little or no cooling will occur.

In a practical system, heat can be extracted from the hot side of the module by a forced air blower or a circulating liquid. In both cases conventional heatsinks and miniature plumbing components can be used.

Commercial thermoelectric modules have more than the single junction shown in Fig. 8-28. Figure 8-29, for example, shows an assortment of miniature Frigichip modules made by Melcor/Materials Electronic Products Corp. These modules may have from four to 66 individual couples. They can produce a hot side-cold side temperature difference of 67.5°C.

Figure 8-30 shows how two module arrays can be stacked or cascaded to achieve even lower temperatures. A two-stage

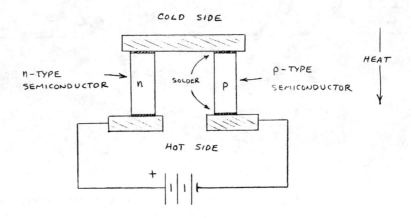

Fig. 8-28. A single-junction semiconductor thermoelectric couple.

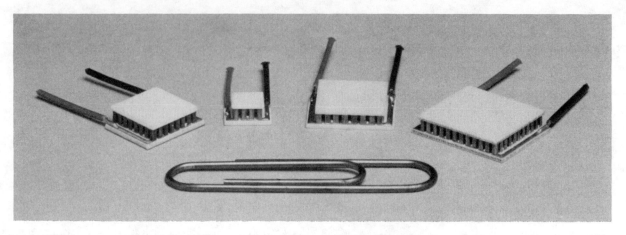

Fig. 8-29. Assorted thermoelectric modules. (Courtesy Melcor)

cooler can achieve a temperature differential of 85°C or more. Three- and four-stage coolers can achieve temperature differentials of 105°C and 125°C or more, respectively. An eight-stage module designed to cool an infrared detector has achieved a temperature drop of 171°C (308°F) below room temperature!

The thermoelectric module in Fig. 8-25, the unit I used to make an ice cube, is a single-stage device having 71 couples. It provides a maximum temperature drop of 60°C or more. It has a maximum current rating of 6 amperes and a maximum forward voltage of 10 volts. The unit sells in single quantities for $31.20. The price drops to $17.80 in quantities of a thousand. These prices are subject to change.

Applications

Thermoelectric modules are used in a surprisingly diverse array of applications. Modules such as those shown in Figs. 8-25, 8-29, and 8-30 are used to make solid-state refrigeration and heating systems. Thermoelectric cooling/heating assemblies equipped with a blower fan are available. Such an assembly can be used to make a portable refrigerator that can double as a food warmer when the power connections are reversed. Several such systems designed to be powered by the 12-volt supply of trucks and cars have been marketed.

Thermoelectric ice makers, baby bottle cooler/warmer units, aquarium coolers, drinking water coolers, and even room air

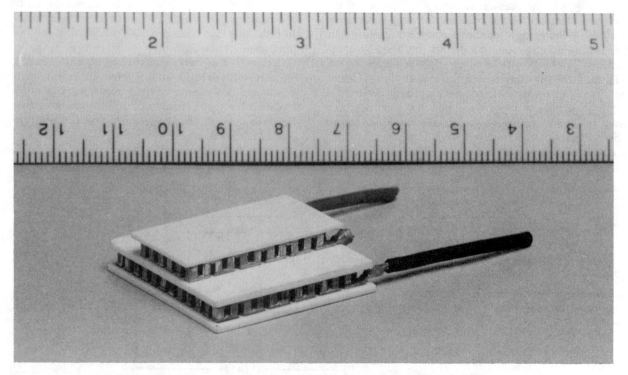

Fig. 8-30. A two-stage thermoelectric cooler. (Courtesy Melcor)

conditioners have also been developed. Westinghouse's Advanced Energy Systems Division developed a noise-free, 5-ton cooling capacity air conditioning system for the Navy's *USS Dolphin* submarine. This system consists of ten modules, each measuring 23 × 21 × 4.25 inches and incorporating 120 couples.

Thermoelectric modules also have many applications in engineering and research. For example, thermoelectric coolers extract excess heat from computer cabinets and microwave waveguides. They also are used to cool laser diodes, far-infrared detectors, CCD imaging arrays, avalanche photodiodes, and photomultiplier tubes. Medical researchers and chemists use thermoelectric coolers to chill and thus immobilize objects and substances being observed with a microscope.

Though still limited to specialized applications, thermoelectric power generators show considerable promise. Several companies in various countries have developed 10 to 100 watt thermoelectric generators fueled by propane, gasoline, or kerosene burners. A system designed to power communication systems in remote regions of the Soviet Union delivers 200 watts when fueled with 4 to 5 pounds of firewood per hour. Smaller thermoelectric generators installed in the chimneys of kerosene lamps provide power for radio receivers in remote Russian homes.

Many kinds of compact nuclear power generators use thermoelectric modules to convert to electricity the heat produced by radioactive decay. Such generators power remote lighthouse beacons in England, unmanned weather monitoring systems floating at sea and installed at remote sites near the north and south poles, and the electrical system of various kinds of satellites and space probes. Thermoelectric modules can also convert sunlight into electricity.

Finally, a thermoelectric module can convert alternating current into direct current. The alternating current is passed through a heating element attached to one side of a module. If the opposite side is kept at a cooler temperature, the module will generate a ripple-free direct current.

For More Information

You can find out much more about thermoelectronics by researching the subject at a good library. Specific articles you may find particularly helpful have appeared in *EDN* ("Thermoelectric Coolers Tackle Jobs Heat Sinks Can't," Jim McDermott, May 20, 1980, pp. 111-117) and *Electronics* ("Thermoelectric Heat Pumps Cool Packages Electronically," Dale Zeskind, July 31, 1980, pp. 109-113). The *Encyclopedia Britannica* has a very thorough article on the subject.

Thermoelectric module manufacturers have published helpful literature about their field. By far the most complete reference guide is Cambion's *Thermoelectric Handbook*. Specification sheets from Cambion, a division of Midland-Ross Corp., and other manufacturers include helpful design, installation, and application information.

Cambion's address is listed in the Appendix. Prices for the Cambion modules given here are contingent upon a minimum order of $100. Cambion distributors do not necessarily impose an order minimum.

Melcor is another good source for information about ther-

moelectric modules. Their applications information is very understandable and well illustrated. Best of all are the performance curves for each of the firm's modules.

Melcor's modules range in cost from $8.85 to $48 each in small quantities. The company requires a minimum billing of $25.

Before ordering thermoelectric modules, you should first request specification sheets and pricing information from the various manufacturers. In addition to Cambion and Melcor, thermoelectric module manufacturers include Borg-Warner Thermoelectrics and Marlow Industries, Inc.

An Experimental Security Alarm

A high crime rate is an unfortunate byproduct of our "modern" civilization. Illegal entry of homes, offices, and businesses are particularly common. The burglary epidemic isn't limited to the cities. Even in the rural area where I reside, several neighbors have been victimized by burglars

With these unpleasant thoughts in mind, let's look at an experimental do-it-yourself security alarm system you can assemble. The system is battery-powered and can identify which of up to eight sensor switches has been triggered. You can use the system with an existing network of normally closed sensor switches. Or you can install your own sensor switch network.

Even if you don't need a security alarm system, you might find the circuit to be described of some interest. That's because it uses an LM3914 dot/bargraph display driver. This chip has many important applications, and the circuit given here might give you some good ideas.

Intrusion Alarm Sensor Switches

Over the past decade, many kinds of sophisticated sensors for detecting intruders have been developed. Nevertheless, various kinds of mechanical switches are still the most common sensor devices for intrusion alarm systems. Generally, these switches are normally closed so they can be used in a closed-circuit alarm configuration.

There are many types of mechanical sensor switches for intrusion alarms. The simplest is a strip of self-adhesive aluminum foil installed around the perimeter of a glass window. If the window is broken, the foil breaks and triggers an alarm. Another simple sensor is a pressure-activated floor switch placed at a doorway or other entrance point.

One of the most common mechanical sensor switches is the magnetic reed switch. This device consists of two cantilevered metal strips installed inside a sealed glass tube. In the normally closed version of this switch, the two strips remain in electrical contact so long as a magnet is placed nearby. If the magnet is moved more than an inch or so away from the switch, one of the strips pulls away from the other, thus opening the circuit. Normally open magnet switches are also available.

Magnet sensor switches are usually installed at entry points like doors and windows. If the door or window is partially opened, the switch opens (or closes), thus triggering an alarm. Figure 8-31 shows a typical installation on a door. The magnet is installed on the door itself, and the sensor switch is installed nearby at a fixed location on the door frame.

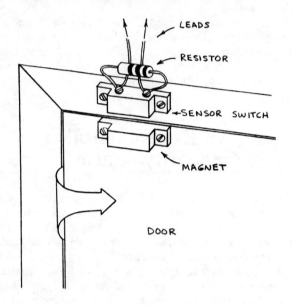

Fig. 8-31. Typical installation of magnet sensor switch.

Some sensor switches are vibration sensitive and will trigger an alarm before actual penetration of a protected structure has occurred. Vibration sensors are often used to protect cars and windows. Others use a simple plunger to detect the movement of a door or window.

A Resistor/Switch Sensor Network

A security alarm system generally incorporates a network of several sensor switches, each of which can trigger a central alarm. This poses the problem of determining which sensor switch has been activated. One way to solve this problem is to run individual connection wires from each sensor switch to the alarm console. A panel of pilot lamps or LEDs can then be used to indicate which switch has been triggered. This method, however, may require considerable wiring and may be difficult to implement in a home or business.

An alternative method with which I've recently experimented is illustrated in Fig. 8-32. Here, a series network of standard, normally closed sensor switches is connected to a multimeter. Each switch is bypassed by a resistor having a different value. Normally, all the switches are closed and the resistance of the network is only that of the wiring, perhaps a few ohms at most. However, if one of the switches is opened, its resistor is placed into the circuit and the multimeter indicates its value. If the resistors have values of 100, 300, and 500 ohms, then the meter can indicate any combination of switches that have been opened as shown in Table 8-1.

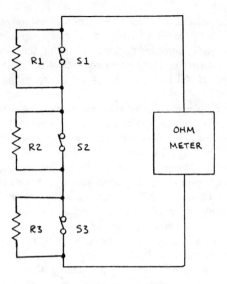

Fig. 8-32. Basic resistor/switch sensor network.

Table 8-1. Resistances Indicated by Combinations of Switches

Resistance	Switch(es)
100	1
300	2
500	3
400	1&2
600	1&3
800	2&3
900	ALL

I don't know if this method is an original idea, but it works quite well. Indeed, I've used it to design a sophisticated intrusion alarm system that uses a personal computer as a programmable controller. The computer permits the system to be switched on at any desired time and it prints on its screen information about which sensor has been triggered.

A Security Alarm with Resistor/Switch Sensors

Though a computerized intrusion alarm system has important advantages, most home and personal computers require household current and therefore such an alarm system will only function when power is available. For this reason, I have designed an experimental, battery-powered circuit that can be used with resistor-bypassed sensor switches. Figure 8-33 shows the circuit diagram. Figure 8-34 shows how the system is connected to a network of up to eight resistor-bypassed sensor switches.

The circuit is powered by a pair of 6-volt lantern batteries connected in series. An LM317T voltage regulator is used to provide a steady output voltage of about 8.5 volts. This guarantees that the circuit will function properly until the battery

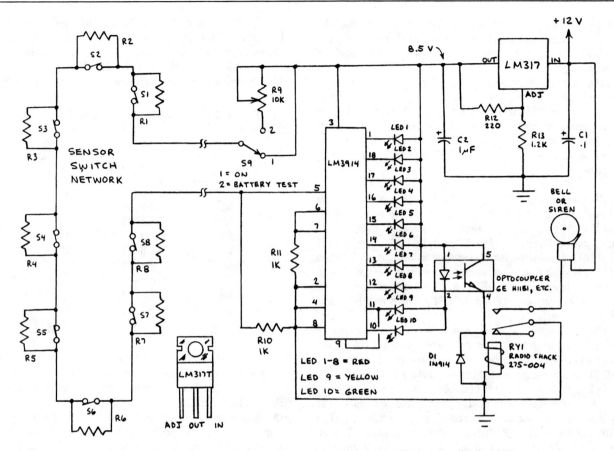

Fig. 8-33. Intrusion alarm system with individual sensor switch LEDs.

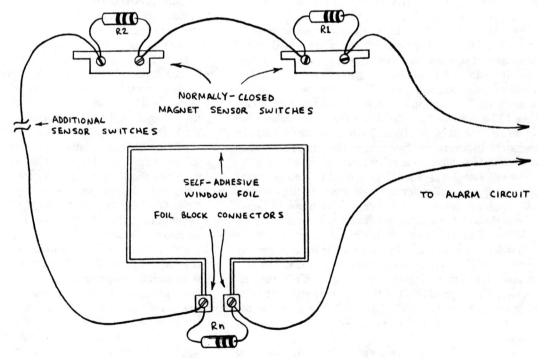

Fig. 8-34. Practical resistor/switch security alarm sensor network.

voltage falls to about 10 volts or so. The output voltage of the LM317T is determined by R13. C1 is included in the event the supply leads from the lantern batteries exceed a length of about 6 inches.

The electronic nerve center of the circuit is an LM3914 dot/bargraph display driver chip. This unique chip can be considered an analog-to-digital converter that indicates the magnitude of an input voltage by means of a row of ten LEDs. The LEDs can function in a bargraph or moving dot mode. The LM3914 is very easy to use and requires a minimum number of external components. Moreover, the output current delivered to the LEDs is controlled by means of a single resistor, R11.

Referring to Fig. 8-33, the LM3914 is operated in a moving dot mode by connecting pin 9 to pin 11. As the voltage at pin 5 on the LM3914 increases, the LEDs connected to the chip's ten output pins switch on and off until the highest order LED (LED 10) is illuminated. A very small amount of overlap between outputs, about a millivolt or so, is designed into the LM3914 so that at no time are all the LEDs off. This prevents an annoying flicker effect that occurs when the LEDs switch rapidly on and off without overlap.

R11 controls the brightness of the LEDs. The current through R11 is approximately a tenth the current through the LEDs. When R11 is 1000 ohms and the supply voltage at pin 3 of the LM3914 is 8.5 volts, the LED current is about 11 milliamperes.

In Fig. 8-33, a network of resistor-bypassed sensor switches is connected in series with R10 to form a voltage divider. The top half of the divider (R1/S1) is connected to +8.5 volts. The bottom of the divider (R10) is connected to ground. When all the sensor switches are closed and S9 is at position 1, the positive supply voltage from the LM317T regulator is applied directly to the input of the LM3914 and the highest order LED (LED10) is illuminated.

Note that the anodes of LEDs 9 and 10 are connected to the positive supply voltage through the internal LED in the optocoupler. Therefore, when either of these LEDs is illuminated, the optocoupler LED also receives forward current. This switches on the internal photo-darlington transistor in the optocoupler which, in turn, applies current to the coil of relay RY1. When RY1's armature is pulled down, the external alarm bell (or siren) is switched off. Diode D1 is connected across the relay's coil to absorb the high-voltage surge generated when the current flow through the coil is switched off.

Though I used a General Electric H11B1 photo-darlington optocoupler, any similar device should work. Devices with an ordinary phototransistor might work, but the photo-darlington will provide better sensitivity and more reliable triggering.

When one of the sensor switches is opened, presumably by an intruder, the resistance of its bypass resistor is placed between R10 and the positive supply from the LM317T. The voltage appearing at the junction of this resistor and R10 then appears at pin 5 of the LM3914. If the values of resistors R1–R8 are properly selected, then the LED having the same number as the resistor will glow when its respective sensor switch is opened.

I measured the resistance range over which each LED switched on in a test version of the circuit in Fig. 8-33. First,

I connected a resistance substitution box between pin 5 of the LM3914 and the positive supply. Then I found the maximum and minimum resistance values over which each respective LED remained illuminated (R1=LED1, R2=LED2 . . . R8=LED8). The results of those measurements are given in Table 8-2.

Table 8-2. Resistance Values for Fig. 8-33

Resistor	Range (kΩ)	Typical Value (kΩ)
R1	36.2–73.0	47
R2	23.9–36.0	33
R3	17.6–23.7	22
R4	13.9–17.4	15
R5	11.5–13.8	12
R6	9.7–11.4	10
R7	8.4– 9.5	6.8 + 2.2
R8	7.3– 8.2	6.8 + 1
R9	6.5– 7.2	10-kΩ Pot
R10	0 – 6.4	Switches

The numbers shown as "typical values" in Table 8-1 are suggested standard resistance values that fall intermediate within the range of each measurement result. To test the circuit with these values, I installed on a plastic breadboard resistors having the values given above and then shorted each resistor(s) with a wire to simulate a normally closed switch. When the shorting wire across R1 was opened, thereby simulating the opening of S1 in Fig. 8-33, LED1 glowed. Likewise, when the wire across R2 was opened, LED2 glowed. All the LEDs glowed in turn as the shorting wires across their respective sensor switch resistors were opened.

By now you may be wondering about the function of LED9. Refer to Fig. 8-33 and you'll note that position 2 of S9 is designated Battery Test. When S9 is at position 2, potentiometer R9 is connected between pin 5 of the LM3914 (through the sensor switch network) and the positive supply voltage. When R9 is adjusted to have a resistance of between 6.5 kΩ and 7.2 kΩ, LED9 will glow when S9 is at position 2.

To use LED9 as a battery test indicator, place switch S9 in position 2 and adjust R9 until both LED9 and LED10 are glowing. Then slightly back off on R9 until only LED10 is glowing. Now, when the total voltage from the two 6-volt lantern batteries falls below about 10 volts, LED9 will glow when S9 is switched to position 2. Otherwise, LED10 will glow. It may be necessary to experiment with R9's setting for reliable results.

Why a battery test mode? Remember that the LM3914 input is voltage sensitive. If the battery voltage falls below about 10 volts, then the output from the LM317T regulator chip will also fall. This means the voltage from the divider formed by the sensor resistors and R10 will be altered. Consequently, the relationship between the sensor resistors and indicator LEDs will be changed so that a different combination of LEDs will glow when the various sensor switches are opened.

Incidentally, when a battery test is made by switching S9 to position 2, the relay will automatically drop out for an instant or so. This will cause a brief burst from the alarm bell or siren, thereby testing it as well as the battery status.

Be sure to take advantage of multicolored LEDs when assembling the circuit in Fig. 8-33. For the prototype version of the circuit, I used red LEDs for LEDs 1–8. I used a green LED for LED10 to indicate the circuit was properly functioning. And I used a yellow LED for LED9 to indicate a low battery condition.

What Can Go Wrong?

I'm unaware of a foolproof security alarm system. Though the experimental circuit described here has several built-in safeguards, it is not perfect. Let's look at some of its vulnerabilities from the burglar's perspective.

First, the system will work properly only when one sensor switch has been opened. If two switches are opened, the alarm will sound, but only one LED will glow. And it might not indicate which of the sensors has been triggered. If enough switches are opened, all the LEDs will be extinguished, but the alarm will sound.

The system does not include a lock-in provision. In other words, the alarm will stop sounding as soon as an opened sensor switch has been closed or jumped by a clip lead.

If any of the sensor network's wiring is installed outside the protective area, an unwise procedure, a naive burglar might be tempted to clip the wire. Should that occur, all the LEDs will be extinguished and the alarm will sound.

A smart burglar will realize that most alarm systems use normally closed sensor switches and that cutting a sensor wire will trigger the alarm. Therefore, he might attempt to bypass, with a clip lead, the sensor switch at the entrance he wants to penetrate. For this reason it's absolutely essential to install the sensor switches inside the protected area, preferably in an inconspicuous location.

Then there's the alarm bell or siren. If it's mounted outside so the neighbors can hear it when the system has been penetrated, are its connection wires exposed? If so, all your work can be defeated in an instant by a single clip of a wire cutter. Even if the wiring is protected or concealed, the alarm device itself can be defeated. Some bells and sirens can be muffled or even silenced by injecting plastic foam inside their housings. Keep this in mind when you select an outside location for an alarm device.

As for cutting off the outside power, the system shown here is battery powered. But if the batteries haven't been checked recently, the wrong indicator LED may glow when a sensor switch is triggered. Worse, if the batteries' reserve power has dropped to a very low level, the heavy current drain of most alarm bells and sirens will quickly exhaust the remaining power.

Fortunately, security alarm systems are like insurance. You may never need them but when you do they more than pay for themselves. Unlike insurance, however, security alarms can cause major problems when they malfunction. Some communities have been so deluged with false alarms from home and business security systems that they have passed ordinances which fine repeat offenders.

In other words, it's essential that a burglar alarm system, particularly a do-it-yourself version, be carefully designed and installed to minimize the possibility of false alarms. In particular, sensor switches should be installed securely. Door and window magnet sensor switches should be mounted close to their respective magnets to avoid false triggering that might occur should the wind cause slight movement. Likewise, vibration sensors should not be adjusted to be so sensitive they respond to wind or loud sounds.

The wire connecting the network of sensor switches can also cause problems. For example, it can pick up stray electrical signals or even lightning, thereby causing an alarm malfunction.

For all these reasons, always consider the installation of a security alarm system a serious matter. All control circuitry should be well built and installed in dust-free enclosures. The batteries should be checked frequently. Each sensor switch should be periodically triggered to determine if the alarm works properly. Be sure to brush away dust and cobwebs from the switch terminals to prevent possible malfunctioning of the system.

Going Further

An obvious drawback of the alarm system shown in Fig. 8-33 is the lack of a timer to allow you to enter and exit a protected area without triggering the alarm. For this reason, it's necessary to activate the system with a concealed switch located outside the protected structure. The switch can short the sensor switch at the entry-exit point. Or it can be placed between the batteries and the circuit.

One way to eliminate the need for an external switch is to add one or two 555 timer circuits. A timer-driven relay whose contacts are placed between the batteries and the circuit will automatically switch the circuit on after a predetermined interval. This will allow you to switch on the alarm and then leave through any exit before the system is activated.

A second timer-driven relay can be connected between the alarm bell or siren and RY1 in Fig. 8-33. This will allow time to enter the protected structure and switch off the system before the alarm sounds. Unfortunately, this timer may allow a fast working intruder to enter through a "protected" entrance and bypass the sensor switch with a clip lead before the alarm sounds.

Still another timer can be used to keep the alarm switched on for a fixed interval even after a triggered sensor switch has been closed. Keep in mind that some communities impose fines if an alarm remains on more than 10 or 15 minutes.

As you can see, building and installing intrusion alarm systems is both serious and tricky. An excellent review article about commercial burglar alarm systems was published in the October 1984 issue of *Consumer Reports* (pp. 568–571, 606). This article observes that living with a burglar alarm can be a nuisance, and that good locks and a smoke detector should have a higher priority. Many books have been written on the subject, and you can probably find one at a good library or book store. Meanwhile, here's hoping the system you build will always work, never give a false alarm, and never be needed.

A Followup

Earlier I observed that "I don't know if this method is an original idea, but it works quite well." After this was published in *Modern Electronics*, Sherwood M. Kidder, president of Northern Instruments, sent this interesting letter: "I really had a surprise when I recently purchased a copy of *Modern Electronics* and saw your article on the security alarm system, because it is quite similar to the one I developed in 1979. My system does provide for some of the shortcomings that you describe. The resistive link and comparator were described in a patent application, No.49807, June 1, 1979. . . . In your article you questioned the originality of the system and I just wanted to let you know that a similar system has been developed."

Incidentally, you can find a detailed explanation of how to design a simple computer-controlled intrusion alarm based on the resistive burglar alarm sensor concept in *Forrest Mims's Computer Projects* (Osborne/McGraw-Hill, 1985, pp. 227–242).

Experimenting with Small DC Motors

Small dc motors are used in robotics, remote control, toys, tape recorders, servos, plotters, and printers. They can even be used as self-generating transducers in devices that measure rotation rates and wind speed.

There was a time when experimenters who needed a small dc motor had to make their own. Today miniature dc motors suitable for many of the foregoing applications can often be purchased for less than a dollar. Much higher quality motors are available for more money. Some include built-in reduction gears and speed-control governors.

Motor Suppliers

Small dc motors are available from most hobby shops, particularly those that specialize in radio-controlled model planes, cars, and boats. Many of these stores also sell the miniature motors that are used in radio-controlled servos. These motors may cost as much as $10 or more, but they are generally of very high quality.

Electronics stores also sell small dc motors. They may also be a good source of defective tape recorders, calculators, and computer printers from which you can salvage motors. I've salvaged good quality motors from broken cassette recorders and radio-controlled toy cars picked up at the "as is" table at Radio Shack stores.

Motors are also available from mail order suppliers. The Edmund Scientific industry and education catalog lists many different kinds of small dc motors. For example, catalog number H31,827 is a pair of 1.5- to 6-volt motors connected to a common shaft and interfaced by a magnetic clutch. One application for this motor is a drive mechanism for small toys and robots. Several motors sold by Edmund include built-in reduction gears.

Another small order supplier is Ace R/C, Inc. Ace sells many kinds of radio-controlled servos and the motors and gears they use. For example, Cat. No. MR012 is an ultraminiature 10-ohm motor that measures only 12 millimeters in diameter.

If your motor requirement cannot be satisfied by any of these

sources, you may have to contact a motor manufacturer or a supplier that specializes in motors. For example, Portescap US is a Swiss company that manufactures a broad line of precision dc motors. Rapidsyn and Warner Electric Brake & Clutch Company both make a complete series of dc stepping motors. Additional suppliers and manufacturers of small dc motors are listed in various electronics industry catalogs.

Motor Control Circuits

Many kinds of low-voltage dc motors can be directly controlled by simple transistorized or integrated circuits. A common example is the drive motor in inexpensive radio-controlled cars. When controlling a motor with a semiconductor device, it's important to avoid exceeding the device's power and voltage ratings. Often a heatsink will be required to couple excess heat into the surrounding air.

CMOS circuits and power MOSFET devices are often used in motor control circuits. Be sure to follow appropriate handling and soldering precautions when using these devices. Excessive voltage and static electricity can cause permanent damage to such components.

Motor Reversers

Figure 8-35 shows how a pair of single-pole, double-throw (spdt) switches can be used to reverse the direction of rotation of a dc motor. The switches permit the polarity of the applied current, hence the direction of rotation, to be reversed.

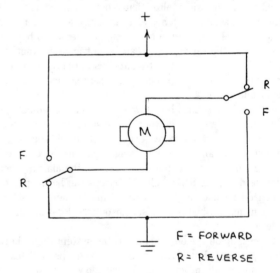

Fig. 8-35. Simple motor reverser.

Generally the switching arrangement in Fig. 8-36 is used to implement a motor reverser. Here the two spdt switches of Fig. 8-36 are replaced by a single double-pole, double-throw switch. This permits the circuit to be controlled by a single switch.

The switches in Figs. 8-35 and 8-36 can be replaced by power MOSFET transistors, thereby permitting the rotation direction of a small dc motor to be controlled by an external logic signal. Figure 8-37 shows one circuit I've devised for this

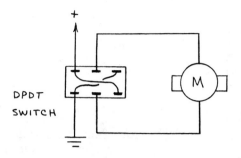

***Fig. 8-36. Reversing motor rotation
direction with a dpdt switch.***

purpose. A pair of gates in a CMOS 4011 quad NAND gate provides the steering logic necessary to switch the MOSFETs on and off. Bipolar power transistors can be used in this kind of circuit, but MOSFETs are a better choice since they are more easily interfaced with external logic circuits.

Referring to Fig. 8-37, when the input is low MOSFETs Q1 and Q2 are switched on and Q3 and Q4 are switched off. Therefore the A terminal of the motor receives a positive bias and the motor's armature rotates accordingly. When the input changes from low to high, MOSFETs Q1 and Q2 are switched off and

Q3 and Q4 are switched on. The A terminal of the motor then receives a negative bias. Now that the polarity of the voltage applied to the motor is reversed, the armature's direction of rotation is also reversed.

If you build and experiment with this circuit, you may notice that changes in the motor's direction of rotation can be implemented merely by momentarily applying a high or low to the input and then allowing the input to "float." When this occurs the circuit appears to remember the status of the last input signal.

In real applications you should never allow the input of a CMOS gate to float. This is because the input lead acts as a miniature antenna that can pick up stray electrical signals. In other words, a floating input may appear to change states spontaneously in response to electrical noise. Even with no noise present a floating CMOS gate may spontaneously change states as the charge previously stored within it gradually leaks away into the surrounding air.

Motor Speed Controller

It's possible to alter the rotation speed of a small dc motor by rapidly interrupting the continuous current normally applied to the motor. This can be readily accomplished by connecting the motor to a simple pulse generator. Increasing the pulse rate will increase the rotation speed.

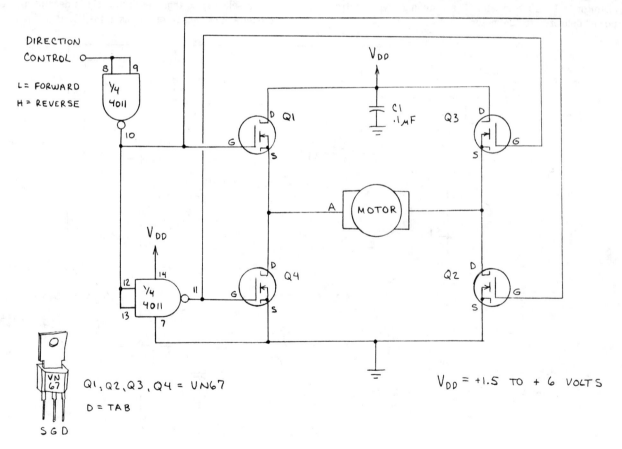

Fig. 8-37. Solid-state dc motor reverser.

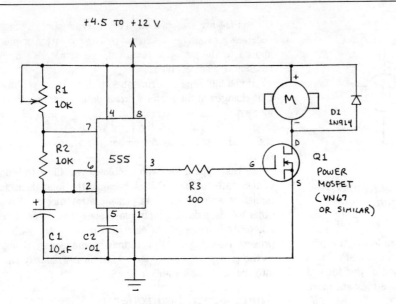

Fig. 8-38. Simple dc motor speed controller circuit.

For best results, the pulse generator should operate at a 50-percent duty cycle. Many different circuits can be used, including simple two-transistor oscillators, cross-coupled gate multivibrators, op-amp pulse generators, and timer ICs. In most cases the pulse generator drives a switching device such as a power transistor or MOSFET transistor. The switching device controls the power delivered to the motor.

Figure 8-38 is a straightforward motor speed controller designed around a 555 timer IC connected as an astable multivibrator. In operation, R1 and C1 control the circuit's pulse rate. Since R1 is a potentiometer, the pulse rate, hence the motor's rotation rate, can be easily varied by altering R1's resistance. R2 is selected so that the duty cycle is near 50 percent. The circuit's operating frequency is $1.44/(R1 + 2R2)C1$.

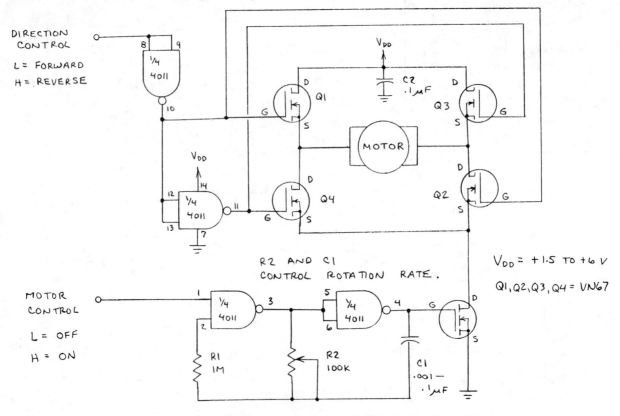

Fig. 8-39. Small motor direction and speed controller.

Pulses from pin 3 of the 555 are applied to the gate of Q1, a power MOSFET transistor such as the VN67. When Q1 is switched on, current is applied to the motor. The pulse rate determines the speed of the motor's rotation.

Note that the motor terminals are bypassed by D1, a reverse-biased diode. Since the diode is connected in the backwards direction, it has no effect upon the circuit. Instead, its function is to slow the rotation of the motor between pulses. After a current pulse applied to the motor ends, the motor's armature will continue to rotate until its angular momentum is overcome by the friction of the brushes and bushings. During this period, the motor functions as a generator. Since D1 is connected as a short circuit directly across the motor's terminals, it places a heavy load on the motor when it is self-generating a current. This imparts a braking force on the armature that can substantially slow its angular momentum.

You may wish to experiment with D1. Its presence may have a substantial effect on some motors and very little on others. You can also experiment with Q1. Though I have used several kinds of power MOSFETs for Q1, an npn power transistor will also work. In any case, be sure the transistor you select is adequately rated and, if necessary, heat sunk.

How well this circuit functions is greatly determined by the motor. It can drive some high quality motors, those with low-friction bearings and brushes, at very slow rotation rates (a few tens of rpm). Cheaper motors can be operated down to a rotation rate of a few hundred rpm. I have had best results with motors fitted with a built-in gear reduction system. The reduction gearing provides reliable slow rotation speeds while allowing the motor to be operated at a higher speed where the effect of the speed controller circuit is more dramatic.

A Direction and Speed Controller Circuit

Figure 8-39 shows a circuit I've designed that controls both the direction and speed of rotation of a small dc motor. The direction-control portion of the circuit is identical to the circuit in Fig. 8-37. The speed-control portion of the circuit is formed by using the two unused gates in the 4011 as a cross-coupled multivibrator. The multivibrator has a duty cycle near 50 percent. Its output is applied to Q5, a power MOSFET connected in series with the direction-controller portion of the circuit. Therefore, current is applied to the motor only when Q1 is switched on by the multivibrator. The operating frequency of the multivibrator is determined by R2 and C1. The frequency is about 1/(2.2RC).

The speed control portion of the circuit in Fig. 8-39 can be replaced by the 555 timer pulse generator in Fig. 8-38. The 555 circuit is somewhat easier to adjust and it can be connected directly to the gate of Q5 in Fig. 8-39. However, the circuit in Fig. 8-39 is much simpler, particularly since it uses all four gates in a single 4011.

Going Further

The circuits given here can be easily modified. For instance, replacing R2 in Fig. 8-39 with a light-sensitive photoresistor will allow the motor speed control portion of that circuit to be controlled by means of a flashlight. Of course both light and radio signals can be used to change the direction of a motor and to switch motors on and off.

Circuit Assembly Tips

- Solderless Breadboards for Experimenters
- How To Assemble Miniature Circuits
- Electrical Shock

Circuit Assembly Tips

Solderless Breadboards for Experimenters

The invention of solderless breadboards has played an essential role in the evolution of modern solid-state electronics. Before such breadboards became available, experimental and prototype circuits had to be laboriously assembled using point-to-point wiring and soldered connections. Needless to say, circuit changes were difficult and time consuming.

In the late 1960s, Barry Instrument Corporation introduced the Springboard, a plastic base containing 120 rectangular slots. In each slot was a 0.3-inch long spring. Component leads and wires were inserted into the springs by means of an awl-like tool that opened a space in any desired spring.

I assembled hundreds of transistor circuits much faster than ever before with the help of a Springboard. The board still comes in very handy for prototyping circuits that have oversize leads such as heavy duty rectifiers and SCRs. Unfortunately, the Springboard is not suitable for use with much smaller pins and leads of integrated circuits.

Wire-Wrappable Panels

Eventually several clever solderless breadboards were developed specifically for ICs. Among these are expensive wire-wrapping panels. Available in numerous styles and configurations, wire-wrappable panels have rows of IC sockets on one surface. The second surface contains matching rows of square cross-section pins extending from the sockets. Circuits are assembled by installing the required chips and making the necessary interconnections with wrapping wire.

Wire-wrappable panels and boards are available from many companies. Among those whose products are sold to experimenters and hobbyists are Vector Electronics, Inc. and Cambion (Electronic Connectors Division of Midland-Ross Corp.). Both these companies make a wide assortment of preassembled wire-wrapping boards as well as perforated boards into which compatible wire-wrappable terminals and sockets can easily be installed.

The Electronics Products Division of the 3M Company has developed a particularly interesting do-it-yourself breadboarding system. The system uses unique dual-contact IC sockets which will accept an IC pin from one side and a plug strip pin from the other side. The sockets are placed on a perforated card and

secured in place by plug strips inserted into the back of the socket from the opposite side of the card. Interconnections are made with the help of an insertion tool that forces insulated wire into the U-shaped slots in the ends of the plug strip terminals.

3M has combined an assortment of dual-contact IC sockets and plug strips in its Scotchflex 3303 Breadboard Kit. This kit also includes a wire insertion tool, socket extraction tool, a cleverly designed plug strip break-off tool, and a spool of connection wire.

You can easily custom design your own breadboards using hardware from Vector, Cambion, 3M, and other companies. Or you can improvise by installing rows of wire-wrappable IC sockets in a standard perfboard having copper solder pads at each hole (Radio Shack 276-152 or similar). Solder the socket's corner pins to the copper pads and the board is ready to use. Alternatively, you can use a standard copperless board and secure the sockets to the board with a thin bead of cyanoacrylate cement.

Wire-wrappable panels are excellent for prototyping complex digital circuits of a repetitive nature such as large capacity solid-state memories. They are therefore very popular with computer enthusiasts. Because of the difficulty of making changes, however, they are unsuited for developing and experimenting with most analog/linear circuits and experimental logic circuits. Indeed, assembled wire-wrapped boards can be used in permanent applications.

Plastic Solderless Breadboards

The plastic solderless breadboard has become an indispensable item on every electronic experimenter's workbench. AP Products, Inc. originated the concept of a solderless, plug-in breadboard. In addition to AP Products, major solderless breadboard makers include Vector Electronics, Interplex Electronics, and E&L Instruments.

The solderless breadboard elements made by Interplex Electronics are similar in configuration to competing modules. Each breadboard element includes parallel arrays of five common tie points. Inside the five socket holes at each row of tie points is a replaceable spring chip having precision formed, nickel silver contacts. The result is five physically independent but electrically common socket hole tie points that accept wire leads having a diameter of from 0.015 to 0.033 inch. The contact resistance of each connection point is under 0.005 ohm.

Some solderless breadboards are configured as modules that include two or more rows of bus connection points across both sides of the breadboard. Others require separate plastic bus strips. Most modules snap or clip together to form arrays of modules.

Most manufacturers of solderless breadboards make various breadboard assemblies or prototyping systems. Interplex Electronics, for example, makes the Proto-Board 103. Measuring 6 × 9 inches, the PB-103 includes eight separate plastic breadboards installed on an aluminum panel having rubber feet. Four power supply binding posts are included. The 450 common terminals of the PB-103 have a total of 2250 connection points (socket holes). Other breadboard assemblies can have fewer or considerably more.

Prototyping systems that include built-in power supplies and other circuit design aids are made by several companies. Among those made by Interplex Electronics, for example, is the CD-1 CMOS/TTL Designer. This compact console-style prototyping tool, which is designed around a single solderless breadboard, includes a 5-volt fixed supply, a variable 3- to 15-volt supply, an adjustable frequency (1-Hz, 10-Hz, 100-Hz, 1-kHz, 10-kHz, and 100-kHz) logic clock, four CMOS/TTL compatible LED indicators, and various switches, binding posts, and connectors. Though sophisticated prototyping systems like the CD-1 are expensive, they provide an unprecedented degree of circuit design flexibility.

A Modular Breadboarding System

AP Products produces the Hobby-Blox family of compatible solderless breadboard products designed specifically for experimenters. Ideally suited for use by experimenters, the Hobby-Blox system is designed around a sturdy plastic tray into which various color-coded bus strips, solderless breadboard strips, and spacers can be installed. The tray will also accept a binding post strip and an exceptionally well designed 9-volt battery holder.

All the Hobby-Blox breadboards include row and column index markings. Two types are designed specifically for ICs or digital readouts. A third is designed for discrete components. A 6-position LED strip is also available.

The most notable feature of the Hobby-Blox system is its flexibility. Breadboard strips, spacers, and other members of the family can be installed in any desired arrangement. One or more additional trays can be added to the main tray with the help of an interlocking bus strip containing 60 connection points or a simple tray extender clip.

Even more important is the ability to install vertical panels that extend upward from the tray as shown in Fig. 9-1. These include a speaker panel with preformed holes (Fig. 9-1, left) and a control panel with 6 holes for installing switches, potentiometers and indicator LEDs or lamps. Blank panels are available for custom designs. All these panels are mounted by inserting their three extensions into the slots in a spacer/support strip. The blank panels are ideal for mounting joysticks, fiber optic connectors, piezoalerters, optoelectronic components, and other specialized devices and components.

A tray shorter than the main tray can be vertically mounted on the main tray by means of an adaptor strip. This tray will accept all the Hobby-Blox modules and is therefore ideal for LED displays. It's also well suited for assembling optoelectronic circuits that require careful location or spacing of such components as infrared emitting diodes and photodetectors.

Hobby-Blox Applications

Of the many applications for the Hobby-Blox system, two I've found particularly handy are compact, self-contained op amp and CMOS prototyping systems. Figure 9-1, for example, shows one way to make an op-amp prototyping system. Dual polarity, self-contained 9-volt supplies are provided as is a speaker. The optional binding posts can be used when the system is powered by an external supply. A control panel with three potentiometers (10 kΩ, 100 kΩ, and 1 megohm) is also included. With a single 741 op amp and a dozen or so resistors, capacitors, diodes, and LEDs, this prototyping system can be used to assemble scores of different op amp and comparator circuits. It can easily be expanded by adding one or more additional trays with the help of extender clips or a bus strip.

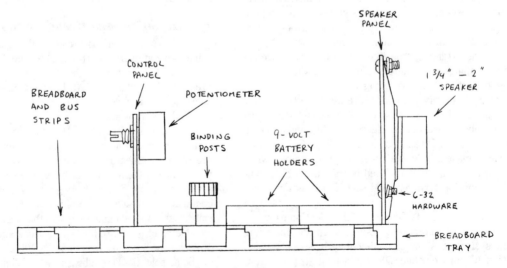

Fig. 9-1. A Hobby-Blox op-amp prototyping system.

The vertically mounted blank panel and short tray make possible many very unique circuit applications. Figure 9-2, for example, shows an experimental optoelectronic detection circuit that will trigger a camera or an electronic strobe when an object enters the space between an infrared emitting diode and a phototransistor. This arrangement will permit you to automatically make clearly focused photographs of flying insects and projectiles since the camera will be triggered only when the object to be photographed is directly between the LED and phototransistor.

In operation, an LED transmitter delivers a train of fast risetime pulses to the sensitive surface of the receiver's phototransistor. Q4 and the 555 form a missing pulse detector that is reset each time a pulse from the transmitter is received by the phototransistor. When an object blocks the path between the transmitter's LED and the receiver's phototransistor, the 555 completes its timing cycle and the output at pin 3 changes from high to low.

The receiver output can be connected directly to the trigger inputs of some cameras, but an inverter and a common ground may be necessary. Other cameras may require a relay (pin 3 to +9 volts) or SCR (pin 3 to gate) interface.

The system can be adjusted to ignore very fast moving objects by increasing the values of R3 and C2 in the receiver and slowing the frequency of the transmitter by means of R1.

You can quickly assemble the circuit in Fig. 9-2 using Hobby-Blox modules. One possible arrangement is to assemble the transmitter and receiver on breadboard strips installed in short trays. The assembled trays should be installed facing one another in a main tray. The battery should be installed in a holder located between the two facing circuits. Any potentiometers and switches should be installed on one or two control panels located anywhere but between the two facing circuits. One possibility is to use extender clips to add a short tray that can contain any required control panels.

If you plan to photograph insects, place a flower or an appropriate bait between the facing circuits. Depending upon the time of year and your location, you may be able to attract flies, moths, beetles, and other insects. Be sure to shield the phototransistor from external light with a length of heat shrinkable tubing.

Going Further

Solderless breadboards can be used for finished circuits, but this can become rather expensive if you build lots of different projects. Interplex Electronics has developed a simple but effective method for quickly transferring a breadboard circuit onto an etched circuit board. This concept is based upon an etched circuit board that exactly duplicates the array of terminal holes in a plastic breadboard. A pad of paper versions of the circuit board allows experimenters to make new designs and layouts on paper and to keep a record of successful designs.

Interplex Electronics calls this clever concept The Experimenter System. It's but one way to convert your circuits into permanent versions. Thanks to some very creative engineers, there are many effective methods for making permanent soldered and wire-wrapped circuits. If you're still building circuits using perforated boards and point-to-point soldered connections, you should take time to investigate the many circuit assembly aids and hardware available.

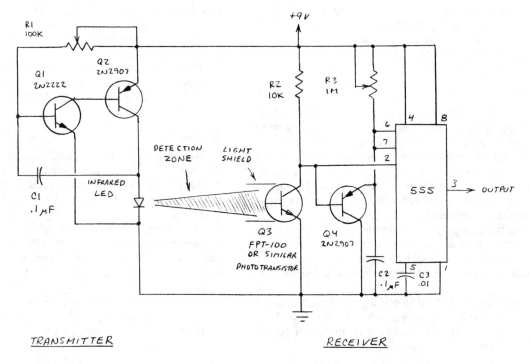

Fig. 9-2. An optoelectronic trigger for a camera or electronic flash.

How To Assemble Miniature Circuits

Since I began building transistor radios in the mid-1950s, I've been fascinated by miniaturized electronic circuits. Even in this era of tiny high-tech products like pocket computers and televisions, digital watches, and credit card-size radios and calculators, there are still plenty of circuits left for experimenters to miniaturize.

The purpose of this discussion is to describe some of the tips and techniques I have used over the years to assemble miniaturized electronic circuits.

Do-It-Yourself Circuit Assembly

Solderless breadboards have made the design, evaluation, and testing of do-it-yourself electronic circuits simple, fast, and convenient. Moreover, modifications are easily made before a breadboard circuit is transformed into a permanent version.

Unfortunately, the act of fabricating a permanent circuit remains relatively time consuming.

Many articles and books have covered circuit assembly methods in depth. If your personal electronics library doesn't include any of these sources, your circuit assembly skills will be greatly enhanced if you visit a good library and look through some of the many publications that cover this subject.

Miniature Electronic Circuits

The tools and techniques for building miniature circuits are often very different from those used to assemble conventional circuits. And sometimes the assembly of miniature circuits can be particularly difficult. Though the active components of such circuits are usually very small, it can be difficult for the average experimenter to find miniature switches, potentiometers, battery holders, and enclosures.

Tools

The most important tool for assembling miniature circuits is a good quality, low-wattage soldering iron. Generally a small chisel or needlepoint tip is best for miniature circuit assembly. The iron's tip should be well tinned and kept free of excess solder and dross during soldering. The best way to keep the tip clean is to pull it across a damp sponge, preferably one specifically designed for this purpose. For best results, be sure to use a small-diameter, rosin-core solder.

A magnifying glass can prove very helpful. With it you can find tiny solder balls and bits of wire that might cause short circuits. And you can locate solder bridges that can easily occur when soldering closely spaced component leads and printed circuit lands.

Pointed tweezers are ideal for holding small components and removing bits of loose wire from crowded circuit boards. They are also handy for twisting wrapping wire around component leads prior to soldering the wire in place.

Pliers and wire cutters are required for virtually every electronic construction project. For best results when assembling miniature circuits, use the smallest needle nose pliers and diagonal cutters you can find.

Screwdrivers are also a must. A set of both standard and crosspoint drivers is indispensable for most miniature projects.

Finally, an effective method for drilling small holes is required. You can use a small drill for this purpose. Or you may find that most holes can be formed by twirling the blade of a hobby knife into the material to be drilled.

Depending on your needs, you may wish to add additional tools to your miniaturization toolkit. For instance, a can of compressed air will help you to blow away the bits of wire and other debris that often hide under the components of a newly assembled circuit board. Soldering probes can be used to remove accumulated solder rosin. A nibbling tool can be used to cut pieces of circuit board to size. And a reamer can be used to enlarge holes.

Enclosures

A wide variety of miniature enclosures designed specifically for electronic projects is available from electronics parts suppliers. Radio Shack, for instance, sells several such cases. Their catalog number 270-220 is the smallest of a series of sturdy all-plastic cases that includes internal slots of circuit boards. Radio Shack's catalog number 270-291 is a somewhat larger enclosure that comes with a pc board having 483 solder-ringed holes, two front-panel labels, snap-in rubber feet, and hardware. Many other parts suppliers also sell enclosures.

As for smaller circuits, like most other experimenters I've used a wide assortment of pill bottles, fishing tackle boxes, tie-tac boxes, and the like. A few years ago, however, I discovered a line of compact plastic boxes that are ideal for miniature projects. These boxes have removable lids and are available in clear or tinted (red, pink, yellow, and green) plastic. These boxes are sold by craft shops and specialty stores. I do not know the name of the manufacturer.

These boxes come in three sizes, each of which is slightly more than 0.7-inch thick. The largest is 2-inches square and the smallest is 1-inch square. The third is a rectangle measuring 1 $\times$ 2 inches.

Batteries

When transistors first became available to experimenters, the choice of a battery to power a miniature do-it-yourself circuit was very limited. For really small circuits, mercury hearing aid batteries could be used. Otherwise, N or AAA penlight cells were used.

Because the choice of batteries was so limited, some experimenters made their own power cells. One common technique was to wrap a small piece of copper with a section of paper towel that had been soaked in a salt water solution and allowed to dry. The towel was then wrapped with a layer of zinc or aluminum foil. The cell was activated by placing a drop or two of water on an exposed portion of the salt-impregnated paper.

In the late 1950s experimenters turned to both selenium and silicon solar cells as a power source. In those days, silicon solar cells were very expensive. Solar cells are very thin and even very small cells supply sufficient power to operate simple transistor radios, oscillators, and the like. The widespread use of low-power components, particularly CMOS integrated circuits, has made solar cells an even more viable power source for miniature electronic circuits. Moreover, solar cells are now much cheaper and more readily available.

Battery Holders

Though dozens of miniature batteries are available to electronic experimenters, the choice of battery holders is much more limited. And some battery holders designed for miniature button cells are much larger than the cell they are designed to hold.

After many years experimenting with do-it-yourself battery holders, I've settled on two basic approaches. Both designs are smaller than commercial battery holders. Both incorporate a built-in on-off switch. And both are built into a small housing with room to spare for a circuit and other components.

The simplest of these do-it-yourself battery holders is shown in Fig. 9-3. This design is well-suited for circuits installed in small plastic boxes like the kind described before. Its key ingredient is a subminiature push-button or toggle switch and a solder lug.

You can assemble this battery holder by first bending a solder lug as shown in Fig. 9-3. Place the large hole of the lug over the switch's threaded neck. Then bend one of the switch terminals toward the lug and gently force it through the small hole in the end of the lug. Solder the lug to the switch terminal.

Next, solder a length of hookup wire to a small spring. Use a spring salvaged from a commercial battery holder. Or take a battery holder to a hardware store and ask where you can buy lengths of spring similar to that used in the holder.

Finally, place the spring and battery in the intended portion of the box as shown in Fig. 9-3. Press the switch lug assembly against the free end of the battery until the tension feels right, and then mark the side of the box directly under the switch's push button or toggle. Complete the battery holder by drilling a hole in the box and installing the switch.

Incidentally, you can omit the switch from this battery holder if you prefer. Just attach the bent solder lug directly to the inside of the plastic case with 4–40 or 6–32 hardware. Be sure to solder a length of hookup wire to the lug before installing it. Otherwise you might melt the side of the box with your soldering iron.

Figure 9-4 shows the second do-it-yourself battery holder. This holder is much more difficult to make, but it is well suited for ultracompact circuits installed in short lengths of tubing. I originally developed this holder for use in an experimental infrared travel aid for the blind housed in two 3.5-inch lengths of brass tubing installed on eyeglass frames.

Construction and assembly details for the second holder are shown in Fig. 9-4. Though the holders I have built all follow this general design, you can modify the basic concept to best suit your needs or the materials you have on hand. For instance, the subminiature slide switch can be replaced by a toggle switch if you prefer. You can also use different kinds and sizes of tubing.

You should be prepared to spend a good deal of time making a battery holder like the one in Fig. 9-4. You will also need access to miniature files and cutting tools to make the necessary slots and holes. Use caution when making the battery holder since it's easy to injure a finger.

Make sure you have all the materials before beginning work. You'll need two lengths of telescoping brass (best) or aluminum tubing. The bulkheads can be cut from solid acrylic rod (best) or wood dowels. Be sure the bulkheads fit snugly inside the smaller of the two metal tubes.

I use metal tubing since it is conductive and provides the connection between the on-off switch and the circuit. This means the circuit installed inside the tube should be well insulated or placed inside an insulating sleeve to avoid accidental short circuits.

Circuit Boards

For the utmost in miniaturization, etched or printed circuit boards are almost always necessary. Were it not for a pair of tiny etched circuit boards, I would never have been able to install the receiver circuitry, lens, and battery for an infrared travel aid for the blind inside a tube measuring only .5 × 3.5 inches.

Many books and articles have described in detail the various ways etched and printed circuits can be fabricated. For best results, use very thin copper-clad board, the kind that can be cut with scissors. If this kind of board is unavailable from local electronic parts dealers, ask for advice about possible suppliers. You might also check with a nearby university or technical school that offers electronics courses.

Should fabricating your own etched circuit boards prove too

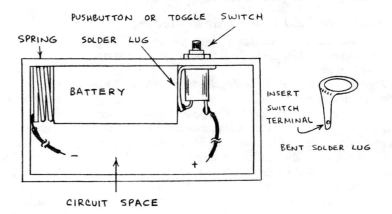

Fig. 9-3. Miniature battery holder with built-in switch.

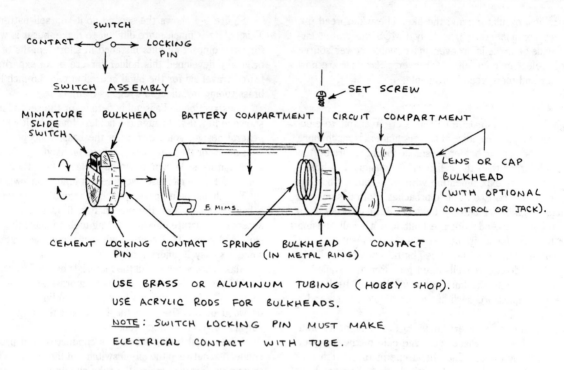

Fig. 9-4. Tubular housing for miniature circuits.

time consuming or inconvenient, it's possible to assemble respectably compact circuits using standard point-to-point wiring. The key here is to use a perforated board which has preetched copper rings, preferably pretinned with solder, around each hole. The leads and pins from the various components can be connected to one another by wrapping wire. With a little planning, many connections can be more easily made by placing leads and pins to be connected in adjacent holes. Bend the pin from one component over the solder ring surrounding the adjacent pin or lead and solder in place.

You can reduce the thickness of circuits made in this fashion by stroking the bottom side of the board across a file. Use care to avoid removing too much solder. Afterwards you must remove all solder filings with a brush and a few puffs of compressed air. Otherwise stray solder particles may cause a short circuit.

A Miniature LED Pulse Transmitter

The best way to become familiar with the miniaturization tips and techniques described here is to build a fully functional miniature circuit.

Figure 9-5 shows the circuit diagram for a miniature LED pulse transmitter you can assemble with room to spare inside a .7 × 1 × 2-inch plastic housing like those described before. This circuit can be used as an optical transmitter for a short-range remote control unit, break-beam object detector, or intrusion alarm. It can also be used as a source for an optical fiber continuity tester. And it's very handy as a workbench source of fast optical pulses for testing various kinds of lightwave receiver systems.

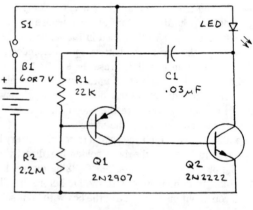

NOTE: R2 CONTROLS PULSE RATE.

Fig. 9-5. Miniature LED pulse transmitter circuit.

Referring to Fig. 9-5, the circuit is a two-transistor multivibrator that delivers a stream of high-current pulses to a high-efficiency red light-emitting diode. Any red LED can be used. For best results, however, use one of the high-efficiency LEDs such as Stanley's H1K (1 candela) or H2K (2 candela). If you can't find either of these super bright LEDs, a good substitute is Radio Shack's Cat. No. 276-066. This LED delivers 300 millicandelas. Though not as bright as the Stanley units, it's much brighter than standard LEDs, it's reasonably priced and readily available.

Begin assembly of the circuit by installing the oscillator components on a piece of perforated circuit board measuring 3/8 × 1 inch (4 holes by 10 holes). Try to orient the components so that connections can be made with only a minimal use of wrapping wire.

Next, refer to the pictorial view of the completed circuit in Fig. 9-6 and assemble the battery holder in accordance with the procedure outlined before. Though I did not incorporate the switch into the holder, you may prefer to do so.

Drill holes in the box to receive the on-off switch and the LED. Then install the circuit and the switch, using care to avoid breaking any of the connection wires. Install a 6- or 7-volt battery and flip the on-off switch to the on (closed) position. The LED should glow brightly with a slightly discernible flicker.

You can transform the flickering light from the LED into an audible tone by pointing the LED at the detector of an optical receiver. You can easily make such a receiver by connecting a silicon solar cell, photodiode, or phototransistor to the input of a small amplifier. For details about various types of lightwave receivers (and transmitters) see *The Forrest Mims Circuit Scrapbook* (McGraw-Hill, 1983).

Electrical Shock

Do you remember the first time you received an electrical shock? Aside from the tingles received from static electrical discharges on cold winter days, my first shock was received at the age of ten when I foolishly touched an exposed copper wire extending from an uncompleted circuit box in a new building.

Knowing that the wire might be "hot," I squatted down on the bare concrete floor, braced myself and quickly brushed my right index finger against the shiny copper. Without these minimal "precautions" I would probably have been electrocuted, for the jolt from the wire was so powerful I clearly remember the incident nearly thirty years later.

Although my crude experiment was exceedingly foolish, the electrician responsible for leaving exposed a potentially lethal hot wire was equally culpable. Unfortunately, many of us who are so aware of the dangers of electrical shock are the very ones who are the most careless. Recently this fact was brought vividly to mind by the tragic loss to accidental electrocution of two talented and respected electrical engineers, Byrd Brunemeier and Gordon Brill.

Byrd H. Brunemeier was an engineer for Far East Broadcasting Company, an organization which broadcasts Christian programming to Asian countries. Since 1949 he had helped install and service transmitter facilities on various Pacific islands. Once he erected a 308-foot transmitter tower, even climbing riggings during typhoon weather. He personally removed thousands of unexploded World War Two shells from a transmitter site on Saipan. Though he survived these and countless other dangers, he did not survive an electrical shock received July 27, 1983 while servicing a new transmitter installation on Saipan.

Gordon Brill was the marketing manager for Plasma Kinetics, a California company that makes metal-vapor lasers. Widely known in the laser industry for his careful attention to customer service, Brill had worked for several laser firms before joining

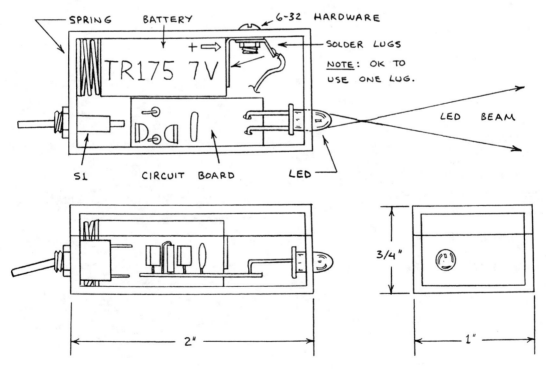

Fig. 9-6. Construction of miniature LED pulse transmitter.

Plasma Kinetics. While servicing a copper-vapor laser at the University of Dayton on June 30, 1983, Brill manually bypassed an interlock designed to keep power from the laser when a service panel was removed. He was instantly electrocuted when he accidentally contacted the laser's power supply.

The tragic loss of these two multitalented, dedicated engineers should be a lesson to all of us who work with or around electronic devices. As far as I'm concerned, any line powered circuit has the potential for delivering a dangerous or even lethal shock and should therefore be treated with respect.

Even battery powered circuits can be dangerous. Portable electronic flash units for cameras, for example, store several hundred volts in a capacitor of a few hundred microfarads. The resultant energy is sufficient to vaporize a small wire or melt the tip of a screwdriver shorted across the capacitor's terminals!

How many times have you and I defeated an interlock or otherwise been negligent when working around high voltage? The loss of Byrd Brunemeier and Gordon Brill should be a lesson to all of us who enjoy experimenting with electronics.

APPENDIX
Company Addresses

The following is a list of the addresses for manufacturers, suppliers, publishers, and others mentioned in this book.

A. C. Interface, Inc.
17911 Sampson Lane
Huntington Beach, CA 92647

Academy of Model Aeronautics
1810 Samuel Morse Dr.
Reston, VA 22090

Ace R/C, Inc.
P.O. Box 511
116 W. 19th St.
Higginsville, MO 64037

Advanced Fiberoptics Corp.
637 Hayden Road
Tempe, AZ 85281

Advanced Micro Devices
901 Thompson Place
Sunnyvale, CA 94086

AEG Telefunken
Route 22, Orr Dr.
Somerville, NJ 08876

American National Standards Institute
1430 Broadway
New York, NY 10018

AP Products/3M
1359 West Jackson St.
Painesville, OH 44077

Bell Industries
J. W. Miller Div.
19070 Reyes Ave
P.O. Box 5825
Rancho Dominguez, CA 90224

Borg-Warner Thermoelectrics
3570 N. Avondale Ave.
Chicago, IL 60018

Digi-Key
P.O. Box 677
Thief River Falls, MN 56701

Direct Safety Co.
7815 S. 46th St.
Phoenix, AZ 85040

Dolan-Jenner Industries, Inc.
P.O. Box 1020
Woburn, MA 01801

E & L Instruments
61 First St.
Derby, CT 06418

Edmund Scientific
101 E. Gloucester Pike
Barrington, NJ 08007

Electronic Connector Division
Midland-Ross Corp.
Electronic Connecter Div.
1 Alewife Place
Cambridge, MA 02140

Electronic Design and Packaging Co.
17425 Ecorse Road
Allen Park, MI 48101

Exar
759 Palamar Ave.
Sunnyvale, CA 94088

Fairchild
464 Ellis St.
Mountain View, CA 94042

Fenwal Electronics
63 Fountain St.
Framingham, MA 01701

General Electric
Semiconductor Products Department
Auburn, NY 13021

Gregson Holdings, Ltd.
382 Blackpool Road
Preston, Lancashire, PR2 2DS England

Hall-Mark Electronics
11333 Pagemill
P.O. Box 222035
Dallas, TX 75222

Harris Corporation
P.O. Box 883
Melbourne, FL 32901

Hayden Book Co.
4300 W. 62nd St.
Indianapolis, IN 46268

Hobby Shack
18480 Bandilier Circle
Fountain Valley, CA 92708

Interlink Electronics, Inc.
535 E. Montecito
Santa Barbara, CA 93103

Interplex Electronics Corp.
P.O. Box 1942
70 Fulton Terrace
New Haven, CT 06509

Intersil
10710 N. Tantau Ave.
Cupertino, CA 95014

Jameco
1355 Shoreway Road
Belmont, CA 94002

Kalmback Books
1027 North Seventh Ave.
Milwaukee, WI 53233

Keystone Carbon Co.
Thermistor Division
1935 State St.
St. Marys, PA 15857

Laser Institute of America
5151 Monroe St.
Toledo, OH 43623

Marlow Industries
1021 S. Jupiter Road
Garland, TX 75042

Melcor/Materials Electronic Products Corp.
992 Spruce Street
Trenton, NJ 08648

McGraw-Hill Book Co.
1221 Avenue of the Americas
New York, NY 10020

Metrologic Instruments, Inc.
143 Harding Ave.
Bellmawr, NJ 08031

Mitel Semiconductor
Suite M
2321 Morena Blvd.
San Diego, CA 92110

Motorola Semiconductor
P.O. Box 20912
Phoenix, AZ 85036

Murata Erie North America, Inc.
2200 Lake Park Drive
Smyrna, GA 30080

National Center for Devices and Radiological Health
   (NCDRH)
8757 Georgia Ave.
Silver Spring, MD 20910

National Semiconductor
2900 Semiconductor Dr.
Santa Clara, CA 95051

Newark Electronics
4801 N. Ravenswood St.
Chicago,IL 60640

Nuclear Products
P.O. Box 5178
El Monte, CA 91734

Omega Engineering, Inc.
One Omega Dr.
P.O. Box 4047
Stamford, CT 06907

Opcoa
330 Talmadge Road
Edison, NJ 00817

Pioneer Electronics

PCI Displays
1145 Sonora Court
Sunnyvale, CA 94086

Piezo Electric Products
186 Massachusetts Ave.
Cambridge, MA 02139

Polaroid Corp.
Ultrasonics Components Group
119 Windsor St.
Cambridge, MA 02139

Polytec Optronics, Inc.
Unit 108
22651 Lambert St.
El Toro, CA 92630

Portescap US
730 Fifth Ave.
New York, NY 10019

Precision Monolithic Industries
1500 Space Park Dr.
Santa Clara, CA 95050

Radio Shack
A Division of Tandy Corp.
One Tandy Center
Ft. Worth, TX 76102

Rapidsyn
Industrial Power Transmission Div.
11901 Burke St.
Santa Fe Springs, CA 90670

RCA
Route 202
Somerville, NJ 08876

Raytheon
350 Ellis St.
Mountain View, CA 94042

Richmond Technology Inc.
P.O. Box 1129
Redlands, CA 92373

Rockwell Associates, Inc.
P.O. Box 43010
Cincinnati, OH 45243

Howard W. Sams & Co.
4300 W. 62nd St.
Indianapolis, IN 46268

Sanyo Electric Ltd.
Hashiridani, Hirakata
Osaka, Japan

S.E. International
P.O. Box 39
Summertown, TN 38483

Sharp Electronics Corp.
10 Sharp Plaza
Paramus, NJ 07652

Siemens AG
Frankfurter Ring 152
D-8000, Munich 46
German Federal Republic

Siemens Components Inc.
186 Wood Ave.
Iselin, NJ 08830

Signetics
P.O. Box 409
Sunnyvale, CA 94986

Silicon Systems, Inc.
14351 Myford Road
Tustin, CA 92680

Siliconix, Inc.
2201 Laurelwood Road
Santa Clara, CA 95054

Statek Corporation
512 N.Main
Orange, CA 92668

Teltone Corp.
P.O. Box 657
Kirkland, WA 98033

Texas Instruments
P.O. Box 5012
Dallas, TX 75222

Thermometrics, Inc.
808 U.S. Highway #1
Edison, NJ 08817

Warner Electric Brake & Clutch Co.
449 Gardner St.
South Beloit, IL 61080

The West Bend Company
P.O. Box 1976
West Bend, WI 53095

3M Corp.
3M Center
St. Paul, MN 55101

U.S. Department of Commerce
Patent and Trademark Office
Washington, DC 20231

Vector Electronics,Inc.
12460 Gladstone Ave.
Sylmar, CA 91342

Venitron Corporation
Piezoelectric Division
232 Forbes Road
Bedford, OH 44146

Vernitech
300 Marcus Blvd.
Deer Park, NY 11729

Watson Industries
3041 Melby Road
Eau Claire, WI 54703

MORE
FROM
SAMS

☐ Electronic Telephone Projects (2nd Edition) *Anthony J. Caristi*

Create a touch-tone dialer, conferencer, computer memory, electronic ringer, and 18 other creative projects that turn a telephone into a favorite household tool. This new edition contains seven completely new projects, clearly detailed with step-by-step construction details and procedures. Included for each project is a photo of the unit, schematic, printed-circuit board pattern, parts layout, and a list of all parts required. Introductory chapters outline telephone system operations and the general construction techniques you will need to build these projects.
ISBN: 0-672-22485-2, $10.95

☐ Fiber Optics Communications, Experiments, and Projects *Waldo T. Boyd*

Another Blacksburg tutorial teaching new technology through experimentation. This book teaches light beam communication fundamentals, introduces the simple electronic devices used, and shows how to participate in transmitting and receiving voice and music by means of light traveling along slender glass fibers.
ISBN: 0-672-21834-8, $15.95

☐ First Book of Modern Electronics Fun Projects *Art Salsberg*

Novice and seasoned electronics buffs will enjoy these 20 fun and practical projects. Electronics hobbyists are introduced to many project building areas including making circuit boards, audio/video projects, telephone electronics projects, security projects, building test instruments, computer projects, and home electronics projects. The necessary tools for each project accompany the step-by-step instructions, illustrations, photos, and circuit drawings.
ISBN: 0-672-22503-4, $12.95

☐ Fun Way Into Electronics *Dick Smith*

This three volume series features 50 introductory projects for beginning electronics enthusiasts. Beginning with Volume 1 and continuing through the series, each project is designed as an instructional building block allowing the beginner to progress to more sophisticated projects. Each book features easy-to-understand, concise construction methods and descriptions, providing a rewarding learning program.

Volume 1: Includes 20 introductory projects including basic materials and tools, component descriptions, component codes, guide to successful projects, component listing, and other projects.
ISBN: 0-672-22548-4, $9.95

Volume 2: Provides 20 projects covering topics such as soldering onto a professional printed circuit board, using a multimeter, reading circuit diagrams, basic circuit laws, and milestones in electronics.
ISBN: 0-672-22549-2, $9.95

Volume 3: Covers advanced projects such as investigating integrated circuits, constructing PC boards, building a mini synthesizer and mini stereo amplifier, and understanding the binary stystem.
ISBN: 0-672-22550-6, $9.95

☐ Electronics: Circuits and Systems *Swaminathan Madhu*

Written specifically for engineers and scientists with non-electrical engineering degrees, this reference book promotes a basic understanding of electronic devices, circuits, and systems. The author highlights analog and digital systems, practical applications, signals, circuit devices, digital logic systems, and communications systems. In a concise, easy-to-understand style, he also provides completed examples, drill problems, and summary sheets containing formulas, graphics, and relationships. An invaluable self-study manual.
ISBN: 0-672-21984-0, $39.95

☐ Gallium Arsenide Technology *David K. Ferry, Editor-in-Chief*

This comprehensive introduction to the structure and properties of this wonder compound also explores its application in analog and digital technology and examines the new band-gap engineering and the uses of gallium arsenide.
ISBN: 0-672-22375-9, $44.95

☐ Solid-State Relay Handbook with Applications *Anthony Bishop*

This comprehensive reference work treats SSRs on a wide range of technical levels. Particularly useful are the applications of SSRs with microprocessor-based equipment for industrial machines and the use of SSRs in interfacing with microcomputers.
ISBN: 0-672-22475-5, $19.95

☐ Transistor Fundamentals, Volume 2 *Training and Retraining, Inc., Charles A. Pike*

This introductory text explains transistor principles, voltage, current resistance, inductance, capacitance, and circuitry. It provides all information you'll need to develop a firm understanding of solid-state electronics and troubleshooting techniques.
ISBN: 0-672-20642-0, $9.95

MORE
FROM
SAMS

☐ Understanding Solid State Electronics (4th Edition)
William E. Hafford and Gene W. McWhorter
This book explains complex concepts such as electricity, semiconductor theory, how electronic circuits make decisions, and how integrated circuits are made. It helps you develop a basic knowledge of semiconductors and solid-state electronics. A glossary simplifies technical terms.
ISBN: 0-672-27012-9, $17.95

☐ Understanding Digital Electronics (2nd Edition)   *Gene W. McWhorter*
Learn why digital circuits are used. Discover how AND, OR, and NOT digital circuits make decisions, store information, and convert information into electronic language. Find out how digital integrated circuits are made and how they are used in microwave ovens, gasoline pumps, video games, and cash registers.
ISBN: 0-672-27013-7, $17.95

☐ Understanding Digital Logic Circuits
Robert G. Middleton
Designed for the service technician engaged in radio, television, or audio troubleshooting and repair, this book painlessly expands the technician's expertise into digital electronics.
ISBN: 0-672-21867-4, $18.95

☐ Active-Filter Cookbook   *Don Lancaster*
Need an active filter, but don't want to take the time to design it? Don Lancaster presents a catalog of predesigned filters which he encourages you to borrow and adapt to your needs. The book teaches you how to construct high-pass, low-pass, and band-pass filters having Bessel, Chebyshev, or Butterworth response. It can also be used as a reference for analysis and synthesis techniques.
ISBN: 0-672-21168-8, $15.95

☐ Computer-Aided Logic Design
Robert M. McDermott
An excellent reference for electronics engineers who use computers to develop and verify the operation of electronic designs. The author uses practical, everyday examples such as burglar alarms and traffic light controllers to explain both the theory and the technique of electronic design. CAD topics include common types of logic gates, logic minimization, sequential logic, counters, self-timed systems, and tri-state logic applications. Packed with practical information, this is a valuable source book for the growing CAD field.
ISBN: 0-672-22436-4, $25.95

☐ Design of Op-Amp Circuits with Experiments   *Howard M. Berlin*
An experimental approach to the understanding of op amp circuits. Thirty-five experiments illustrate the design and operation of linear amplifiers, differentiators and converters, voltage and current converters, and active filters.
ISBN: 0-672-21537-3, $12.95

☐ Design of Phase-Locked Loop Circuits with Experiments   *Howard M. Berlin*
Learn more about TTL and CMOS devices. This book contains a wide range of lab-type experiments which reinforce the textual introduction to the theory, design, and implementation of phase-locked loop circuits using these technologies.
ISBN: 0-672-21545-4, $12.95

☐ Handbook of Electronics Tables and Formulas (6th Edition)
Howard W. Sams Engineering Staff
Stay abreast of the rapidly changing electronics industry with this new edition containing computer programs (written for Commodore 64®, with conversion information for Apple®, Radio Shack, and IBM®) for calculating many electrical and electronic equations and formulas. The easy-to-access format contains formulas and laws, constants and standards, symbols and codes, service and installation data, design data, and mathematical tables and formulas.
ISBN: 0-672-22469-0, $19.95

☐ Reference Data for Engineers: Radio, Electronics, Computer, and Communications (7th Edition)
Edward C. Jordan, Editor-in-Chief
Previously a limited private edition, now an internationally accepted handbook for engineers. Includes over 1300 pages of data compiled by more than 70 engineers, scientists, educators and other eminent specialists in a wide range of disciplines. Presents information essential to engineers, covering such topics as: digital, analog, and optical communications; lasers; logic design; computer organization and programming, and computer communications networks. An indispensable reference tool for all technical professionals.
ISBN: 0-672-21563-2, $69.95

MORE FROM SAMS

☐ **Security Dictionary**
Richard Hofmeister and David Prince
This reference work brings together definitions and descriptions of video equipment, computer hardware and software, ultrasonics, fiber optics, biometric ID, infrared sensors, and microwaves as they apply to the security business.
ISBN: 0-672-22020-2, $8.95

☐ **Introduction to Digital Communications Switching**   *John P. Ronayne*
Here is a detailed introduction to the concepts and principles of communications switching and communications transmission. This technically rigorous book explores the essential topics: pulse code modulation (PCM), error sources and prevention, digital exchanges, and control. Sweeping in its scope, it discusses the present realities of the digital network, with references to the Open Systems Interconnection model (OSI), and suggests the promising future uses of digital switching.
ISBN: 0-672-22498-4, $23.95

☐ **How to Read Schematics (4th Edition)**
Donald E. Herrington
More than 100,000 copies in print! This update of a standard reference features expanded coverage of logic diagrams and a chapter on flowcharts. Beginning with a general discussion of electronic diagrams, the book systematically covers the various components that comprise a circuit. It explains logic symbols and their use in digital circuits, interprets sample schematics, analyzes the operation of a radio receiver, and explains the various kinds of logic gates. Review questions end each chapter.
ISBN: 0-672-22457-7, $14.95

☐ **Semiconductor Device Technology**
Malcom E. Goodge
This text explains fundamental principles of semiconductor technology, then discusses the practical operation and performance of commercial diodes, FETs, bipolar transistors, specialized switching, and optical devices. It shows in detail how planar fabrication takes place and thoroughly covers design, manufacture, and application of monolithic and film-type ICs. Contains tutorial questions with answers, information on network modeling, terminology, preferred component values, device numbering, and coding.
ISBN: 0-672-22074-1, $34.95

☐ **Basic Electricity and an Introduction to Electronics (3rd Edition)**
Howard W. Sams Engineering Staff
Extensive two-color illustrations and frequent questions and answers enhance this introduction to electronics. The mathematics of electrical calculations are clearly presented, including Ohm's law, Watt's law, and Kirchhoff's laws. Other topics include cells and batteries, magnetism, alternating current, measurement and control, and electrical distribution.
ISBN: 0-672-20932-2, $11.95

☐ **IC Timer Cookbook (2nd Edition)**
Walter C. Jung
You can learn lots of ways to use the IC timer in this second edition which includes many new IC devices. Ready to use applications are presented in practical working circuits. All circuits and component relationships are clearly defined and documented.
ISBN: 0-672-21932-8, $17.95

☐ **IC Op-Amp Cookbook (3rd Edition)**
Walter G. Jung
Hobbyists and design engineers will be especially pleased at this new edition of the industry reference standard on the practical use of IC op amps. This book has earned respect in the industry by its comprehensive coverage of the practical uses of IC op amps, including design approaches and hundreds of working examples. The third edition has been updated to include the latest IC devices, such as chopper stabilized, drift-trimmed BIFETS. The section on instrumentation amps reflects the most recent advances in the field.
ISBN: 0-672-22453-4, $21.95

MORE
FROM
SAMS